CONTENTS

Part I: Theoretical Foundation

POLARIZED ELECTRONS IN SURFACE PHYSICS

POLARIZED
ELECTRONS IN
SURFACE PHYSICS

R. Feder

World Scientific

Published by

World Scientific Publishing Co. Pte. Ltd.
P. O. Box 128, Farrer Road, Singapore 9128

Library of Congress Cataloging-in-Publication Data is available.

Polarized Electrons in Surface Physics

ISBN 9971-978-49-0
    9971-978-50-4 pbk

Printed in Singapore by Kim Hup Lee Printing Co. Pte. Ltd.

Foreword

Surface physics, dealing with physical (and thence also chemical) properties and processes related to the solid/vacuum boundary, has developed into a vast field of still growing scientific and technological importance. (An extensive survey and references may be found in four volumes recently edited by King and Woodruff (1981-1984)). Electrons are essential for surface physics in two respects: firstly, bound electrons are a constituent part of any surface system, essential not only for the very existence of the system (its local bonding and geometry), but also for its vibrational, magnetic and chemical properties; secondly, free electrons provide - in a wide variety of scattering and emission techniques - powerful means of studying surface properties. Particularly interesting and useful phenomena occur, if the ensemble of electrons under consideration is polarized, i.e. if the number of electrons with spin parallel to a preferential direction differs from the number with spin antiparallel. For electrons bound to the surface system, this is associated with ferro- or ferrimagnetic ordering. In electron spectroscopy techniques, one is dealing with a beam of polarized electrons which is incident on the surface or emerges from it. (A general introduction to polarized free electrons and their use in atomic, solid state and high-energy physics is given in a monograph by Kessler (1976 and, updated, 1985)). Interacting with the surface system via exchange or spin-orbit coupling, polarized electron beams have - due to substantial advances in producing them ("sources") and analysing their spin polarization ("detectors") - within the past decade established themselves as a unique tool for studying magnetic, electronic and even geometrical surface properties.

In view of the achievements already made, the rapidly growing interest in and the future promise of this field, the time appears ripe for a comprehensive presentation, which initiates the non-specialist (with a general physics background at the graduate level) and reviews the current state of the art. This is the aim of the present book. It consists of a coherent sequence of fourteen chapters written by top level experts, who have significantly promoted progress in the respective sub-areas of the field.

The book is organized in two main parts: (I) theoretical foundation, (II) experiments and (experimental and theoretical) results.

Part I (Chapters 1 to 4) introduces fundamental concepts and theoretical approaches. Chapter 1 is devoted to the electronic and magnetic structure of clean and adsorbate-covered surfaces at temperatures well below their ferromagnetic transition temperature. After setting the theoretical framework of local spin-density-functional formalism and thin film approximation, the currently most fruitful first-principles theory for calculating the spin-polarized electronic ground state is introduced, and numerical results for a variety of typical transition metal surfaces and overlayers are presented and discussed. While ferromagnetism of the infinite solid ("bulk") at low temperatures is well understood (in particular itinerant Stoner model for 3d transition metals and Heisenberg Model for rare-earth systems) (cf. standard textbooks on Solid State Physics, e.g. Ashcroft and Mermin (1976), Callaway (1976), Harrison (1970)), transition metal ferromagnetism near the Curie temperature is a subject of very recent controversy and progress. Since an understanding of the "bulk" is a prerequisite for understanding the surface, and since polarized electrons (in photoemission) have provided valuable new insight, key concepts (spin fluctuations, local moments, short-range magnetic order) and theoretical state of the art are therefore included in the present book (Chapter 2). This is naturally followed by a survey of surface ferromagnetism near the Curie temperature (Chapter 3), i.e. "critical behaviour", including scaling theory, renormalization group and Monte Carlo simulations. Chapters 1 to 3 having dealt with the spin-polarized (ferromagnetic) structure of surface systems "by themselves", Chapter 4 addresses the interaction of polarized free electrons with magnetic and non-magnetic surface systems (semi-infinite solid with clean or adsorbate-covered surface), whicn is fundamental to a wide variety of electron scattering and emission techniques for investigating surface properties. Observable spin polarization effects may arise from ferromagnetic exchange interaction and/ or from spin-orbit coupling, which are both formally incorporated in a one-electron Dirac Hamiltonian containing an effective magnetic field. The theory of elastic spin-polarized low-energy electron diffraction (LEED) is presented in some detail, firstly because of its intrinsic importance and secondly because the "LEED state" (or its time reverse) is an essential ingredient for quantitative theories of other methods like photoemission, inverse photoemission and inelastic electron scattering, which are subsequently discussed. General results due to symmetry properties are presented, and principles of deducing, with the aid of theory, surface properties from experimental data are explained.

Part II (Chapters 5 to 14) deals with experimental techniques, experimental results and physical information obtained by comparing experimental data with their theoretical counterparts. Chapter 5 introduces the essential experimental tools: sources of polarized electrons, spin polarization detectors (polarimeters), synchrotron radiation (linearly and circularly polarized) and photon detectors. Elastic spin-polarized low-energy electron diffraction is presented for non-magnetic and for ferromagnetic surfaces in Chapters 6 and 7, respectively. In particular, the determination of the surface geometry, of the layer-dependent-magnetization at low temperatures and of the ferromagnetic critical behaviour of surfaces is illustrated. The subsequent two Chapters are devoted to spin-dependent electron-electron collision processes. The techniques, which are presented, include in particular high-resolution electron energy loss spectroscopy from ferromagnets, culminating in a "triple scattering" experiment involving both a polarized primary beam and spin analysis of inelastically scattered electrons (Chapter 8), and secondary electron emission (especially Auger emission and the very-low-energy "cascade") (Chapter 9). In addition to revealing details of the electron-electron interaction and ferromagne-

tic surface and bulk properties, these studies have led to the techno-
logically important development of a magnetic scanning electron microscope. Chapters 10 to 12 are devoted to spin-resolved photoemission
due to radiation in the (vacuum) ultraviolet range (photon energies up
to about 70 eV). Chapter 10 focuses on semiconductor surfaces, for
which spin-orbit coupling together with circular light polarization
produces highly polarized photoelectrons, which carry information on
the bulk band structure, on doping with impurities and - due to a most
recently discovered spin precession effect in non-centrosymmetric crystals - on the spatial extent of the band-bending region near the surface. For non-magnetic metal surfaces, spin-, angle- and energy-resolved photoemission experiments performed with circularly polarized
ultraviolet synchrotron radiation permit a direct observation of the
symmetry types of the occupied states and promise detailed information
on the electronic structure of adsorbed overlayers (Chapter 11). For
ferromagnets (Chapter 12), spin-resolved photoemission by linearly
polarized or unpolarized light reveals the majority- and minority-spin
(quasi-particle) bulk and surface band structures. While photoemission
observes the occupied electronic states, its inverse, bremsstrahlung
induced by polarized electrons (Chapter 13), provides complementary
information on the unoccupied states (in particular in the vicinity
of the Fermi level). The retrieval of physical information from experimental photoemission and bremsstrahlung data by means of theoretical
model calculations is illustrated in Chapter 14 for ferromagnetic Fe
and Ni. In particular, the determination of short-range magnetic order
near the Curie temperature is demonstrated, and some light is shed on
the influence of chemisorption on surface magnetism. The final Chapter 15 gives a synopsis and an outlook on future prospects for polarized electrons both in fundamental surface physics research and in
technological applications.

The organization of Part II is such that physical properties of
specific materials are presented in conjunction with the polarized-
electron method by which they were revealed. An alternative classifica-

tion of the results according to materials is indicated in the following "cross reference" table, which may also serve as a Reader's guide
to Chapters 6 to 14.

Material Method	ferromagnetic			non-magnetic	
	metals 3d	4f	non- metals	metals	semi- conductors
elastic scattering	7	7		6	
inelastic scattering	8	8		8	
secondary emission	9		9	9	
photoemission	12,14	12	12	11	10
bremsstrahlung	13,14				

Table: Chapters, in which results for different types of material
as obtained by polarized-electron methods are presented.

Having outlined the scope and contents of the present book, it
seems pertinent to briefly mention some related methods for studying
ferromagnetic surfaces, which have not been included. About a decade
ago, a substantial research effort was devoted to field emission, i.e.
to extracting polarized electrons from ferromagnets by applying a
strong electric field (for reviews and references cf. Kessler 1976
and 1985, Campagna et al. 1976, Feuchtwang et al. 1978, Celotta and
Pierce 1980). The development of spin-, angle- and energy-resolved
photoemission (cf. Chapters 10-12) has, however, superseded field emission as a magnetic surface diagnostic technique, and the polarized-
electron source based on field emission from an EuS-coated W tip has
- despite its merits of about 90 % polarization and high brightness -
not survived in the competition against the presently most widely
used and even commercially available GaAs-photoemission source (cf.
Chapter 5). Polarized field emission has therefore - to our knowledge -

not been pursued further since about 1980, and belongs to the history of surface physics.

Since we concentrate on techniques involving free polarized electrons, tunneling between superconductors and ferromagnets (cf. e.g. Tedrow et al 1982, Feuchtwang et al. 1978), which yields information on the spin polarization of electron states very close to the Fermi energy, has not been included. For the same reason, Mössbauer spectroscopy (cf. review by Keune 1985) is not represented nor is another interesting nuclear-physics method known as "Electron Capture Spectroscopy" (cf. review by Rau 1982) , in which deuterons impinging at grazing angles on a ferromagnetic surface pick up conduction electrons and thus carry information on surface magnetism. Another recent technique, which yet has to prove its quantitative merits, employs a polarized beam of low-energy positrons and measures the spin dependence of the positronium formation rate at the magnetic surface (Gidley et al. 1982) (cf. also a monograph on positron studies: ed. Mills and Canter 1985). Further, an atomic-physics method, in which spin-polarized metastable atoms are ionized at a ferromagnetic surface and subsequently neutralized in a polarization-dependent conduction-band Auger process, appears promising (Onnelion et al. 1984).

In conclusion of this Foreword, may we express a personal thought? The authors hope that the book is not only useful and enjoyable to you, dear Reader, but that you may also share some of their enthusiasm about this flourishing area of surface physics.

Roland Feder

References

Ashcroft N W and Mermin N D 1976 Solid State Physics (Holt, Rinehart and Winston, New York)

Callaway J 1976 Quantum Theory of the Solid State (Academic Press, New York)

Campagna M, Pierce D T, Meier F, Sattler K and Siegmann H C 1976 Adv. El. and El. Physics $\underline{41}$ 113

Celotta R J and Pierce D T 1980 in Adv. in Atomic and Molecular Physics $\underline{16}$ 101

Feuchtwang T E, Cutler P H and Schmit J 1978 Surface Sci. $\underline{75}$ 401

Gidley D W, Koymen A R and Capehart T W 1982 Phys. Rev. Lett. $\underline{49}$ 1779

Harrison W A 1970 Solid State Theory (McGraw Hill, New York)

Kessler J 1976/1985 Polarized Electrons (Springer, Berlin Heidelberg New York)

Keune W 1985 in Proceedings of the International Conference on Applications of the Mößbauer Effect, Leuven, Sept. 1985, in "Hyperfine Interactions" (North Holland, Amsterdam)

King D A and Woodruff D P 1981-85 (ed) The Chemical Physics of Solid Surfaces and Heterogeneous Catalysis (North Holland, Amsterdam)

Mills A and Canter K F (ed) 1985 Positron Studies of Solids, Surfaces and Atoms (World Scientific Publishing Co, Singapore 1985)

Onellion M, Hart M W, Dunning F B and Walters G K 1984 Phys. Rev. Lett. $\underline{52}$ 380

Rau C 1982 J. Magn. Magn. Mater. $\underline{30}$ 141

Tedrow P M, Moodera J S and Meservey R 1982 Solid State Commun. $\underline{44}$ 587

Part I
Theoretical Foundation

Chapter 1

Electronic and Magnetic Structure of Solid Surfaces

A.J. Freeman and C.L. Fu

Department of Physics and Astronomy
Northwestern University
Evanston, Illinois 60201, USA

and

S. Ohnishi

NEC Corporation, 1-1 Miyazaki 4-chome,
Miyamae-ku, Kawasaki 213, Japan

and

M. Weinert

Department of Physics,
Brookhaven National Laboratory
Upton, New York 11973, USA

1.1 Introduction

The study of magnetism at surfaces and interfaces has added new excitement to this largely unexplored frontier field. Much of this theoretical and experimental effort has been addressed to transition metal materials and represents part of a general increase in interest in the study and understanding of these materials on the microstructure scale. The unique chemical and physical properties of transition metals play an especially important role in determining the observable phenomena in such reduced symmetry systems as small particles (or clusters of atoms), surfaces, interfaces, and modulated structures. A driving force in this exciting area is that new insights into their electronic structure and magnetism are expected because (i) the existence and role of surface and interface states and (ii) the reduced coordination and symmetry lead to important property differences with respect to bulk systems. In particular, the possibility of magnetically dead layers (Lieberman et al. 1969; 1970) at the surface of ferromagnetic Fe and Ni invoked considerable discussion and stimulated the development of theoretical methods for describing surface electronic structures.

Early on, a number of studies (Dempsey et al. 1975, 1976a, 1976b; Dempsey and Kleinman, 1977, 1978; Arlinghaus et al. 1980a, 1980b, 1979; Jepsen et al. 1978; Krakauer et al. 1979; Feibelman and Hamann, 1980) by the finite slab approximation to the semi-infinite solid showed that the electronic structures of the surface are well described by this finite slab model. Further, with the success of the local spin density functional theory in describing magnetism in bulk systems, it became of great interest for theorists to extend bulk methods to treat surface magnetism by means of ab initio self-consistent finite slab calculations. In this regard, it was fortunate for the theoretical treatment of these problems that the theory of itinerant electron magnetism has been considerably advanced in recent years by the success of local density and local spin-density ab initio self-consistent band theory calculations in providing a quantitative understanding of many ground state properties of the ferromagnetic transition metals, iron, cobalt, and nickel (Moruzzi et al. 1978; Connolly, 1967; Wang and Callaway, 1977; Anderson et al. 1979). These calculations have been remarkably successful in obtaining good agreement with such experimental quantities as magnetization, neutron form factors, hyperfine fields, lattice parameter, bulk modulus, cohesive energies, and the Fermi-surface properties. This is particularly impressive, considering that all many-electron effects are included only through an effective one-electron local potential. This achievement is a major confirmation of the utility of Hohenberg-Kohn-Sham (1964, 1965) (von Barth and Hedin, 1972; Gunnarsson et al. 1972; Rajagopal and Callaway, 1973) local-density-functional theory which provides the formal justification for using the single-particle picture to determine ground-state properties. While the ground-state properties are now quantitatively understood on this basis (Moruzzi et al. 1978) this is not true, unfortunately, for the elementary excitations and temperature-dependent effects of itinerant-electron ferromagnet (Edwards, 1980), although there have been significant advances in these areas in recent years (Prange and Korenman, 1979; Feldkamp and Davis, 1979; Liebsch, 1979; Kleinman, 1979; Treglia et al. 1980a, 1980b; Liebsch, 1981; Kleinman and Mednick, 1981).

From the experimental side, the renewed interest in surface and interfaces has been stimulated, in part, by recent developments in experimental methods which have provided unique information about spin polarization in magnetic materials (Busch et al. 1971; Eib and Alvarado, 1976; Eastman et al. 1978; Bergmann, 1978; Sato and Hirakawa, 1975; Landolt and Campagna, 1977; Paraskevopoulos et al. 1977; Eichner et al. 1977; Smith et al. 1977). Much of this will be discussed in the chapters which follow. Here it needs to be pointed out that the past few years have witnessed major advances in the use of angle-resolved photoemission (Eastman et al. 1978, 1980; Himpsel et al. 1979; Eberhardt and Plummer, 1980) and such novel techniques as electron capture spectroscopy (Rau, 1980), spin-polarized low-energy electron diffraction (LEED) (Celotta et al. 1979; Feder, 1981), and inverse photoemission (Himpsel and Fausten, 1982; Woodruff et al. 1982). New and remarkably detailed information on the ferromagnetic transition metals (especially from photoemission) have thus provided a well-charted area for testing the limitations of the single-particle picture and our understanding of the important many-body processes.

Many of the experimental methods such as photoemission, however, are intrinsically surface sensitive, and the experimentally observed features may involve bulk effects and/or surface effects such as the possible existence of magnetically "dead" surface layers (Liebermann et al. 1970; Pierce and Siegmann, 1974; Bergmann, 1978; Meservey et al. 1980) and magnetic surface states (Plummer and Eberhardt, 1979; Erskine, 1980; Dempsey and Kleinman, 1977). These surface effects are interrelated as demonstrated by the recent photoemission observation (Eberhardt et al. 1980) of exchange-split surface states (with temperature-dependent exchange splitting) on the Ni(110) surface. Unless these states are well localized in the surface layer, however, the possibility of "dead layers" for this surface could not be ruled out. As shown by Krakauer et al. (1983), these states are indeed extremely localized in the surface layer. Taken together, the good agreement between experiment (Eberhardt et al. 1980) and these theoretical results for this surface state provide very strong evidence for the presence of surface magnetism on Ni(110). The absence of dead layers is also supported by recent spin-polarized LEED experiments (Celotta et al., 1979; Feder, 1981) on the Ni(100) surface. The photoelectron-spin-polarization reversal (Dempsey and Kleinman, 1977; Kleinman, 1981; Eib and Alvarado, 1976; Kisker et al. 1979; Gudat et al. 1980; Kisker et al. 1980; Moore and Pendry, 1978) observed just above threshold in Ni is another example where surface effects (surface states and unbound evanescent states) have been suggested (Dempsey and Kleinman, 1977; Kleinman, 1981) to play a decisive role.

There has, therefore, been great interest in the past few years in the electronic structure of ferromagnetic metal surfaces. Exchange-split surface states have been mapped out on the Ni(100) surface by Plummer and Eberhardt (1979) and by Erskine (1980). Some of these states had been predicted previously to exist in the non-self-consistent calculations of Dempsey and Kleinman (1977). As mentioned,

prominent exchange-split surface states were also observed by
Eberhardt et al. (1980) on the Ni(100) surface and used to rule out
magnetically dead surface layers, and these states have been found for
the first time in the linearized-augmented-plane-wave (LAPW)
calculations (Kraukaer et al, 1983).

A related area of interest is the magnetic properties of very thin
ferromagnetic overlayers on nonmagnetic substrates (Rau, 1980;
Liebermann et al. 1970; Pierce and Siegmann, 1974; Bergmann, 1978;
Meservey et al. 1980). Liebermann et al. (1970) observed magnetically
dead layers for less than about 2.5 layers of Ni deposited on a Cu
substrate. Anomalous Hall-effect measurements by Bergmann (1978)
confirmed the existence of dead layers, but the spin-polarized
photoemission measurements of Pierce and Siegmann (1974) suggest that
Ni becomes ferromagnetic for overlayer thickness as low as a
monolayer. Similarly, electron-capture spectroscopy measurements (Rau,
1980) find that a Ni monolayer on Cu is not magnetically dead but has
a reduced moment. Self-consistent thin-film calculations of Ni
overlayers on a Cu(100) substrate (Wang et al. 1982) have found that
even one monolayer of Ni on Cu is not magnetically dead (the moment is
reduced to $0.39\mu_B$).

Compared to the number of bulk calculations, there have been
relatively few such studies for surface and other reduced symmetry
systems. This chapter concerns itself with describing the present
state of ab initio-self-consistent local spin density determinations
of the electronic structure and magnetism at surface and interfaces.
Of necessity - because of space limitations - we will restrict
ourselves essentially to the work in which we have been involved.
(Abundant references to other work are given in the papers reviewed
here. In this way, the reader will have what we hope is a coherent
overview of developments in this field.

1.2 Theoretical Framework

This section describes the theoretical framework which supports
all the studies reviewed here. Any theoretical approach must address

two basic problems. Within the Born-Oppenheimer approximation, the
first (and more difficult problem) is the treatment of many-body
electron-electron interactions which are so essential for the
description of magnetism. The second problem is to devise a structural
model of the surface for which realistic calculations can be
performed. By realistic, we mean calculations comparable to state-of-
the-art bulk calculations.

1.2.1 Local Spin Density Functional Theory

One of the most important advances in the determination of the
electronic and magnetic structure of materials was the development of
density functional theory (Hohenberg and Kohn (1964); Kohn and Sham
(1965); and Rajagopal (1980)) and its spin-polarized extensions
(Rajagopal (1980); von Barth and Hedin (1972); Gunnarson et al.
(1972)). The basic result is the proof by Hohenberg and Kohn (1964)
that the ground state energy of a many-body system is a unique
functional of the density, $n(\vec{r})$, and is a minimum when evaluated for
the true ground state density. Kohn and Sham (1965) then showed how an
equivalent one-particle equation could be set up which, in principle,
includes all correlations. Here we will give a brief review of the
theory of relativistic spin-polarization (Weinert and Freeman (1983b))
within the local spin-density framework. This development has the
standard local spin density results as the non-relativistic limit, but
is also valid for high Z systems and has important numerical
consequences for quantities such as contact hyperfine fields.

Let us consider the coupling of Dirac particles of mass m and
charge e to the electromagnetic field (we will be in the radiation
gauge $\nabla \cdot A = 0$)(Bjorken and Drell (1964, 1965)).

$$\mathcal{H} = \mathcal{H}_0 + \frac{1}{c} \int d\vec{r} \, \mathcal{J}_\mu A^\mu \tag{1}$$

where $\hat{\mathcal{H}}_0$ is the Hamiltonian in the absence of external fields and the
four-current operator and four-potential are given by

$$\mathcal{J}^\mu = (c\hat{\rho}, \vec{\hat{J}}) = ec \, \hat{\bar{\psi}} \, (\hat{r}) \gamma^\mu \hat{\psi}(\vec{r}) \tag{2}$$

$$A^{\mu} = (\Phi, \vec{A}_{ext}) \tag{3}$$

(Our notation follows the convention of Bjorken and Drell (1964 and 1965.) Rajagopal and Callaway (1973), MacDonald and Vosko (1979), and Rajagopal (1978) have shown that the Hohenberg-Kohn (1964) theorems can be generalized to include relativistic effects. Moreoever, these authors have shown that one can obtain Kohn-Sham single particle equations of the form (MacDonald and Vosko (1979))

$$\left\{ c\vec{\alpha} \cdot (\vec{p} - \frac{e}{c} \vec{A}_{eff}) + \beta mc^2 + eV_{eff}(\vec{r}) \right\} \phi_i(\vec{r}) = \epsilon_i \phi_i(\vec{r}) \tag{4}$$

where the effective potentials are given by

$$V_{eff}(\vec{r}) = \Phi(\vec{r}) + e \int \frac{n(\vec{r}\,')d\vec{r}\,'}{|\vec{r} - \vec{r}\,'|} + \frac{\delta E_{xc}[J_{\mu}]}{\delta J_0(\vec{r})} \tag{5}$$

$$\vec{A}_{eff}(\vec{r}) = \vec{A}_{ext} \frac{\delta E_{xc}[J_{\mu}]}{\delta \vec{J}(\vec{r})} \tag{6}$$

and $n(\vec{r})V$ is the number density. The exchange-correlation energy functional $E_{xc}[J_{\mu}]$ contains magnetic effects through its dependence on the spatial components of the current. If we are interested in spin effects, this approach is not appropriate since spin and kinetic effects are not separable. Following MacDonald and Vosko (1979), we take the non-relativistic viewpoint that the external fields (in analogy with non-relativistic spin-density functional theory) couple only to the particle and spin densities:

$$\left\{ c\vec{\alpha} \cdot (\vec{p} - \frac{e}{c} \vec{A}_{eff}) + \beta mc^2 + eV_{eff}(\vec{r}) \right\} \phi_i(\vec{r}) = \epsilon_i \phi_i(\vec{r}) \tag{7}$$

where μ_B is the Bohr magneton and

$$\sigma^{\mu\nu} = \frac{1}{2} [\gamma^{\mu}, \gamma^{\nu}] \tag{8}$$

$$F^{\mu\nu} = \partial^{\mu}A^{\mu} - \partial^{\nu}A^{\nu} \tag{9}$$

If we consider A^{μ} and $F^{\mu\nu}$ to be given classical objects with $F^{\mu\nu}$ having only spatial components, we have

$$\hat{\mathcal{H}}_{ext} = e \int d\vec{r} : \hat{\psi}(\vec{r}) \gamma_0 \hat{\psi}(\vec{r}) : \phi(\vec{r}) - \mu_B \int d\vec{r} : \hat{\psi}(\vec{r}) \vec{\sigma} \hat{\psi}(\vec{r}) : \vec{B}$$

If we now define the magnetization density operator $\hat{\vec{m}}(\vec{r}) = \mu_B \hat{\psi} \vec{\sigma} \hat{\psi}$, we can write

$$\hat{\mathcal{H}}_{ext} = \int d\vec{r} \left(\rho(\vec{r})\Phi(\vec{r}) - \vec{m}(\vec{r})\cdot\vec{B} \right) \tag{10}$$

The first term contains the usual minimal electromagnetic coupling while the second term represents a coupling to the magnetic dipole moment only. This Hamiltonain leads to single-particle equations of the form

$$\left\{ c\vec{\alpha}\cdot\vec{p} + \beta mc^2 + eV_{eff}(\vec{r}) - \mu_B \vec{\Sigma}\cdot\vec{U}_{eff}(\vec{r}) \right\} \phi_i(\vec{r}) = \varepsilon_i\phi_i(\vec{r}) \tag{11}$$

where $V_{eff}(r)$ is given by Eq. (5) and the spin density operator $\vec{\Sigma}$ and effective magnetic potential are given by

$$\vec{\Sigma} = \begin{pmatrix} \vec{\sigma} & 0 \\ 0 & -\vec{\sigma} \end{pmatrix} \tag{12}$$

$$\vec{U}_{eff} = \vec{B} + \frac{\delta E_{xc}}{\delta\vec{m}(\vec{r})} \tag{13}$$

and $\vec{\sigma}$ denotes the usual 2x2 Pauli spinors. The number density $n(\vec{r})$ and magnetization density $\vec{m}(\vec{r})$ are given by

$$n(\vec{r}) = \sum_i \phi_i^\dagger(\vec{r})\phi_i(\vec{r})$$

$$\vec{m}(\vec{r}) = \mu_B \sum_i \phi_i^\dagger(\vec{r}) \vec{\Sigma} \phi_i(\vec{r}) \tag{14}$$

where the sums are over all occupied states.

If we take the non-relativistic limit of (11) retaining the first relativistic correction, we obtain the familiar Pauli-like equation for a magnetic field coupling to the spins only:

$$\left\{ \left[\frac{p^2}{2m} - \frac{p^4}{8m^3c^2} \right] - \mu_B \vec{\sigma}\cdot\left[\vec{B} - \frac{1}{2mc}(\nabla V \times \vec{p}) \right] \right.$$

$$\left. + \left[eV + \frac{\hbar^2 e}{8m^2c^2} \nabla^2 V \right] \right\} \psi = (\varepsilon - mc^2)\psi \tag{15}$$

In this equation, $\vec{B}$ and V are the effective magnetic fields and
potentials which include the effects of exchange-correlation.

The set of self-consistent equations (5), (11)-(14) in principle
yields the correct charge and magnetization densities. The total
energy for the charge and magnetization densities is as a sum of
kinetic, potential and exchange-correlation terms:

$$E[n,m] = T_s[n] + U[n] + E_{xc}[J_\mu] \qquad (16)$$

where $T_s[n]$ is the kinetic energy of non-interacting particles of
density n:

$$T_s[n] = \sum \int d\vec{r}\; \phi^\dagger(\vec{r})\; K_{op}\phi(\vec{r}) \qquad (17)$$

and K_{op} is the kinetic energy operator in the single particle
equation. If one knew the exact E_{xc}, one would have a solution of the
many-electron system. Unfortunately these functionals are not known;
instead one must make approximations. The most common one is the local
spin density approximation (LSDA) (Kohn and Sham (1965); Rajagopal
(1980)). Here, one uses the exchange-correlation energy, E_{xc}, for the
homogeneous electron gas of density n to give an approximation to E_{xc}
and $\delta E_{xc}/\delta J_\mu$.

The great advantage of the density functional method is that given
some approximate form of E_{xc}, the many-body problem has been reduced
to a single-particle problem. It should be kept in mind, however, that
when the densities and the total energy are observables, the
eigenvalues and eigenfunctions of the one-particle equation are not
directly related to the quasiparticles observed in, for example,
photoemission.

1.2.2 Thin-Slab Approximation

Our focus in this review is the treatment of extended surfaces
with or without ordered overlayers of absorbed atoms. Probably the
most successful structural model uses a thin-slab to simulate both
surface and bulk effects on an equal footing. Calculations in this
model are performed for a finite thickness slab which is infinitely

periodic in the plane parallel to the surface. Typically, slabs 5-13 atomic-layers thick are used. The features of this model have been discussed at length by Appelbaum and Hamann (1976). To insure accurate results, the slab should be thick enough so that the electronic structure in the interior of the slab resembles closely the expected bulk structure. Thicker slabs also reduce size-effect energy splittings between surface states which are localized on the upper and lower surfaces. For metals with well localized surface states (such as transition metals) these requirements are satisfied for slabs as thin as five-layers. If it is necessary to identify more extended surface states, thicker slabs must be used. The desirability of thicker slabs for the above reasons must be balanced, however, by the practical consideration that the magnitude of the calculation (measured by the amount of computer memory and computer time required) increases rapidly with increasing slab thickness (roughly between N^2 and N^3, where N is the number of layers in the slab).

1.3 Approach and Methodology

1.3.1 FLAPW Method for Thin Films

Currently, one of the most successful structural models for ab initio surface calculations is the single slab (or thin film) geometry. A film thickness of five to ten atomic layers is usually sufficient to obtain bulk-like properties in the center of the film and consequently true surface phenomena on the two film/vacuum interfaces. As discussed in the previous Section, density-functional theory (Hohenberg and Kohn, 1964; Kohn and Sham, 1965) provides an elegant and powerful framework to describe the electronic structure of condensed systems (e.g. bulk crystals, surfaces, interfaces). In its local approximation density-functional theory leads to Schrödinger-like one-particle equations (Kohn-Sham equations) containing an effective potential energy operator which is determined by the self-consistent charge distribution. Thus, the local density-functional one-particle equations have to be solved iteratively.

One of the most precise and powerful schemes to solve the local (spin) density (LSD) one-particle equations for the film geometry is the all-electron full-potential linearized-augmented-plane-wave (FLAPW) method (Wimmer et al., 1981). The basic idea in this variational method is the partition of real space into three different regions, namely spheres around the nuclei, vacuum regions on both sides of the film, and the remaining interstitial region. In each of these regions the "natural" form of the variational basis functions is adopted, i.e. plane waves in the interstitial region, a product of radial functions and spherical harmonics inside the spheres, and in the vacuum a product of functions which depend only on the coordinate normal to the film and 2-dimensional plane waves. Each of these basis functions is continuous in value and derivative across the various boundaries. This is possible because inside the spheres (and analogously in the vacuum) two radial functions for each ℓ-value are used, namely the solution of the radial Schrödinger equation for the correct potential and its energy derivative.

In the FLAPW method no shape approximations are made to the charge density and the potential. Both the charge density and the effective one-electron potential are represented by the same analytical expansions described above, i.e. a Fourier representation in the interstitial region, an expansion in spherical harmonics inside the spheres, and in the vacuum 2-dimensional Fourier series in a set of planes parallel to the surfaces. The generality of the potential requires a method to solve Poisson's equation for a density and potential without shape approximations. This is achieved by a new scheme which goes beyond the Ewald method (Weinert, 1981). The key idea in this new scheme is the observation that the potential outside a sphere but only on its multi-pole moments. Now, Poisson's equation is solved straight-forwardly when the charge density is given in a Fourier representation. Because of the sharp structure of the charge density in the core region (including the nuclear charge), a Fourier expansion of the total density would be extremely slowly convergent. However, since the potential outside the sphere depends on the charge inside only through the multipole moments, the true charge density can be replaced by a smooth density which has a rapidly converging Fourier series and the same multipole moments as the true density. With this replacement of the density inside the spheres we have a Fourier expansion of a charge density which gives the correct potential outside and also on the sphere boundaries. To find the potential inside the sphere we are faced in a final step with a standard boundary-value problem of classical electrostatics which can be solved from the original charge densities inside the spheres and the potential on the sphere boundaries by a Green's function method.

Thus, the FLAPW method allows a fully self-consistent solution of the LSD one-particle equations for the film geometry and yields charge densities and spin densities close to the LSD limit. Beside the total charge density, the key quantity in density-functional theory is the total energy corresponding to the ground state charge density. Recently, we have presented a new scheme (Weinert et al., 1982) to calculate accurate and stable all-electron density functional total energies and have applied it within the FLAPW method. The capability of total energy calculations for various geometrical arrangements provides us with a powerful theoretical tool to study the energetics and, at least in principle, the dynamics of surfaces and overlayers.

We will now highlight the key formulas of the FLAPW method without making use of special symmetries such as inversion symmetry or mirror reflection on the central plane of the film. In practice these symmetries are of great advantage, since the inversion symmetry makes the Hamiltonian and overlap matrix real and the mirror reflection breaks these matrices into two blocks according to even and odd states.

For a thin film geometry and within the LDF approach, the wavefunctions for each state are solutions of the one-particle equations

$$\left[-\nabla^2 + V_{eff}(\vec{r}) \right] \psi_i(\vec{k}, \vec{r}) = \epsilon_i(\vec{k}) \ \psi_i(\vec{k}, \vec{r}) \tag{18}$$

where $\vec{k}$ is a vector of the two-dimensional first Brillouin zone and i is a band index. The effective potential, V_{eff}, is given as the sum of the electrostatic Coulomb potential, related to the charge density by Poisson's equation, and the local exchange-correlation potential as obtained from many-body theory. In Eq. (18) Rydberg atomic units are used ($\hbar^2/2m=1$, $e^2 = 2$). In the FLAPW method the wave function of each state is expanded variationally in the reciprocal lattice

$$\psi_1(\vec{k}) = \sum_j c_{ij} \phi (\vec{k} + \vec{G}_j) \tag{19}$$

where each of the basis functions is an augmented-plane-wave given by

$$\phi(\vec{K}_j) = \begin{cases} (\Omega)^{-1/2} \exp (i\vec{K}_j \cdot \vec{r}) & \epsilon \text{ interstitial} \tag{20a} \\[2mm] \sum_j [A_L(\vec{K}_j)u_\ell (E_\ell,r) + B_L(\vec{K}_j)\dot{u}_\ell (E_\ell,r)] Y_L(\hat{r}) & \text{for } \vec{r} \epsilon \text{ sphere} \tag{20b} \\[2mm] \sum_q \left[A_q(\vec{K}_j)u_q (E_v,z) + B_q (\vec{K}_j)\dot{u}_q (E_v,z) \right] \exp\left[i(\vec{k}+\vec{K}_q)\vec{r} \right] & \text{for } \vec{r} \epsilon \text{ vacuum} \tag{20c} \end{cases}$$

with $\vec{K}_j=\vec{k}+\vec{G}_j$. $\vec{G}_j$ is a vector of a three-dimensional reciprocal lattices defined in terms of the auxiliary periodicity domain $\hat{D}$ (Fig. 1). The reason for choosing $\hat{D}$ larger than D is simply to gain a greater variational freedom in the basis functions. (For $\hat{D}=D$ the charge density would have at the vacuum boundary an artificial zero slope in the direction perpendicular to the surfaces). Ω is the volume of the unit cell between the vacuum boundaries at $\pm$ D/2. Here $u_\ell(E_\ell,r)$ are solutions of the radial Schrödinger equation obtained with the actual spherical part of the effective potential inside a sphere for a fixed energy E_ℓ and $\dot{u}_\ell(E_\ell,r)$ is the energy derivative of this radial function. The coefficients $A_L(\vec{K}_j)$ and $B_L(\vec{K}_j)$ are determined by the requirements that the plane wave (20a) is continued smoothly in value and derivative across the sphere boundaries. Similarly in the vacuum, $u_q(E_v,z)$ are solutions of the equation

$$\left[-\frac{\delta^2}{\delta z^2} + V(z) - E_v + (\vec{k}+\vec{K}_q)^2 \right] u_q (E_v,z) = 0 \tag{21}$$

where V(z) is the component of the effective potential in the vacuum which depends only on the distance perpendicular to the surface. E_v is an energy parameter for the vacuum analogously to the parameters E_ℓ inside the muffin-tin spheres, $\dot{u}_q(E_v,z)$ is the energy derivative of the function $u_q(E_v,z)$. $\vec{K}_q$ denotes a two-dimensional (i.e. parallel to the surface) reciprocal lattice vector. The matching coefficients $A_q(\vec{K}_j)$ and $B_q(\vec{K}_j)$ are determined by the continuity conditions of $\phi(\vec{K}_j)$

across the vacuum boundaries at $\pm D/2$.

In the FLAPW method for thin films, the electronic charge density is represented in each of the three spacial regions by the "natural" representation, namely

$$\rho(\vec{r}) = \begin{cases} \sum_i \rho_j \exp{(i\vec{G}_j \cdot r)} & \text{for } \vec{r} \in \text{interstitial} & \text{(22a)} \\ \sum_L \rho_L(r) \, Y_L(\hat{r}) & \text{for } \vec{r} \in \text{sphere} & \text{(22b)} \\ \sum_q \rho_q(z) \exp{(i\vec{K}_q \cdot \vec{r})} & \text{for } \vec{r} \in \text{vacuum} & \text{(22c)} \end{cases}$$

The electrostatic potential is obtained from the electronic charge density and the nuclear charges by solving Poisson's equation using the technique described by Weinert (1981) as implemented into the FLAPW method (Wimmer et al., 1981). The exchange-correlation potential is calculated from the local electronic charge density by a least squares fitting technique where the root-mean-square deviation is usually about 1 mRy in the interstitial region and better than 0.1 mRy inside the spheres and in the vacuum. Finally, the effective one-electron potential (as the sum of the electrostatic Coulomb potential and the exchange-correlation potential) is represented in the form completely analogous to the charge density as given by Eqs. (22a-c).

The LD one-particle equations (18) are now solved iteratively. A starting density in the form (22) is constructed from a superposition of self-consistent atomic densities. From this density the corresponding potential is calculated which defines the effective one-particle operator in Eq. (18). Using the expansion (19) and the explicit form (20) of the basis functions, the coefficients c_{ij} of (19) are obtained via a Rayleigh-Ritz variational procedure. These coefficients now define the film-wavefunctions $\psi_i(\vec{k})$ which in turn yield, according to Fermi-Dirac statistics, a new charge density

$$\rho^-(\vec{r}) = e^2 \sum_{occ} \int_{BZ} \psi_i^*(\vec{k},\vec{r}) \, \psi_i(\vec{k},\vec{r}) d^3k \qquad (23)$$

where the summation runs over all occupied states. The density of the

core electrons is obtained by solving fully-relativistically a free
atom-like problem using the effective potential. This
completes one iteration cycle. The new density is fed back and self-
consistency is achieved when $\rho^{\sim} = \rho$, i.e. when the output-density is

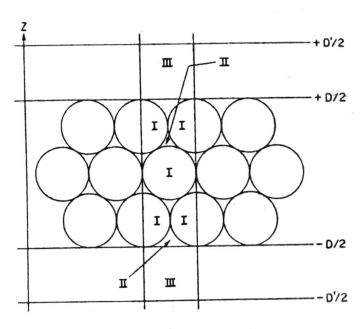

Figure 1. Thin film geometry as used in the FLAPW method. I, II, and
III are the spheres, the interstitial, and the vacuum regions,
respectively. Note that the vacuum regions start at $\pm D/2$. D' provides
an auxilliary periodicity domain as discussed in the text. [After
Krakauer et al. (1979)]

equal to the input density. In practice, self-consistency is assumed
when the potentials corresponding to the input and output densities
differ on the average by less than about 1 mRy. The self-consistency
procedure is accelerated by using an attenuated feedback, i.e. the new
input density is a mixture of, say, 95% of the input density and 5%
output density of the previous iteration. Faster convergence is
achieved by employing a more sophisticated scheme (Andersen, 1965)
involving the input and output densities of two previous iterations.

1.3.2 Energetics of Surfaces: All-electron Total Energy Approach

The intense experimental interest in surface problems such as chemisorption, surface reconstruction and relaxation, and dynamics has produced a wealth of data, much of it still not understood. Complicating the theoretical understanding is the fact that often the important structural parameters needed are not available experimentally. In order to solve this problem theoretically, one can use the general principle of minimization of the total energy to determine the stability of a system. Density functional theory (Hohenberg and Kohn, 1964; Kohn and Sham, 1965) provides an elegant framework in which the total energy of solid-state systems can be obtained for any geometrical configuration of the nuclei. With the advent of accurate methods to solve the local-density (LD) one-particle equations, there has been increasing interest (Janak, 1974; Zunger and Freeman, 1977; Moruzzi et al., 1978; Ferrante and Smith, 1978; Ihm et al., 1979; Wendel et al., 1979) to use these methods to determine the total energy and related properties, such as equilibrium phases, lattice constants, and force constants of both bulk solids and surfaces. For magnetism, it is often the only way to determine theoretically the magnetic structure which has the lowest energy and so is the ground state of the system.

The major difficulty in any straightforward application of the total-energy expressions involves numerical problems arising from the cancellation between the very large kinetic and potential-energy contributions (Janak, 1974). The problem obviously becomes more severe for heavier atoms since the (chemically inactive) core electrons are responsible for the largest part of the total energy. To avoid this problem, one successful approach has been to remove the core electrons from the problem as is done in the pseudopotential method (Ihm et al., 1979). Within an all-electron approach, using the muffin-tin approximation, Moruzzi et al. (1978) and Janak (1974) have obtained an algebraic cancellation of part of the core contributions in the expressions for the total energy and pressure.

Weinert et al. (1982) go beyond these treatments and consider the

total energy using an all-electron, general potential approach. A key
feature of this new approach is the high accuracy that results from an
explicit cancellation of the Coulomb singularities in the kinetic and
potential-energy terms arising from the nuclear charge. As an example
of the applicability of this method to solid-state systems, Weinert
et al. (1982) have implemented it in the full-potential linearized
augmented-plane-wave (FLAPW) method (Wimmer et al., 1981) for thin
films. In their earliest applications, results were presented for
characteristic problems: (i) the equilibrium distance in a monolayer
of covalently bonded graphite for which comparisons of the calculated
equilibrium structural properties and cohesive energy can be made
with experiment (Weinert et al, 1982), (ii) the relaxation and
reconstruction of the W(001) surface (Fu et al. 1984a, 1985b), and (iii)
surface energies of the W(001) and V(001) surfaces (Fu et al. 1985a).

1.4 Magnetism of Transition Metal Surfaces

1.4.1 Ferromagnetic Fe(001) Surface

It is natural to begin our discussion of magnetic surfaces with
the classic ferromagnetic, bcc iron which has a bulk magnetic moment
of $2.12\mu_B$ (Danan et al., 1968). Let us now consider the ferromagnetic
Fe(001) surface. The study of the magnetism of Fe surfaces has a
relatively long history, beginning with the observation of magnetic
"dead" layers (Liebermann et al., 1969; Liebermann, et al., 1970). This
early result is now believed to be due to impurities; the present
experimental and theoretical consensus is that there is a ~ 30%
increase in the magnetic moment at the clean surface. In an earlier
theoretical study of surface states, surface magnetization, and
electron spin polarization of the Fe(001) surface, Wang and Freeman
(1981) used an LCAO thin film method and found an enhancement of the
surface magnetism and strong Friedel-type oscillations in the spin
density. In this LCAO calculation a small variational basis set for the
wave functions and a superposition model of spherical charges were
used. Both computational restrictions can cloud the significance of the

results, particularly for delicate quantities such as spin densities. Therefore Ohnishi et al. (1983) undertook a re-examination of the electronic and magnetic surface properties of Fe(001) using the highly accurate FLAPW method.

As in the earlier LCAO calculation, the surface is represented by a 7-layer slab and the spin-polarized exchange-correlation potential given by von Barth and Hedin (1972) is used. The FLAPW result of a surface induced enhancement of the magnetic moment agrees with that of the earlier LCAO calculation. However, in this new calculation no significant Friedel oscillation is found. The FLAPW results for the magnetic moments going from the bulk-like center of the 7-layer film to the surface as 2.27, 2.39, 2.35 and 2.96μ_B, i.e. the moment at the surface is increased by 30% compared to the center.

A study of the density of states (DOS) decomposed into atomic (i.e. layer) and ℓ-like components gives insight into the mechanism of the surface induced enhancement of the magnetism. The DOS in the central layer of the 7- layer Fe(001) film (lower panels of Fig. 2) is very close to the DOS of bulk bcc Fe (compare e.g. with Moruzzi et al. (1978), p. 170): the DOS shows a three-peak structure, so typical for the bcc structure, with a pronounced minimum below the highest peak. The d-band for the majority spin has a small unoccupied part whereas for the minority DOS the Fermi energy falls into the characteristic minimum leaving about 30% of the minority d-band unoccupied. The resulting spin-imbalance is reflected in the large magnetic moment of bcc iron. The DOS for the surface layer (top panels of Fig. 2) is dramatically changed compared with the bulk-like DOS of the central layer. Due to the reduced symmetry and fewer number of nearest neighbors in the surface layers the characteristic three-peak bcc structure is lost and the d-band is narrowed. As a consequence, the majority d-band is now almost completely filled. For the minority DOS we observe (upper right panel in Fig. 2) a peak just at the Fermi energy with its center of gravity slightly above E_F. The states leading to this peak in the surface DOS fall in the minimum of the bulk-like DOS (lower right panel in Fig. 2) and can be identified as surface states. These surface states are also present in the majority DOS and are shifted to lower energies by an exchange-splitting of about 2 eV. The net result of the surface-induced d-band narrowing and the occurrence of surface states is a larger spin-imbalance corresponding to an enhancement of the magnetic moment in the surface.

We now consider the spin density of the Fe(001) surface which is

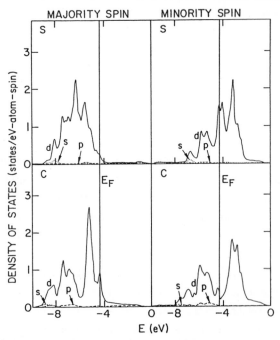

Figure 2. Atomic- and ℓ-projected densities of states for the center (lower panels) and surface (upper panels) atoms in a 7-layer Fe(001) slab. [After Ohnishi et al. (1983)]

shown for the (110) plane perpendicular to the surface in Fig. 3. The greatest part is dominated by positive spin densities with only small pockets of negative spin densities between the atoms. In the bulk-like center the shape of the spin density shows important non-spherical components of t_{2g} symmetry. Surprisingly, this bulk-like shape is found even for the iron atoms just one layer below the surface and only the surface atoms exhibit a different shape of the spin density which is markedly more spherical. The polarization on the vacuum side of the surface atoms is found to be positive.

The main conclusion that we can draw from the FLAPW calculation

(Ohnishi et **al.** 1983) on a 7-layer Fe(001) film is that there is a
large increase in the magnetic moment at the surface (2.98μ_B) compared
to the bulk (2.25μ_B). This increase is due to a band of surface states
below the Fermi energy. (The agreement between experiment (Turner et
al. 1982; Turner and Erskine, 1983) and theory for the surface states
is quite good.) Not only are "dead" layers ruled out, but a simple

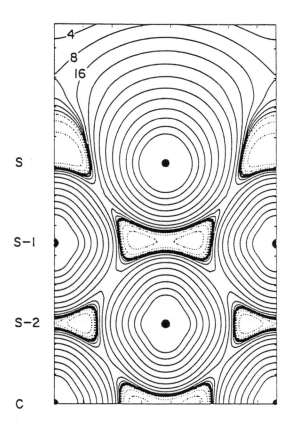

Figure 3. Spin density of a 7-layer Fe(001) slab in the (110) plane
perpendicular to the surface in units of $10^{-4}e/(a.u.)^3$. Dotted
contours indicate negative spin densities. [After Ohnishi et al.
(1983)]

picture of the magnetism of the clean surfaces emerges. In this
picture, the magnetization is a monotonic function of the

dimensionality, with the bulk at one end and the free atom at the other. This simple picture is supported by self-consistent calculations (and experiments) on bulk, surfaces, monolayers, linear chains (Weinert and Freeman, 1983a), and atoms.

Because of the large moment in Fe, it is not surprising that the magnetism also has an effect on the electronic properties. One example is the dependence of the work function ϕ (a mainly electrostatic property) on the allowed magnetic order. A paramagnetic (i.e. non-spin-polarized) film yields ϕ = 4.86 eV, whereas the spin-polarized result is 4.29 eV. (Experimental values of 4.31 (Fomenko, 1966) and ~ 4.4 eV (Turner et al. 1982; Turner and Erskine, 1983). This result is easily understood: Ferromagnetic ordering fills the more compact anti-bonding type states at the expense of bonding states, resulting in a dipole barrier (relative to the paramagnetic surface) such as to reduce ϕ. Obviously this mechanism is important only if there is a large moment, as in the present case.

In Fe, it is possible to measure the hyperfine field via the Mössbauer effect. The standard interpretation (Freeman and Watson, 1964) of the Fe data for bulk systems is that H_{hf} scales with the magnetic moment since the dominant term is the (negative) core polarization. (The bulk valence electrons generally give an additional small negative contribution.) This interpretation was used by Tyson, et al. (1981) to deduce that there was an increase of the magnetic moment at the Fe surface. If we compare the calculated (Ohnishi et al. 1983) H_{hf} for Fe(001) (see Table I), we see a different behavior. While we do have an increase of the core contribution to H_{hf} that scales with the moment (-130 kG/μ_B), the valence contribution is large and _positive_ (+140 kG), resulting in a reduction of H_{hf} at the surface (-250 kG) compared to the bulk (-360 kG).

A complicating factor in a direct comparison is that the theory is for a clean Fe surface, while the experiments (Tyson et al. 1981) were on Fe films coated with Ag to prevent oxidation. In order to understand the Mössbauer results, Ohnishi et al. (1984) performed a set of calculations on Fe(001) films covered with an ordered overlayer of Ag with which to compare the clean Fe results. We will return to a discussion of the Fe film Mössbauer measurements and their interpretation in Sec. 1.5.2 which describes the Fe/Ag(001) magnetic studies.

Table 1. Layer-by-Layer hyperfine field and magnetic moment in the seven-layer Fe(001) film. [After Ohnishi et al., 1985)]

	Moment (μ_B)	Core	Hyperfine field (kG) Conduction electron	Total
S	2.98	−398	+143	−252
S-1	2.35	−306	−89	−395
S-2	2.39	−311	−16	−320
C	2.25	−291	−75	−366

1.4.2 Ferromagnetic Ni(001) Surface

Since the report of magnetically "dead" layers on a Ni(001) surface (Liebermann et al., 1970) this system has attracted a host of experimental and theoretical investigations, partly with contradicting conclusions. It is now established, however, that the clean Ni(001) surface is not magnetically "dead". This result is based on experiments using spin-polarized field emission (Landolt and Campagna, 1977), spin-polarized photoemission (Moore and Pendry, 1978), electron capture spectroscopy (Rau, 1982), and polarized electron diffraction (Feder et al., 1983). It seems that the experimental results showing magnetically "dead" layers have been clouded by difficulties in the sample preparation and characterization, partly due to interface problems.

This wealth of experimental information provided a challenging subject for theoretical studies in terms of accurate ab-initio calculations. The first steps in this direction were made by Dempsey and Kleinman (1977) who reported on results from a parameterized, non-self- consistent Ni surface calculation. Wang and Freeman (1980) presented the first self-consistent study of a nine-layer slab of Ni(001) using the linear-combination of atomic orbitals discrete-variational-method (LCAO-DVM). They found that the surface was not magnetically dead, but that the magnetic spin moment for the surface atoms was reduced compared with the bulk-like center of the film. However, as stated above, these LCAO-DVM calculations have possible limitations in the variational freedom of the basis and the representation of the charge density. Jepsen et al. (1980; 1982) concluded from their five-layer LAPW film calculations that the surface magnetic moment is slightly increased which is in agreement with theoretical/computational studies by Freeman et al. (1982b) and Krakauer et al. (1983).

Recent highly accurate FLAPW studies on a 7-layer Ni(001) slab (Freeman et al., 1982a; Wimmer et al., 1984) shed new light onto the magnetism of the Ni(001) surface: (i) the magnetic moments in atomic-nearest neighbor volumes are found to be (from the central layer to the surface) 0.56, 0.59, 0.60, 0.68μ_B, i.e. there is an enhancement of the magnetic moment of ~20% of the surface compared to bulk; (ii) no Friedel oscillation of the moment going from the surface to the center of the 7-layer film; and (iii) the majority surface state $\bar{H}_3$ lies 0.14 eV below the Fermi energy, i.e. there is no majority-spin d-hole at $\bar{H}$. In all three points the FLAPW results disagree with the early pioneering calculations on a 9-layer Ni(001) film of Wang and Freeman (1980). Differences between the two approaches can be seen in the spin densities (compare Fig. 3 with Fig. 7 of W-F), which in the FLAPW case shows much more negative values between the atoms. The theoretical result for the work function, 5.37 eV, agrees well with the experimental value of 5.22 eV.

The decomposition of the majority and minority charges into ℓ-projected partial charges inside atomic spheres gives insight into the mechanism of the surface induced enhancement of the magnetic moments. The dominant partial charge inside the Ni-spheres has d-character. For the majority spin this d-like charge is increased for the surface atom compared with the interior of the film and the s,p-charge is decreased; in other words for the atoms with a reduced number of nearest neighbors the charges become more d-like. For the minority spin, we find a similar trend, but here also the d-charge decreases in the surface leading to an increased d-moment. The majority d-band is completely filled giving a rather small density of states at the Fermi energy of 16.9 states/Ry whereas the only partially filled minority d-band gives a high density of states at the Fermi energy of 175.8 states/Ry. The exchange-splitting varies from 0.19 eV for a state near the bottom of the d-band with s-admixture to 0.69 eV for a pure d-state about 1 eV below E_F. For the $\bar{M}_3$ surface states with the majority state just below E_F, the unoccupied minority state is found 0.78 eV higher in energy.

The spin densities for the 7-layer Ni(001) slab, shown in Fig. 4 exhibit bulk-like features for all atoms except those at the surface. We observe a localized positive spin density inside atomic spheres of a radius of about 2 a.u. and a pronounced negative spin density in the interstitial region. The spin density inside the atomic spheres includes important non-spherical components originating from wave function of t_{2_g} symmetry. The dominant component of this spin density maps out the top part of the completely filled d-band which has no occupied counterpart in the minority spin system. The shape of the spin density of the surface atoms is quite different from the bulk-like atoms in the interior of the film particularly on the vacuum side where the spin density shows an egg-like shape. The spin density in the vacuum region is slightly negative.

We consider now the spin densities at the nuclei for the 7-layer Ni(001) slab, which give the contact polarization hyperfine field. For all layers the total spin density at the nucleus is negative. The value for the surface layer is reduced by 15% compared to the center of the film. A decomposition of the contact spin density into core and valence contributions shows that the core part scales with the moment in the spheres, i.e. Wimmer et al. (1984) find an increase for the surface whereas the valence part becomes positive for the surface atoms. The ratio of the core polarization at the nucleus and the magnetic moment

28

due to the d- valence electrons is nearly constant.

The FLAPW results indicate a surface induced core level shift to smaller binding energies for the $3p_{3/2}$ states of 0.30 and 0.40 eV for majority and minority spin, respectively. For the $1s_{1/2}$ states the core level shifts are 0.35 eV to smaller binding energies for both spin directions. The surface-induced core level shifts are caused by a global electrostatic shift of the potential to smaller limiting energies. This effect also leads to the formation of the $\bar{M}_3$ surface

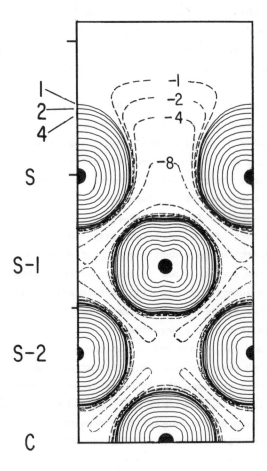

Figure 4. Spin density of a 7-layer Ni(001) slab in the (100) plane perpendicular to the surface in units of $10^{-4}e/(a.u.)^3$. [After Wimmer et al. (1984)]

states which is split off from the top of the d-band and shifted to smaller binding energies by about 0.3 eV. Thus, there is a close correspondence between surface states and core level shifts on the Ni(001) surface (Wimmer, 1984).

It is remarkable that for both Fe and Ni the reduced number of nearest neighbors on the surface leads to an enhancement of the magnetic moment. This trend becomes even more obvious and a simple picture emerges when we include linear chains and free atoms in the consideration: going from bulk to a (001)-surface, then to a linear chain (Freeman and Weinert, 1982c, Weinert and Freeman, 1983a) and finally to the free atom, the moments are 0.56, 0.68, 1.1, and 2.0 for Ni and 2.27, 2.96, 3.3, and 4.0 for Fe. Thus, as the dimensionality is decreased, the magnetic moments approach the values of the free atoms.

1.4.3 On the Possibility of Surface Magnetism V(001)

Interest in the V(001) surface has centered on the possibility of inducing ferromagnetism in its surface layer while retaining paramagnetic behavior in the bulk (Akoh and Tasaki, 1977; Allan, 1979; Grempel and Ying, 1980; and Yokoyama, et al., 1981). Until recently, despite some efforts, no quantitative measurement of the magnetic structure of V(100) has been performed on single-crystalline surfaces. Very recently, preliminary results on the electron-spin polarization measurement using electron-capture spectroscopy by Rau and Schmalzbauer (1984) reveals magnetic ordering of the V(001) surface below 540 K. However, these authors do not exclude the possibility of surface magnetism as being due to surface reconstruction. Much earlier, based on their magnetic susceptibility measurements of vanadium superfine particles, Akoh and Tosaki (1977) suggested the existence of a surface magnetic moment ($\sim 2\mu_B$/surface atom). This observation was supported by theoretical studies using either a tight-binding model (Allan, 1979) or spin-fluctuation theory (Grempel and Ying, 1980), which estimated the size of the magnetic moment as $2.4\mu_B$ per surface atom. Self-consistent spin-polarized calculations on a five-layer V(100) film by Yokoyama et al. (1981), using the LCAO-DVM method was concluded that the appearance of the magnetism on the surface layer was likely but with a magnitude which is relatively small (~ $0.2\mu_B$/surface atom).

The nonmagnetic to magnetic transition of vanadium has aroused considerable theoretical attention in the last decade. Hattox et al. (1973) employed a self-consistent augmented-plane-wave calculation to show that the transition to the ferromagnetic state in <u>bulk</u> vanadium occurred abruptly at a lattice constant which is about 1.25 times larger than the equilibrium value. This behavior has also been used to explain qualitatively the occurrence of the appearance of magnetism in certain vanadium alloys simply from the increased vanadium-vanadium distance. For the V(001) surface, Yokoyama et al. (1981) found that the magnetic moment of surface atoms increases by $0.2\mu_B$/atom in going from a 10% contraction to a 10% expansion of the topmost interlayer spacing (from $0.2\mu_B$ to $0.4\mu_B$). All of these calculations indicate that the tendency to form a magnetic moment is accompanied by the dilation of the lattice spacing. This seems in contrast to the LEED observation by Jensen et al. (1984) for the clean V(001) surface which showed a 7% contraction of the topmost interlayer spacing with respect to the bulk value. Thus, the electronic and magnetic structure of V(100) cannot be considered as complete and understood. It is therefore of some interest to examine in more detail the electronic structures and surface magnetism of V(100) from a first-principles calculation.

Ohnishi et al. (1985) presented results of an all-electron, local-density-functional (LDF) study of the V(001) surface using the total energy all-electron full-potential-linearized-augmented-plane-wave (FLAPW) method discussed above (Wimmer et al., 1981; Weinert et al., 1982). Starting with a V(001) monolayer, self-consistent calculations were carried out for both the paramagnetic and spin-polarized states of films with 3, 5, and 7 layers. Since this is such a challenging case we discuss these results in some detail.

We begin with a presentation of the paramagnetic film results. Table 2 lists their calculated results for the layer-projected DOS at E_F, the exchange-correlation integrals, and the Stoner factors for the monolayer to 7-layer films. The analysis shows that the surface Stoner factor is larger than 1 only for the monolayer system; for the multilayer systems, the paramagnetic state tends to be more stable, owing to the increased coordination number of the surface atoms. Surprisingly, the exchange-correlation integral, I, does not show enhanced values from the center layer even for atoms of a monolayer film. Its value, 0.45 eV, is in agreement with the result obtained for bulk vanadium by Janak (1977) (0.35 eV) using a self-consistent KKR calculation, and by Gunnarsson (1976) (0.4 eV) within the spin-density-functional formalism.

Table 2. Density of states, interaction parameter, I, and Stoner factors of V(001) films [After Ohnishi et al., (1985)].

V(001)		$N(E_F)$ (State/Ry)	$I(Ry)$	$N(E_F)I$
mono-layer	C	64.5	0.034	2.19
3-layer	S	27.4	0.033	0.90
	C	15.5	0.37	0.58
5-layer	S	26.1	0.033	0.58
	S-1	14.6	0.034	0.51
	C	10.1	0.031	0.31
7-layer	S	20.6	0.033	0.68
	S-1	12.5	0.033	0.41
	S-2	13.7	0.033	0.45
	C	18.4	0.031	0.57
bulk[a]		22.0	0.026	0.57

[a]Janak (1977)

Since the Stoner factor analysis indicates the inherent instability of the paramagnetic monolayer film, spin-polarized calculations were carried out to investigate the magnetic moment and the total energy difference between the paramagnetic and ferromagnetic states. For the spin-polarized calculations, an initial external magnetic field is applied to separate energetically the spin-up and spin-down states obtained in the self-consistent paramagnetic calculation. For the monolayer film, a large magnetic moment

(3.09μ_B/atom) is found and the total energy of the ferromagnetic state is 57 mRy lower than the paramagnetic one. However, for multilayer systems, the reduction of the surface DOS at E_F (as shown in Table 2) causes the paramagnetic state to be more stable. For the spin-polarized seven layer film calculations, the self-consistent results with an initial magnetic field of 1 mRy, 5mRy, or as large as 30 mRy, yield negligible induced moments on the surface; the magnetic moment is found to be less than 0.05μ_B within the MT sphere of the surface atom. The calculated total energy difference (ΔE) between the spin-polarized and paramagnetic calculations is found to be less than 0.1 mRy, out of a total energy for a seven layer film of −13263.154 Ry.

The stability of the paramagnetic multilayer V(001) films can be understood from the layer-projected partial density of states (DOS) in the surface (S) and center (C) layers MT sphere as shown in Fig. 5.

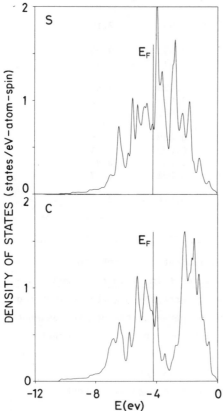

Figure 5. Density of states (per eV-atom-spin) of the surface (S) and center (C) layers but smoothed by a Gaussian broadening function of 0.05 eV full width at half-maximum. [After Ohnishi et al. (1985)]

The DOS of the center layer for a seven layer film is essentially bulk-like with well-separated bonding and antibonding regions. For the surface layer, the sharp surface state DOS peak which is characteristic of the occurrence of surface magnetism in the 3d transition metals is located 0.3 eV above E_F. As a result, the paramagnetic state is stable. For the case of paramagnetic Fe(100), this surface state band lies below E_F and contributes to the enhancement of the surface magnetism; for paramagnetic Cr(001), as we will discuss later, it crosses E_F and causes a ferromagnetic phase transition.

In contrast to the results of a susceptibility measurement on superfine vanadium particles (Akoh and Tasaki, 1977) and earlier theoretical studies (Allan, 1979; Grempel and Ying, 1980; Yokoyama, et al., 1981), no convincing evidence of surface magnetism is found in this FLAPW investigation by Ohnishi et al. (1985) for multilayer films. A comparison with these theoretical studies reveals the difference; the surface exchange-correlation integral employed in the semiempirical tight-binding model of Allan (1979) was 1 eV and 0.8 eV by Grempel and Ying (1980) in their spin-fluctuation approach. These values are a factor of 2 larger than the result (0.45 eV) obtained using the self-consistent first-principles all-electron approach (Ohnishi, 1985). Consequently, a Stoner instability is expected in the earlier calculations (Allan, 1979; Grempel and Ying, 1980) and this leads to surface ferromagnetism.

In order to be conclusive, Ohnishi et al. (1985) also examined the effect of atomic spacings near the surface was examined in more detail. This may be carried out through the FLAPW total energy approach for bulk solids (Jansen and Freeman, 1984) and surfaces (Weinert et al., 1982). Here we focus only on those results which are relevant to the possible occurrence of surface magnetism: The equilibrium bulk lattice constant as determined from a total energy minimization (5.55 a.u.) is 3% smaller than the experimental value. Now it is known that because of the relatively narrow d-bandwidth of 3d transition metals, LDF theory tends to give smaller lattice spacings as is found in the total energy studies for the bulk crystal and, as we shall see, also for the

surface. Starting from the calculated bulk lattice spacing and a
paramagnetic V(001) surface, multilayer relaxation was investigated for
the (1x1) structure by this total energy approach. This calculation
gives a 9% contraction of the topmost interlayer spacing and a 1%
expansion of the second interlayer spacing with respect to the bulk
spacings. These results are in good agreement with the LEED measurement
by Jansen et al. (1984) — a 7% contraction and 1% expansion of
the first and second interlayer spacing respectively. Thus, the
experimental value of the lattice constant used by Ohnishi et al.
(1985) in their unrelaxed calculation has already favored the formation
of the magnetic moment, since such a lattice spacing is dilated
compared with the value obtained from the LDF variational approach.
Furthermore, from the agreement of the calculated relaxation of the
interlayer spacings for the paramagnetic state with experiment, the
possible occurrence of surface magnetism in V(001) appears to be
excluded. This conclusion was further examined by means of a spin-
polarized calculation for a relaxed surface (Ohnishi et al., 1985). The
effect of the surface relaxation reduces the surface DOS at E_F;
however, the overall electronic structure of the surface states does
not show significant changes. As a result, surface magnetism is not
found in all our calculations for multilayer systems.

The interaction between oxygen and vanadium has been a long
discussed problem in the literature. This lies perhaps in the
difficult problem of preparing and maintaining clean surfaces, owing
to diffusion of bulk oxygen to the surface. Recently, it was concluded
(Jensen et al., 1984; Foord et al., 1983) that the V(100)–(5x1)
structure is not a characteristic of the clean surface, but rather is
associated with the presence of significant oxygen concentration in
the surface region. Foord et al. (1983) further proposed that this
(5x1) structure consists of an outermost layer of vanadium atoms of
(100) symmetry and a subsurface layer of adsorbate atoms with (5x1)
translation symmetry. This adsorbate layer presumably increases the
first interlayer spacing between vanadium atoms. This observation is
supported by the kinematical scattering calculation of Davies and
Lambert (1981). Our results showed that the Stoner factor of the

surface atom increases as the number of film layers decreases. The significance of this analysis implies that either (i) an increase of the first interlayer spacing as the surbace layer approaches the monolayer limit or (ii) a reduction of the coordination number of the surface atoms lead to favorable conditions for the formation of surface magnetism for V(100). Consequently, it is not surprising if the surface magnetism of V(100) were induced by subsurface oxygen or by impurity-induced surface reconstructions.

1.4.4 Surface Magnetism of Cr(001)

Intense interest exists in the Cr(001) surface because of (1) the possibility of induced ferromagnetism in its surface layer while retaining antiferromagnetic behavior in the bulk (Ferguson, 1978; Allan, 1979; Matsuo et al., 1980; Rau and Eichner, 1981; Meier et al., 1982; Foord et al., 1983; Klebanoff et al., 1984; Grempel, 1981; Hirashita et al., 1981); (2) interest in the intriguing surface electronic structure of the (001) surface for the Group VIB transition metals in general and possible phenomena associated with the high density-of-states which arises from localized surface-state bands near the Fermi level in particular. For W(001) (Fu et al., 1985b) and Mo(001) the surface state coupling plays a dominant role in the structural phase transition to the c(2x2) structure, whereas for Cr(001) the relatively narrow 3d surface-state band favors a large enhancement of the local spin susceptibility. For these reasons, one might expect ferromagnetic ordering in the surface layer of Cr(001).

The existence of a ferromagnetic surface layer in antiferromagnetic chromium was first reported with the observation of surface magnetoplasma waves via an attenuated total reflection method (Ferguson, 1978) on polycrystalline chromium films. It has also been reported (Matsuo et al., 1980) that in small particles of chromium magnetic ordering persists well above the Néel temperature at 310 K. Recently, an electron capture spectroscopy study by Rau and Eichner (1981) further confirmed the existence of ferromagnetic order below 365 K, but with a c(2x2) reconstructed surface. Meier et al. (1982) found no indication of ferromagnetism in spin-polarized photoemission studies from a chromium surface which, as they noted, included coverage of a nitrogen monolayer on top of the surface. It was later confirmed by LEED analysis (Foord et al., 1983) that the c(2x2)- Cr(001) structure is related to surface impurities such as C,N,O, or S, and that a clean surface exhibits a p(1x1) LEED pattern at room temperature. Very recently, a surface magnetic phase transition was observed on Cr(001) using angle-resolved photoelectron spectroscopy (ARPES) by Klebanoff et al. (1984); from the temperature dependence of the splitting of the surface resonance state, the transition temperature was found to be 780 K, and the surface magnetic moment was estimated to be 2.4μ_B. These findings are supported by the studies of Zajac et al. (1985) on the epitaxy and electronic structure of Cr overlayers deposited on single-crystal Au(001).

Recent theoretical studies on a 7-layer Cr(001) slab (Fu and Freeman, 1985d) using the FLAPW method confirms a large enhancement of the magnetization at the surface. The calculated spin density for the 7- layer Cr(001) film, shown in Fig. 6, exhibits antiferromagnetic coupling between adjacent layers in the (001) direction. The plot reveals highly anisotropic components of the spin density around the atomic sites with an essentially bulk-like feature for the S-2 and C layers. The eruption of the spin density into the vacuum region is entirely positive and indicates a large enhancement of the magnetic moment at the surface. The induced ferromagnetic state has layer projected magnetic moments within each MT sphere from the center to the surface layer given as -0.89, 0.89, 1.29, and 2.49μ_B. This surface induced magnetic moment of 2.49μ_B is comparable with earlier theoretical results obtained from the tight-binding method (2.8μ_B) by Allan (1979), and a DVM-LCAO calculation (2.56μ_B) by Hirashita et al. (1981). In addition, the calculated spin polarization shows large penetration of the enhanced surface magnetization into the bulk. (For bulk Cr, the measured maximum magnetic moment is 0.59μ_B (Shirane and Takei, 1962).)

Consider now the spin densities at the nuclei for the 7-layer Cr(001) slab, which gives the contact polarization hyperfine field

(H_{cf}). The total hyperfine field can be conveniently decomposed into two components for transition metals (Freeman and Watson, 1965): (1) a negative-polarization of the core s electrons due to the d-moment; (2) the contribution from valence s-like electrons. The core part, as is now expected (Freeman and Watson, 1965) even for metals, scales very

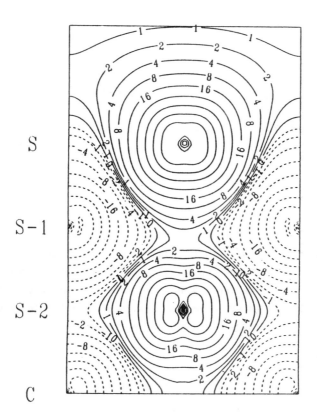

S

S-1

S-2

C

Figure 6. Spin density map on the (110) plane for the seven-layer Cr(001) film in units of $1 \times 10^{-3} e/(a.u.)^3$. Subsequent contour lines differ by a factor of 2. [After Fu and Freeman (1985d)]

precisely with the moment within each MT sphere (cf. Table 3). An increase in magnitude of H_{cf} is found for the surface atom and is predominantly due to the large core contributions (from the magnetic moment enhancement) at the surface. The predicted value, −73 kGauss, for the surface atoms should be experimentally measureable.

Table 3. Electronic spin densities at the nuclei in a seven-layer Cr(001) film in units of $e/(a.u.)^3$. The last column shows the ratio of the core spin density at the nucleus and the magnetic moment in the corresponding MT sphere (in KG)/unpaired spin). [After Fu and Freeman (1985d)].

| | Hyperfine field (KG) | | | |
	core	valence	total	H_{cf}(core)/M
S	−328	256	−72	131
S−1	163	−140	23	126
S−2	−113	75	−38	127
C	113	−115	−2	127

Similar to the case of Fe(001) (Ohnishi et al., 1984a), the surface magnetization reduces the work function by 0.35 eV from that of the paramagnetic state (from 4.40 eV to 4.05 eV). While this theoretical value is 0.4 eV lower than the experimental value obtained by Wilson and Miller (1983) (4.46 eV) it is closer to the photoemission result by Meier et al. (1982) (4.10±0.05 eV). Although both experiments claimed to stand for the work function of the clean Cr(001) surface, the agreement with the latter may be fortuitous owing to the presence of one monolayer of nitrogen and zero spin-polarization on their p(1x1) surface. On the other hand, it should be noted that the often quoted experimental value (4.46 eV) (Wilson and Miller, 1983) refers to a surface contaminated by a submonolayer of impurities (O,S, and N) with a c(2x2) reconstructed structure. Thus, these experimental values cannot be regarded as the work function of a clean Cr(001) surface.

For a 7-layer paramagnetic Cr(001) film, a dramatic change of the

surface LDOS from its bulk counterpart can be seen in Fig. 7. The DOS
at E_F (3 states/eV-atom) suggests ferromagnetic ordering on the
surface, since a Stoner instability can be easily achieved even with
an exchange integral assumed equal to the bulk value, 0.67 eV. [As
discussed earlier for V(001), this sharp surface LDOS peak which is
characteristic of the occurrence of surface magnetism in the 3d
transition metals is located 0.3 eV above E_F in V(001). As a result, the
paramagnetic state is more stable than the ferromagnetic state for
V(001)]. In Fig. 8, the ℓ-decomposed partial LDOS in the surface layer
MT sphere, shown for the ferromagnetic state, reveals a large spin
density imbalance and gives rise to an enhancement of the local atomic
moment to 2.49μ_B per surface atom. In agreement with earlier estimates
(Klebanoff et al., 1984; Allan, 1979), our calculated ferromagnetic
surface exchange splitting is found to be 1.8 eV.

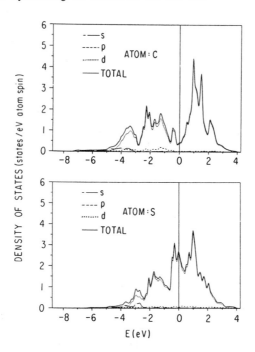

Figure 7. Layer projected and ℓ-decomposed partial density-of-states
in units of states/eV-atom in the paramagnetic state of a seven-layer
Cr(001) film for the surface (S) and center (C) layer. [After Fu and
Freeman (1985d)]

Normal emission ARPES spectra at 300 K by Klebanoff et al. (1984)
display two surface related features with binding energy (BE) = 0.16 eV
and 0.75 eV. Both of these features were identified as due to surface
states. HeI photoemission studies by Zajac et al. (1985) of Cr
overlayers deposited on single-crystal Au(001) also display similar
features: a broad peak with a binding energy of 1 eV and a shoulder
near 0.1 eV. Indeed, in our calculation a majority spin LDOS peak with
BE = 1 eV is clearly exhibited in Fig. 8, and is attributed to the
surface state of $\bar{\Gamma}_5$ symmetry. In addition, a LDOS peak near (slightly
above) E_F for majority spin is also seen in Fig. 8; the occupied
portion of this feature may correspond to the shoulder (0.1 eV BE) in
the photoemission spectra ($\bar{\Gamma}_1$ symmetry).

Since the localized surface states play an important role in the

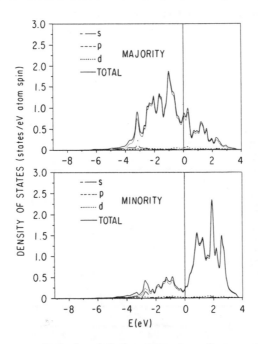

Figure 8. Layer projected and ℓ-decomposed surface partial density-of-
states in the units of states/eV-spin-atom for majority spin, and
minority spin of a seven layer Cr(001) film. [After Fu and Freeman
(1985d)]

enhancement of surface magnetism for 3d transition metals, much attention has been on the possibility of inducing two-dimensional magnetism in a controlled way (such as overlayers, sandwiches, and modulated structures). This will be the subject of Section 1.7.

1.4.5 Induced Magnetism and Knight Shift of Pt(001)

So far, we have discussed ferromagnetic surfaces, Fe(001), Ni(001), and Cr(001), for which we found a surface-induced enhancement of the magnetic moment and investigated the possibility of surface magnetism for V(001). We now consider a case of a paramagnetic transition metal surface and study the response to an external magnetic field. Pt is an interesting candidate, since recent ^{195}Pt NMR results (Yu et al., 1980; Yu and Halperin, 1981; Stokes et al., 1981) on very small particles (d~29-300A) indicate that due to the large surface to volume ratio their electronic and magnetic properties are quite different from bulk Pt metal. The experiments suggest that there is a Knight shift distribution related to the surface and a shift towards positive values.

Although there is no net magnetization in the ground state of Pt, the magnetic properties can be probed by NMR techniques; in particular, one quantity that can be extracted is the Knight shift. The Knight shift K is defined (Slichter, 1980) as the relative shift in the NMR frequency, ω_0 of a nucleus (with gyromagnetic ratio γ) in a dc magnetic field H_0 due to the polarization of the conduction electrons in a metal:

$$\omega_0 = \gamma(1 + K) H_0$$

where the frequency of the bare nucleus is just γH_0. The main contributions to the night shift arise from the orbital paramagnetism and from the Fermi contact terms due to the valence electrons and core polarization.

Recently, several NMR studies on superfine Pt particles have been reported with emphasis on the surface properties (Yu and Halperin, 1981; Rhodes et al., 1982). The NMR spin-echo field gradient results of Yu and Halperin (1981) showed that the characteristic surface region has a thickness of 1.5±0.5 lattice constants independent of the size of the particles. The experiments by Rhodes et al., (1982) on particles with diameters in the 10-100A range showed a distribution of the NMR signal ranging from the bulk signal (K = -3.4%) to a rather broad peak at K ≃ 0% that can be identified with the clean surface.

The Pt films studied by Wang et al. (1983) and Weinert and Freeman (1983b) are non-magnetic in the absence of an external applied field. From studies of the bulk, Pt is found to be a Stoner-type paramagnet. In this model, the magnetization density in small fields is proportional to the paramagnetic density at the Fermi level, $\rho(\underline{r}, \epsilon_F)$. This density (Weinert and Freeman, 1983b) shows a decreased magnetization at the surface compared to the bulk. As would be expected from the density of states (DOS), the magnetization density is mainly d-like. These results have implications for the hyperfine fields which we now discuss.

The relativistically correct Fermi contact term, which is generally the largest contribution to the hyperfine energy, is given by (Fermi, 1930)

$$\Delta E_{hf} = \frac{8\pi}{3} \quad m(\underline{r}=0) \langle \underline{\mu}_e \cdot \underline{\mu}_N \rangle \tag{24}$$

where $\underline{\mu}_e$ ($\underline{\mu}_N$) is the magnetic dipole moment of an electron (nucleus) and $m(\underline{r}=0)$ is the magnetization density at the nucleus. Since the energy of the nuclear moment in an external field, $\underline{H}_{ext}$, is $-\underline{\mu}_N \cdot \underline{H}_{ext}$, it is standard to define a (non-relativistic) effective hyperfine field strength, H_{hf} by

$$H_{hf} = \frac{8\pi}{3} \mu_B m(\underline{r}=0) \tag{25}$$

where μ_B is the Bohr magneton; with $m(\underline{r}=0)$ given in a.u., then H_{hf} is found (in kG) from the conversion 1 a.u. = 524 kG. In general there are two contributions to the contact term for transition metals:(1) the large negative polarization of the core electrons due to the d moment and (2) a polarization of the valence s electrons. From calculations on transition metals, it has been found that the core

polarization per unpaired spin, $H_{hf}(d)$, is roughly a constant
regardless of the local environment (Freeman and Watson, 1964). Hence,
since there is a decreased magnetization at the surface, there should
be a decrease in magnitude of the negative core polarization field. In
the simple Stoner-like picture for a paramagnet such as Pt, the
valence contribution is proportional to the s-DOS at the Fermi level,
which increases at the surface relative to bulk. Combining these two
effects, we expect that the contact hyperfine field at the surface is
more positive than in the bulk. The Knight shift is then related to
the hyperfine energy (or non-relativistically to the hyperfine field)
by

$$K = \frac{E_{hf}(H_{ext})}{\Delta E_N(H_{ext})} = \frac{H_{hf}}{H_{ext}} \qquad (26)$$

Hence, we expect that the change in the Knight shift between the bulk
and the surface, $\Delta K = K(surface)-K(bulk)$, is positive. These results
give a qualitative understanding of the NMR results; in order to get
quantitative results, the magnetic field must be included explicitly,
thereby allowing the charge and spin densities to readjust self-
consistently beyond linear response.

Weinert and Freeman (1983b) carried out self-consistent
calculations for a number of external magnetic fields in the range of
0.1-2.0 (1.0 Ry = 2.35 x 10^{6}G) using both the standard non-relativistic
exchange-correlation potential and the relativistically corrected one.
Even for these high fields, the magnetization scaled approximately with
the field. To obtain values of K in quantitative agreement with
experiment for the bulk, the relativistic corrections are necessary
(without them K is too large by a factor of 3-4). In Table 4 we present
their results for the relativistic (spin-only) valence and core
contributions to K for each layer. The general trends expected from the
Stoner-type model are borne out by the spin-polarized calculations,
i.e. $\Delta K > 0$. If we add the experimental estimate for the orbital
contribution (Clogston et al, 1964; Shaham et al., 1978) of +0.4% to
our spin-only result of -4.1%, we obtain a "bulk" Knight shift of
-3.7%, in very good agreement with the experimental value of -3.4%. By

contrast, K at the surface is approximately zero when the positive orbital and dipolar contributions at the surface are added to our spin-only results. As stated earlier, the core-polarization field per unpaired spin $H_{hf}(d)$ is approximately constant for different local environments. From an analysis of the experimental data (Clogston et al., 1964; Shaham et al., 1978), a value of 1.2×10^6 Gauss for the core polarization per unpaired spin is obtained; our calculated value is 1.1×10^6 Gauss.

Table 4. Calculated Knight shift contact contributions (in %) by layer for a five layer Pt(001) film. [After Weinert and Freeman (1983b)].

	Central	S-1	Surface
valence	1.3	1.3	2.4
core	-5.4	-3.9	-3.0
total	-4.1	-2.6	-0.6
Bulk experiment	-3.4		

From these theoretical results, we clearly see that the resulting near zero value of K at the surface is due to a cancellation of contributions. This prediction is subject to verification by considering the Korringa relationship between the spin-relaxation time T_1 and the sum of the squares of the individual contributions to K: We would expect that T_1 should be characteristic of a metallic environment. This interpretation (Weinert and Freeman, 1983b) of the Knight shift data is strongly supported by the very recent experiments of van der Klink et al. (1984) on small Pt particles.

1.5   Magnetism at Bimetallic Interfaces

1.5.1 Magnetism at the Ni/Cu Interface

One of the best studied interfaces between magnetic and non-

magnetic metals is that of the Ni/Cu system. Because of the good match
of their lattice constants [a(Ni) = 3.526A and a(Cu) = 3.615A]
overlayers and interfaces are readily accessible experimentally.
However, there are still open questions about the magnetism of Ni
overlayers on a Cu surface. Interface phenomena also play a key role in
layered coherent modulated structures (CMS) (Hilliard, 1979), a
promising new kind of synthetic material for which the Ni/Cu system is
the prototype.

Wang et al. (1981a,b; 1982) and Freeman et al. (1982a) have
carried out the first ab-initio determination of the electronic
structure and magnetism of Ni overlayers on Cu(001). They used the
linearized augmented plane wave (LAPW) thin film method to obtain
accurate self-consistent spin-polarized semi-relativistic energy band
solutions for Ni overlayers on a Cu(001) substrate, consisting of a 4-
layer Cu(001) slab plus one or two p(1x1) layers on Ni on either
side, referred to as Ni/Cu and 2Ni/Cu, respectively.

The spatial distribution of the spin density in both cases shows
that the magnetization is localized in the Ni layers. As seen in Fig.
9 the magnetization is essentially zero on the Cu layers. The vacuum
and interstitial region are slightly polarized in the opposite
direction, similar to that reported for the clean Ni(001) film shown in
Fig. 9 An examination of the layer-by-layer magnetic moments
(contributed to by the electrons inside the touching muffin-tin
spheres), indicates that the magnetic moment of the surface Ni layer of
the 2Ni/Cu film increases by about 10% to $0.69_{\mu B}$ compared to the "bulk"
value (Freeman et al., 1982a). The moment of the interface layer Ni(I)
of the 2Ni/Cu film decreases by 24%, and the Ni layer of the Ni/Cu
system decreases by 37% (to $0.39_{\mu B}$) compared to the calculated bulk
value. From an orbital angular momentum decomposition, we find that
the contribution to the moments arises almost completely from the

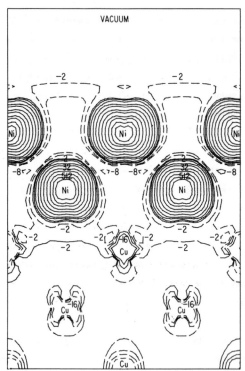

Figure 9. Spin density of 2 layers of Ni on both sides of a Cu(001) 5-layer film. The units are $10^{-4}e/(a.u.)^3$. [After Wang et al. (1982)]

d-like component, similar to that in the clean Ni film.

An analysis of the theoretical results leads to two observations which relate to the problem of surface magnetism. First, the surface and interface affect the total number of electrons so that in general the surface atoms have fewer electrons and the Ni atoms in contact with the Cu substrate have more electrons due to charge transfer from Cu to Ni. Secondly, the change in the total number of electrons arises almost completely from the minority spin electrons, and this leads to the decrease of the moment with an increase of the total of number of electrons.

The influence of charge transfer and the change of bonding on the position of the 3d majority and minority spin bands are not independent; their splitting will change through the exchange/correlation. However, the exchange splitting (ΔE) is not the same for different states. For the Ni/Cu film, the exchange splitting is 0.39 and 0.35 eV and $\overline{\Gamma}_4$ and $\overline{\Gamma}_5$, 0.28 and 0.37 eV for $\overline{M}_1$ and $\overline{M}_3$, and 0.38 eV for $\overline{X}_2$. The moment of the various layers is found to be proportional to the exchange splitting. However, the ratio $\Delta E/\mu$, the so-called Stoner-Hubbard parameter, remains almost unchanged, and also close to the value for the isolated monolayer and bulk Ni.

One of the interesting results of this analysis is the correlation of the magnetic moment with the number of p-like electrons. At the surface this dehybridization of the s, p, and d electrons acts to increase the magnetic moment and is related to the d-band narrowing seen there. This dehybridization is also related to the simultaneous removal of p-electrons from the muffin-tin (MT) spheres (where the d-electrons are mainly localized) as they spill out into the vacuum region. It is not surprising that the number of p-electrons in the MT region is correlated with the degree of dehybridization; in the free Ni atom, the p-orbitals are completely unoccupied. There is a remarkable correlation between the total s and p charge of both spins, $q_s + q_p$ (μ_s and μ_p are essentially zero) and the magnetic moment. Since q_s is relatively unchanging ($\sim\pm0.03$ electrons) this is, indeed, a correlation with q_p. In the unsupported Ni monolayers, where the electrostatic shift mechanism is absent, this dehybridization accounts for the large increase of μ_d compared to the bulk value. In all cases, the total charge in each vacuum region (there are two per slab) is equal (to within 0.01 electrons) to the loss of p-electrons from the MT spheres.

In the 2Ni/Cu slabs, other effects are seen; charge transfer (about 0.1 electrons) into the d-bands of the interfacial Ni layer reduces the magnetic moment to $0.48\mu_B$ in this layer. Since this atom still has a coordination number of twelve, the dehybridization should not take place, as is also seen. By contrast, the surface Ni atoms of the 2Ni/Cu slab show some dehybridization, and could also have an upward electrostatic shift. Indeed, this Ni(S) atom does show a $3p_{3/2}$ core-level shift of 0.26 eV to reduced binding energy relative to the Ni(I) atom. Both these effects are consistent with the increased moment of $0.69\mu_B$.

Finally, it is interesting to compare these results with those for

NiCu modulated structures obtained by Jarlborg and Freeman (1980).
Their model uses a unit cell for the modulated CuNi structure along
[100], and contains, in all, 8 layers with no local distortion
considered. The result for 1 or 2 Ni layers (Ni_1Cu_7 and Ni_2Cu_6) show a
strong quenching of the moment, but no "dead" layers. More Ni layers
(Ni_3Cu_5 and Ni_4Cu_4) also give some reduction, but the variation of
moments between different layers is negligible. Thus, they find
reduced magnetic moments on Ni when it is layered with Cu but no
"magnetically dead" Ni layers when Ni is deposited onto a Cu(100)
substrate.

1.5.2 Magnetism of the Fe(001) Surface Overlayered with Ag

We described and discussed in Sec. 1.4.1, the surface magnetism of
the Fe(001) surface and the calculated hyperfine fields. As mentioned
there, the Mössbauer experiments of Tyson et al. (1981) were performed
on Fe films coated with Ag in order to prevent oxidation. Ohnishi et
al. (1984) studied the Ag/Fe(001) system in order to make direct
comparison with experiment. Figure 10 compares their calculated spin
densities, ρ_S, for the Fe surface and the Ag/Fe film. The spin density
of the clean surface is highly anisotropic and has regions of both
positive and negative spin density, consistent with that obtained by
neutron scattering (Shull and Mook, 1966) for bulk Fe. The spin density
in the vacuum region is almost entirely positive and indicates a large
increase in the magnetic moment at the surface. For the Ag-covered
surface, the spin density for the layers below the Ag/Fe interface is
the same as for the clean surface. However, at the interface, the spin
density is strongly modified even though the charge density is already
rather bulk-like. In contrast to the clean surface, the spin density in
the vacuum region is now mainly negative due to the exchange
polarization of the Ag sp-electrons towards the unpaired spins of the
Fe. The magnetic moment of the Fe at the interface has a moment of
$2.5\mu_B$ (cf. Table 5), which is still a substantial (~ 10%) increase
compared to bulk and in excellent agreement with recent direct
magnetization measurements (Bayreuther and Lugert, 1983) of Fe(100)-

Ag(111) interfaces.

Fe(001) Ag/Fe(001)

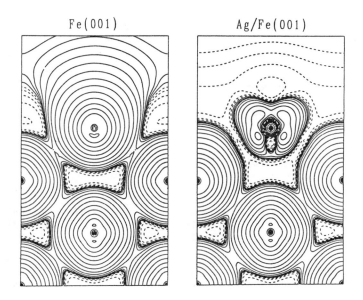

Figure 10. Total spin densities for Fe(001) and Ag/Fe(001). Solid (dotted) lines mark the zero and positive (negative) 2^n ($n = 0,1,2...$) contours in units of 10^{-4} electrons/a.u.^{3}. [After Ohnishi et al. (1984)]

Now let us return to H_{hf}. For atoms below the surface, the indirect polarization (Anderson and Clogston, 1961) should be dominant for the valence electrons, yielding a negative contribution. Then the standard interpretation relating the measured H_{hf} to the magnetization should be valid. At the surface, however, the s density is more nearly atomic-like and hence we have a positive valence contribution to H_{hf} since the direct polarization in an atom dominates. In the sense that the bands narrow at the surface and that the magnetic moments are more atomic-like, this is a consistent picture. Moreoever, we now see that the standard interpretation of the data is not valid at the surface.

Table 5. Magnetic moments (in μB) in muffin-tin spheres and the total conduction, and core contributions to the hyperfine field (in kG) nor each layer of the clean and Ag-covered 7-layer Fe(001) films. (For the Ag/Fe film, the surface atom is Ag) [after Ohnishi et al. (1984)].

| | Fe(001) | | Ag/Fe(001) | |
	moment	H_{hf}	moment	H_{hf}
Surface	2.98	−252	0.08	−587
		+143		−553
		−395		− 34
S−1	2.35	−395	2.52	−335
		− 89		− 7
		−306		−328
S−2	2.39	−327	2.37	−339
		− 16		− 31
		−311		−308
Center	2.25	−366	2.27	−359
		− 75		− 64
		−291		−295

This interpretation is born out by the calculated H_{hf} for both the clean and Ag-covered surfaces (see Table 5). The Fe H_{hf} at the interface acts much like that of the bulk atom, i.e. the core contribution scales with the moment and the valence contribution is small and negative, and hence the results of Tyson et al. (1981) are consistent with an increased Fe moment at the Ag interface. These results not only resolve the apparent contradiction between theory and experiment, but also demonstrate that surfaces and interfaces are different, and individually interesting, systems.

1.6 Effects of Adsorbates on Magnetic Surfaces: p(1x1)H/Ni(001)

The catalytic properties of Ni in hydrogenation reactions make the study of the H/Ni system not merely of academic interest, but also of practical importance. From a fundamental viewpoint, the H/Ni system is of interest since it is one of the simplest systems in which to study chemisorption on a magnetic substrate. The electronic and magnetic structure of this system was studied by Weinert and Davenport (1985). The surface was modeled by a seven layer Ni(001) film with an ordered p(1x1) monolayer of H on each side in the four-fold hollow. (The p(1x1)

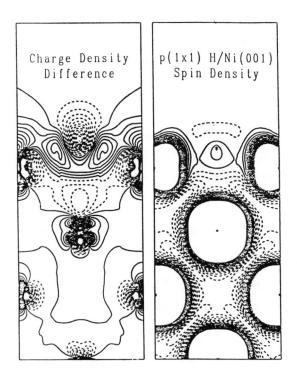

Figure 11. Charge density different between H/Ni(001) and the superposition of Ni(001) + atomic H (contour spacing of 10^{-3} electrons/a.u.^{3}) and the H/Ni(001) spin density (contour spacing 10^{-4}a.u.^{3}). Negative contours are dotted and the position of the H is marked by *. [After Weinert and Davenport (1985)]

is the only ordered overlayer observed in He diffraction (Rieder and Wilsch, 1983).) The H-Ni bondlength was varied between 1.78A and 1.85A and the equilibrium properties were determined both from the total energy (Weinert et al., 1982) and the force (Weinert, unpublished). The calculated bondlength, d_{Ni-H}, is 1.80A (0.69 a.u. above the outermost Ni plane) and is shorter than that of bulk nickel hydridge (1.86A). [As is typical of LDA calculations (Weinert et al., 1982), the cohesive energy of 3.2 eV/H atom (including zero point motion) is overbound by ~0.5 eV compared to experiment (Christmann, et al., 1974).] The calculated value of the bondlength is supported by their calculated vibrational frequency of 82 meV, which is in good agreement with the experimental value (Andersson, 1978), of 74 meV, and the theoretical work of Upton and Goddard (1979) and Umrigar and Wilkins (1985). (The slightly shorter d_{Ni-H} found in these calculations is consistent with their neglect of magnetic ordering.)

In order to display the polarization of the density that occurs due to H adsorption, the difference between the H/Ni(001) density and the superposition of the densities appropriate to Ni(001) and atomic H is given in Fig. 11The polarization of the density, as expected, is localized to the surface region and shows the importance of correctly including the non-spherical contributions to the density and potential. There is a clear enhancement of the charge along the Ni-H bond indicating substantial covalency. This bonding is the essential reason for the reduction in the magnetic moment at the surface-namely the formation of a Ni-H bond pairs electrons.

The spin density is strongly reduced at the surface, yielding a moment of ~$0.2_{\mu B}$/Ni atom, while below the surface it rapidly approaches bulk-like behavior. The spin density around the H site (marked by a *) is highly non-spherical with regions of both positive and negative densities which integrate to a near zero value about the H. At the H nucleus itself, the spin density is positive-- unlike the negative contact term found for H in the bulk (Jepsen et al., 1980)--because of the reduced symmetry at the surface.

From the reduced moment at the surface and the changes in the electronic structure at the surface, one would expect that those single particle states localized in the surface region should be modified. The reason that the magnetic surface state (Plummer and Eberhardt, 1979) at $\bar{M}$, which is composed of $d_{x^2-y^2}$ orbitals (in the coordinate system of the calculation), is unaffected by H is simply that this state and the H level belong to different irreducible representations. A simple way to see this is to note that the H sits between the lobes of this state. That the exchange splitting is not reduced is due to the localized nature of the planar $d_{x^2-y^2}$ orbitals of this state and its position near E_F: The calculated exchange integral of this state is so much larger than that of bulk Ni (Janak, 1977) that the Stoner criterion for the surface is satisfied (Weinert and Davenport, unpublished). This implies that the surface will remain ferromagnetic, albeit with a much reduced moment.,

Through direct interaction, on the other hand, the H induces a localized non-exchange split state below the Ni bonds. The relative position of this band below the Ni continuum increases with decreasing d_{Ni-H}, as expected for the bonding combination of Ni and H states; the corresponding anti-bonding combination is clearly seen at $\overline{M}$ 4.5 eV above E_F. At $\overline{\Gamma}$ the H induced states are a mixture of H and Ni s, while at $\overline{M}$ they are H s – Ni $3d_{xy}$ hybrids, clearly showing that the bonding with H involves both Ni s and d orbitals.

The increased moment on the clean Ni surface can be viewed (Weinert and Davenport, 1985) as a consequence of the band narrowing and an upward electrostatic shift which occurs to maintain charge neutrality, resulting in a decrease of the minority population. (Without a shift, the minority and total charge would increase.) For the H-covered surface; the bands still narrow, but because of N_i-H hybridization, charge neutrality is maintained without an electrostatic shift or charge transfer. Hence, the decreased magnetic moment can be thought of as a minority band-filling effect. However, this is not the same as the simple picture in which the extra band resulting from the H-Ni interaction appears above E_F and the extra electron just fills the top of the Ni d-band which results in no moment at the surface. Instead there are significant changes to the band structure and density of states in the d-bands on H adsorption. This mechanism is supported by the calculated 3s initial state core level shifts: there is no minority shift while the majority shift, as always for a strong ferromagnet, is determined by the exchange splitting. This latter shift of 0.5 eV and the hybridization of the H

s with the filled majority Ni d-band act to decrease the majority d
character inside the surface Ni sphere by 0.13 electrons. Combined
with an increase of 0.25 minority d electrons due to band narrowing,
this results in a decrease in the moment to $\sim 0.2_{\mu B}$. These results
(Weinert and Davenport, 1985) give a rather simple physical picture of
the reduction of the surface moment and should apply to other
adsorbates on Ni as well.

All of these results apply to zero temperature. At finite
temperatures, the surface moment is expected (Weinert and Davenport,
1985) to decrease more rapidly than the bulk. At room temperature, H
adsorbed on a Ni surface would reduce the surface Ni moment to ~ 0.1 of
the T=0 bulk value, giving the appearance of a magnetic "dead" layer.
These results support the suggestion (Gradmann, 1977) that the
original observation (Liebermann, et al., 1970) of "dead" layers was
due to H contamination.

1.7 Two-Dimensional Magnetism of Metallic Overlayers, Interfaces and Superlattices

Since the recognition that localized surface and interface states
offers the possibility of inducing new and exotic phenomena and so
holds out the promise of new device applications, a key question and
goal has been the possibility of inducing 2D magnetism in a controlled
way. The availability of recently developed sophisticated synthesis and
characterization techniques makes observation of such phenomena in
laboratory-made structures such as overlayers, epitaxial sandwiches and
modulated structures (superlattices) a challenging possibility.
Concurrently, as described earlier, all-electron theoretical approaches
have been developed which give highly precise descriptions of the
electronic, magnetic and structural properties of thin films and bulk
solids. The lively interplay of theory and experiment has already
yielded interesting results. Thus, for example, enhanced magnetism at
Fe (Ohnishi et al., 1984), Ni (Wimmer et al., 1984), and Cr (Fu and
Freeman, 1985d) surfaces has been predicted and observed. As regards
interface magnetism in say bimetallic systems, however, it is generally

believed and (so far) observed that in the absence of large negative pressure effects, such as arise from lattice parameter mismatches (Brodsky and Freeman, 1980), a non-magnetic metal would diminish the magnetism, if any, of the other metal.

Recently, Fu et al. (1985c) have carried out extensive FLAPW investigations which predicted enhanced 2D magnetism in a number of transition metal-noble metal systems. Consider first the Cr_m-Au_n system for which a large number of configurations have been investigated for different numbers of atomic layers, m and n. In each case, when the plane of Au(001) is rotated by 45° with respect to that of Cr(001) there is an exact matching of their primitive 2D square nets, because the experimental lattice constant of 2.88 A used for Cr is exactly a factor of $\sqrt{2}$ smaller than that of Au (4.08A); the stacking has atoms in the four fold hollow site of adjacent atomic planes. Now, since the work of Brodsky et al. (1982) has raised the possibility of the formation of fcc Cr when sandwiched by Au, Fu et al. (1985c) first carried out FLAPW total energy (Weinert et al., 1982) determinations of the Cr-Cr, Cr-Au and Au-Au interlayer spacings. They found that: (i) just one layer away from the Cr/Au interface, the Au-fcc structure and the Cr-bcc structure is retained; (ii) the Cr-Au interlayer spacing is close to the average of the bulk bcc Cr and fcc Au spacings – a result long suspected by materials scientists.

Selected results for the magnetic moments calculated within each atomic sphere at the Au/Cr interface are presented in Table 6. As an aid in understanding the underlying physics and to emphasize its 2D nature, we focus first on results for the experimentally unattainable free monolayer Cr(001) film. A large magnetic moment of $4.12\mu_B$ is obtained which is close to the atomic limit and substantially larger than that for bulk antiferromagnetic Cr metal ($0.59\mu_B$). This is simply related to the band narrowing of the d-band due to the reduced coordination numbers and lower symmetry for the monolayer.

Surprisingly, when a monolayer of Cr(001) is deposited onto Au(001), the magnetic moment of the Cr overlayer decreases by only a small amount from the free mono-layer to $3.70\mu_B$. This extremely large moment – the largest value reported for a transition metal other than for Mn - is 50% greater than that of the surface layer in Cr(001) predicted theoretically (2.49μ_B) (Fu and Freeman, 1985d) or derived experimentally (2.4μ_B) (Klebanoff et al., 1984). Also surprising is the finding that this enormously enhanced moment is only moderately reduced (to 3.10μ_B) when the Cr overlayer is itself covered by an Au layer. A very similar result was obtained by Oguchi and Freeman (1985), 2.95μ_B per Cr atom, when the (1x1) Cr/Au coherent modulated structure (superlattice) was studied using the LMTO approach (Andersen, 1975). Since Cr is a notorious getter, the retention of this enhanced 2D magnetization in either the single sandwich or superlattice structures might make its observation much easier. Further, additional theoretical studies show that the Cr magnetic moment is hardly affected by the increase of Au thickness; more layers of Au tend to reduce the antiferromagnetic coupling between the ferromagnetic monolayer sheets in the superlattice geometry (Oguchi and Freeman, 1985). Thus, there is the added degree of freedom of variable wavelength of modulation - and magnetic coupling - which may be important for technical applications including magnetic recording. Significantly, there is also a sizeable magnetic moment induced onto neighboring Au atoms (0.10 - 0.14μ_B) indicating a large magnetic perturbation of the Au induced by the magnetic Cr atoms. As in the case of the Cr atoms, the polarization of the Au atoms is mainly from the d-like electrons; the contribution from the s- and p-like electrons is only 1 - 2%. Only one layer away from the interface, the electronic structure of Au is essentially unperturbed and the bulk structure and properties of Au are retained.

To gain a better understanding of the mechanism behind the giant magnetic moment formation on the Cr sites, Fig. 12 shows the layer density-of-states (LDOS) within the MT sphere of Cr (in the paramagnetic state) for the Cr(001) monolayer (Fig. 12a), the Cr/Au(001) overlayer (Fig. 12b), and the Au/Cr/Au(001) sandwich (Fig. 12c). It is seen, as the number of Au atoms which are nearest-neighbors to a Cr atom increases, that the basic feature of the sharp LDOS peaks about E_F remains similar in all cases, except for some transfer of the LDOS to states lying at lower energy due to the enhanced hybridization between the Au and Cr d-bands. It should be noted, however, that the center of the Au d-band lies too far below E_F (about 4.5 eV) to have an effective interaction with Cr-localized surface (or interface) states in the vicinity of E_F. A calculation by Feibelman and Hamann

Table 6. Theoretical layer-by-layer magnetic moments (in μ_B) for specified cases (with estimated uncertainties $\pm$ 0.03μ_B). S and (S-n) indicate surface and sub-surface layers. The last column shows the spin-polarization energy (in eV). CMS denotes coherent modulated structures. [After Fu et. al. (1985c)].

	Cr	Nearest Au	E(para.) -E(spin-pol.)
Cr monolayer	4.12		1.69
1 Cr/Au(001) overlayer	3.70	0.14	0.78
2 Cr/Au(001)	2.90 (S) -2.30 (S-2)	-0.08	0.60
Au/Cr/Au(001) sandwich	3.10	0.14 (S) 0.13 (S-2)	0.38
(1x1) Au/Cr CMS	2.95	0.10	0.25
	Fe	Nearest Cu,Ag,Au	
Fe monolayer	3.20		1.34
1 Fe/Cu(001)	2.85	0.04	0.70
1 Fe/Ag(001)	2.96	0	1.14
2 Fe/Ag(001)	2.94 (S) 2.63 (S-1)	0.05	1.15
Ag/Fe/Ag	2.80	0	0.88
Au/Fe/Au	2.92	0.08	0.97
	V	Nearest Ag,Au	
1 V/Au(001)	1.75	0.04	0.10
1 V/Ag(001)	1.98	0.06	0.14
2 V/Ag(001)	1.15 (S) <0.05 (S-1)	0	0.08
Ag/V/Ag	0	0	

(1985) has also confirmed that the interface states of Cr/Au at E_F retain the same feature of the surface states for clean Cr(001). Moreover, the broadening effect of the LDOS peak due to the s-d hybridization is weakened by the very low LDOS of Au at E_F (~0.25/eV). For these reasons, the Cr overlayer on Au(001) or the Cr monolayer

sandwiched by Au essentially retains the 2D electronic properties of a
monolayer Cr(001) film. As a result of this electronic isolation, the
high LDOS at E_F leads to a Stoner instability and drives the Cr layer
to be ferromagnetic.

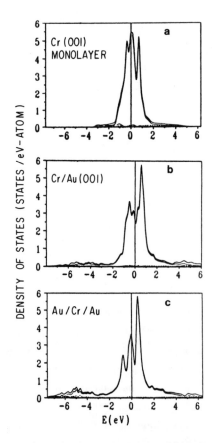

Figure 12. Layer-projected partial density of states in units of
states/eV-atom in the paramagnetic state of Cr(001) monolayer, Cr
monolayer on Au(001) [Cr/Au(001)], and sandwich [Au/Cr/Au(001)].

The magnetic passivity of the Au and the independence of the Cr
magnetic moment with increasing non-magnetic Au layer thickness is in
remarkable contrast with the effects of adding additional magnetic Cr
layers; the magnetic moment of these single-Cr-layer structures is
substantially decreased with increasing number of Cr layers. For

example (Fu et al., 1985c): (i) with two layers of Cr deposited on
Au(001), the magnetic moment of the surface and subsurface Cr atoms
are $2.9\mu_B$ and $-2.3\mu_B$, respectively; (ii) with three layers of Cr
sandwiched on each side by two Au layers, the magnetic moments of the
interface and sub-interface Cr atoms are reduced to $1.75\mu_B$ and
$-1.07\mu_B$; (iii) with five layers of Cr sandwiched on each side by two Au
layers, these sizeable moments are further reduced to $1.55\mu_B$, $0.77\mu_B$
and $0.67\mu_B$ for interface, sub-interface, and center Cr atoms
respectively. (Note that the moment in the center layer is now quite
close to that of bulk Cr.)

Similar investigations (Fu et al., 1985c) have also been performed
for monolayers of Fe and V adsorbed on Ag(001). For Fe there is a close
matching of the lattice constants but a 5% mismatch for V. As in the
case of Cr/Au(001), large magnetic moments, $2.96\mu_B$ (Fe) and $2.0\mu_B$ (V),
are found for the adsorbate monolayers. The magnetic moment of the Fe
overlayer on Ag(001) is remarkably close to the theoretical magnetic
moment (Ohnishi et al., 1984a) of the surface layer of an Fe(001) film
($2.98\mu_B$) – again indicating the lack of interaction with the substrate.

The result for V/Ag(001) is much more surprising since, like the
bulk, the surface layer of V(001) is not magnetic (Ohnishi et al.,
1985). We thus have the remarkable prediction that an overlayer of V on
Ag(001) is magnetically ordered with a sizeable magnetic moment ($2.0\mu_B$)
which is almost as large as the moment of Fe in bulk Fe. [If confirmed,
this will be the first solid material for which elemental vanadium
demonstrates magnetic ordering.] The origin of magnetic ordering is not
negative pressure since, in fact, the matching to the Ag(001) substrate
results from a reduced lattice constant for V by 5%.

For both the Fe and V overlayers, it is found – despite some
hybridization between the d-bands of the adsorbate and substrate – that
the adsorbate localized surface state bands retain their quasi-2D
behavior for these systems. This behavior is further demonstrated for a
monolayer of Fe sandwiched by Ag(001) – the magnetic moment of Fe is
only slightly decreased from that of the overlayer of Fe/Ag(001) to
$2.80\mu_B$. (Similarly, a moment of $2.92\mu_B$ on the Fe site is found for an
Au/Fe/Au(001) sandwich.) However, when a monolayer of V(001) is

sandwiched by Ag(001), the vanadium layer becomes paramagnetic. Apparently, the hybridization with Ag further reduces the DOS of V near E_F and results in a Stoner factor which is just less than one (the exchange correlation parameter for V is 0.4 eV (Ohnishi et al., 1985)).

The sensitivity of the magnetic ordering of V to its metallic environment is further illustrated by calculations with two layers of V on Ag(001) (Fu et al., 1985c); in this case the interface layer atoms remain essentially in the paramagnetic state (zero moment) and the surface layer has a moment of $1.15\mu_B$ which is substantially reduced from that of the single overlayer value ($2.0\mu_B$). This calculation leads to the result that an Ag substrate is more amenable for the magnetism of the V monolayer than is another V layer.

In order to further illustrate that the noble metal substrate is more amenable for the magnetism of a 3d metal monolayer sheet than is another magnetic material, Fu and Freeman (1985e) have investigated the electronic and magnetic properties of Cr/Fe(100) and Fe/Cr(100), as well as combinations of Cr and Fe layers on metal substrates (i.e. Fe/Cr/Ag(100) and Cr/Fe/Au(100)). As shown in Table 7, the Cr and Fe layers tend to couple antiferromagnetically in the ⟨100⟩ direction; further, although there is an enhancement of the Cr moment at the Cr/Fe interface (to $3.10\mu_B$ accompanied by a decrease of the Fe moment) from that of the clean Cr(001) surface, it is $0.6\mu_B$ less than the magnetic moment of Cr on Au(100). They have also carried out calculations to examine the effect of a nonmagnetic simple-metal substrate on the spin-polarization of a Cr monolayer (i.e. Cr/Al(100)). Although a larger d-sp hybridization is realized in this case, no evidence of a magnetic proximity effect is found; the magnetic moment of the Cr monolayer is largely retained ($3.10\mu_B$). Again, these analyses indicate that the isolation of the interface states of the 3d transition metal is a key factor in allowing the formation of "giant" magnetic moments.

There have been some experimental and other theoretical investigations of the magnetic properties of overlayers on non-magnetic substrates. For example: (1) experimental studies on the epitaxity and electronic structures of Cr on the Au(100) (Zajac et al., 1985) indicate the persistence of Cr surface states (which are

Table 7. Theoretical magnetic moments for the transition metal-
transition metal interfaces (see Table 6 for notation).
[After Fu and Freeman (1985e)]

	moment (μ_B)	
Cr/Fe(100)	3.10 (S),	−1.98 (S−1)
	−2.53 (S−2),	−2.35 (S−3)
Fe/Cr(100)	2.40 (S),	−0.83 (S−1)
	0.80 (S−2),	−0.85 (S−3)
Cr/Fe/Au(100)	3.10 (Cr),	−1.96 (Fe)
Fe/Cr/Ag(100)	2.30 (Fe),	−2.40 (Cr)

magnetic) at the interface; (2) photoemission spectra of Fe on Ag(100)
(Binns et al., 1985) show that the monolayer film is ferromagnetic
with a similar local moment as that of bulk Fe; and (3) extensive
studies (Meservey et al., 1981; Moodera and Meservey, 1984; Bergmann,
1983) of magnetic films (e.g., Fe, Co, Gd, etc.) in contact with
simple or noble metal films find no magnetic "dead layers" or give no
evidence of a proximity effect. On the theoretical side, a very
similar result for the magnetization of Fe/Ag(100) was obtained by
Richter et al. (1985) using the local orbital method; however, a
considerably large magnetic moment $(3.6\mu_B)$ for a Cr monolayer on
Fe(100) was found by Victoria and Falicov (1985) using a parameterized
tight-binding scheme.

Finally, we must emphasize that these predictions of strongly
enhanced 2D magnetism, thus far on only a fraction of various
combinations possible, are part of a growing literature which
indicates the range and wealth of magnetic phenomena available for
experimental investigation by a variety of techniques. It also
indicates the new predictive role of all-electron local spin density
theory.

62

Acknowledgements

We are grateful to our colleagues and collaborators (H.J.F. Jansen, H. Krakauer, T. Oguchi, D.S. Wang, and E. Wimmer) who participated in the work reported here. A critical reading of the manuscript by J.-K. Lee is acknowledge with thanks. The excellent work of J.E. Bosak on this manuscript is greatly appreciated. The research reported here was supported by the U.S. National Science Foundation, Office of Naval Research, and Department of Energy.

References

Akoh H and Tasaki A 1977 J. Phys. Soc. Jpn. **4** 791
Allan G 1979 Phys. Rev. B **19** 1447
Andersen D G 1965 J. Assoc. Comp. Math **12** 547
Andersen O K 1975 Phys. Rev. B **12** 3060
Anderson J R, Papaconstantopoulos D A, Boyer L L and Schirber J E 1979
     Phys. Rev. B **20** 3172
Anderson P W and Clogston A M 1961 Bull. Am. Phys. Soc. **6** 124
Andersson S 1978 Chem. Phys. Lett. **55** 185
Appelbaum J A and Hamman D R 1976 Rev. Mod. Phys. **48** 478
Arlinghaus F J, Gay J G and Smith J R 1980a Phys. Rev. B **21** 2055
Arlinghaus F J, Gay J G and Smith J R 1980b Phys. Rev. B **21** 2201
Arlinghaus F J, Gay J G and Smith J R 1979 Phys. Rev. Lett. **42** 332
von Barth J and Hedin L 1972 J. Phys. **C5** 1629
Bayreuther G and Lugert G 1983 J. Mag. Magn. Matls. **35** 50
Bergmann G 1978 Phys. Rev. Lett. **41** 264
Bergmann G 1983 J. Mag. Magn. Matls. **35** 68
Binns C, Stephenson P C, Norris C, Smith G C, Padmore H A, Williams
     G P and Barthes-Labrousse M -G 1985 Surf. Sci. **152** 237
Bjorken J D and Drell S D 1964 in "Relativistic Quantum Mechanics"
     (McGraw-Hill, New York)
Bjorken J D and Drell S D 1965 in "Relativistic Quantum Fields"
     (McGraw-Hill, New York)
Brodsky M B, Marikar P, Friddle R J, Singer L and Sowers C H 1982
     Solid State Commun. **42** 675
Brodsky M B and Freeman A J 1980 Phys. Rev. Lett. **45** 133
Busch G, Campagna M and Siegmann H C 1971 Phys. Rev. B**4** 746
Celotta R J, Pierce D T, Wang G C, Bader S D and Felcher G P 1979
     Phys. Rev. Lett. **43** 728
Christmann K, Schober O, Ertl G and Neumann M 1974 J. Chem. Phys.
     60 4528
Clogston A M, Jaccarino V and Yafet Y 1964 Phys. Rev. **134** A640
Connolly J W D 1967 Phys. Rev. **159** 415
Danan H, Heer A and Meyer A J P 1968 J. Appl. Phys. **39** 669
Davis L C and Feldkamp L A 1979 J. Appl. Phys. **50** 1944
Davies P W and Lambert R M 1981 Surf. Sci. **107** 391
Dempsey D G, Kleinman L and Caruthers E 1975 Phys. Rev. B **12** 2932
Dempsey D G, Kleinman L and Caruthers E 1976a Phys. Rev. B **13** 1489
Dempsey D G, Kleinman L and Caruthers E 1976b Phys. Rev. B **14** 279
Dempsey D G and Kleinman L 1977 Phys. Rev. Lett. **39** 1297
Dempsey D G and Kleinman L 1978 Phys. Rev. B **18** 1270
Eastman D E, Himpsel F J and Knapp J A 1978 Phys. Rev. Lett. **40** 1514
Eastman D E, Himpsel F J and Knapp J A 1980 Phys. Rev. Lett. **44** 95
Eberhardt W and Plummer E W 1980 Phys. Rev. B **21** 3245
Eberhardt W, Plummer E W, Horn K and Erskine J 1980 Phys. Rev. Lett.
     45 273
Edwards D M 1980 J. Mag. Magn. Matls. **15-18** 262
Eib W and Alvarado S F 1976 Phys. Rev. Lett. **37** 444
Eichner S, Rau C and Sizmann R 1977 J. Mag. Magn. Matls. **6** 204
Erskine J L 1980 Phys. Rev. Lett. **45** 1446
Feder R 1981 J. Phys. C **14** 2049
Feder R, Alvarado S F, Tamura E and Kisker E 1983 Surf. Sci. **127** 83

Feibelman P and Hamann D R 1980 Phys. Rev. B 21 1385
Feldkamp L A and Davis L C 1979 Phys. Rev. Lett. 43 151
Ferguson P E 1978 J. Appl. Phys. 49 2203
Fermi E 1930 Z. Physik 60 320
Ferrante J and Smith J R 1978 Phys. Rev. B 19 3911
Fomenko V S 1966 in "Handbook of Thermionic Properties" ed E.G.V.
        Samsanow (Plenum, New York), p. 20
Foord J S, Reed A P C and Lambert R M 1983 Surf. Sci. 129 79
Freeman A J and Watson R E 1964 in "Magnetism" Vol. IIA, eds.
        G.T. Rado and H. Suhl (Academic, New York) p. 167
Freeman A J, Wang D-S and Krakauer H 1982a J. Appl. Phys. 53 1997
Freeman A J, Krakauer H, Ohnishi S, Wang D-S, Weinert M and
        Wimmer E 1982b J. Phys. (Paris) 43 C7-167
Freeman A J and Weinert M 1982c Bull. Amer. Phys. Soc. 27 180
Fu C L, Ohnishi S, Wimmer E and Freeman A J 1984 Phys. Rev.
        Lett. 53 675
Fu C L, Ohnishi S, Jansen H J F and Freeman A J 1985a Phys. Rev.
        B 31 1168
Fu C L, Freeman A J, Wimmer E and Weinert M 1985b Phys. Rev. Lett.
        54 2261
Fu C L, Freeman A J and T Oguchi 1985c Phys. Rev. Lett. 54 2700
Fu C L and Freeman A J 1985d Phys Rev. B (to be published)
Fu C L and Freeman A J 1985e J. Mag. Magn. Matls. (to be published)
Gradman U 1977 J. Mag. Magn. Matls. 6 173
Grempel D R and Ying S C 1980 Phys. Rev. Lett. 45 1018
Grempel D R 1981 Phys. Rev. B 24 3928
Gudat W, Kisker E, Kuhlmann E and Campagna M 1980 Phys. Rev. B 22 3282
Gunnarson O, Lundqvist B I and Lundqvist S 1972 Solid State Commun.
        11 149
Hasegawa H 1979 J. Phys. Soc. Jpn 46 1504
Hattox T M, Conklin, Jr. J B, Slater J C and Trickey S B 1973
        J. Phys. Chem. Solids 34 1627
Hilliard J E 1979 in "Modulated Structures - 1979" eds. J.M. Cowley
        et al. AIP Conference Proceedings No. 532 (American Institute
        of Physics, New York)
Himpsel F J, Knapp J A and Eastman D E 1979 Phys. Rev. B 19 2919
Himpsel F and Fausten Th. 1982 Phys. Rev. B 26 2679
Hirashita N, Yokoyama G, Kambara T and Gondaira K I 1981 J.
        Phys. F 11 2371
Hohenberg P and Kohn W 1964 Phys. Rev. 136 B864
Hubbard J 1979 Phys. Rev. B 19 2626
Ihm J, Zunger A and Cohen M L 1979 J. Phys. C12 4409
Janak J F 1974 Phys. Rev. B 9 3985
Janak J F 1977 Phys. Rev B 16 355
Jansen H J F and Freeman A J 1984 Phys. Rev. B 30 561
Jarlborg T and Freeman A J 1980 Phys. Rev. Lett. 45 653
Jepsen O, Madsen J and Anderson O I 1978 Phys. Rev. B 18 605
Jepsen O, Madsen J and Andersen O K 1980 J. Mag. Magn. Matls.
        15-18 867
Jepsen O, Madsen J and Andersen O K 1982 Phys. Rev. B 26 2790
Jepsen O, Nieminen R M and Madsen J 1980 Solid State Commun.
        34 575
Jensen V, Andersen J N, Nielsen H B and Adams D L 1984 Surf.

Sci. 116 66

Kisker E, Gudat W, Campagna M, Kuhlmann E, Hopster H and Moore
    I D 1979 Phys. Rev. Lett. 43 966

Kisker E, Gudat W, Kuhlmann E, Clauberg R and Campagna M 1980
    Phys. Rev. Lett. 45 2053

Klebanoff L E, Robey S W, Liu G and Shirley D A 1984 Phys.
    Rev B 30 1048

Kleinman L 1979 Phys. Rev. B 19 1295

Kleinman L and Mednick K 1981 Phys. Rev. B 24 6880

Kleinman L 1981 Comments Solid State Phys. 10 29

van der Klink J J, Buttet J and Graetzel M 1984 Phys. Rev. B
    29 6352

Kohn W and Sham L J 1965 Phys. Rev. 140 A1133

Korenman V and Prange R E 1980 Phys. Rev. Lett. 44 1291

Krakauer H, Posternak M and Freeman A J 1979 Phys. Rev. B
    19 1706

Krakauer H, Freeman A J and Wimmer E 1983 Phys. Rev. B 28 610

Landolt M and Campagna M 1977 Phys. Rev. Lett. 38 663

Liebermann L N, Fredkin D R and Shore H B 1969 Phys. Rev. 22 539

Liebermann L N, Clinton J, Edwards P M and Mathon J 1970 Phys.
    Rev. Lett 25 232

Liebsch A 1979 Phys. Rev. Lett. 43 1431

Liebsch A 1981 Phys. Rev. B 23 5203

MacDonald A H and Vosko S H 1979 J. Phys. C12 2977

Matsuo S and Nishida I 1980 J. Phys. Soc. Jpn. 49 1005

Meier F, Pescia D and Schriber T 1982 Phys. Rev. Lett. 48 645

Mermin N D 1965 Phys. Rev. 137 A1441

Meservey R, Tedrow P M and Kalvey V R 1980 Solid State Commun.
    36 969

Meservey R, Tedron P W and Kalvey V R 1981 J. Appl. Phys. 52 1617

Moodera J S and Meservey R 1984 Phys. Rev B 29 2943

Moore I P and Pendry J B 1978 J. Phys. C11 4615

Moruzzi V L, Janak J F and Williams A R 1978 in "Calculated Electronic
    Properties of Metals (Pergamon, New York)

Ohnishi S, Freeman A J and Weinert M 1983 Phys. Rev. B 28 6741

Ohnishi S, Weinert M and Freeman A J 1984 Phys. Rev. B 30 36

Ohnishi S, Fu C L and Freeman A J 1985 J. Mag. Magn. Matls. 50 161

Oguchi T and Freeman A J J. Mag. Magn. Matls. (to be published)

Paraskevopoulos D, Meservey R and Tedrow P W 1977 Phys. Rev. B
    16 4907

Penn D R 1979 Phys. Rev. Lett. 42 921

Pierce D T and Siegmann H C 1974 Phys. Rev. B 9 4035

Plummer E W and Eberhardt W 1979 Phys. Rev. B 20 1444

Prange E and Korenman V 1979 Phys. Rev. B 19 4691

Rajagopal A K and Callaway J 1973 Phys. Rev. B 7 1912

Rajagopal A K 1978 J. Phys. C11 L943

Rajagopal A K 1980 in "Advances in Chemical Physics" Vol. 41,
    eds. I. Prigogine and S.A. Rice (Wiley, New York) p. 59;
    and references therein

Rau C 1980 Comments Solid State Phys. 9 177

Rau C and Eichner S 1981 Phys. Rev. Lett. 47 939

Rau C 1982 J. Mag. Magn. Matls. 30 141

Rau C and Schmalzbauer (unpublished)

Rhodes H E, Wang P K, Stokes H T, Slichter C P and Sinfelt J H
　　1982 Phys. Rev. B 26 3559
Richter R Gay J G and Smith J R 1985 Phys. Rev. Lett. 54 2704
Rieder K H and Wilsch H 1983 Surf. Sci. 131 245
Sato M and Hirakawa K 1975 J. Phys. Soc. Jpn. 39 1467
Shaham M, El-Hanany U and Zamir D 1978 Phys. Rev. B 17 3513
Shirane G and Takei W J 1962 J. Phys. Soc. Jpn. 17 Suppl. BIII, 35
Shull C G and Mock H A 1966 Phys. Rev. Lett. 16 184
Smith R J, Anderson J, Hermanson J and Lapayre G L 1977 Solid
　　State Commun. 21 459
Stokes H T, Rhodes H E, Wang P, Slichter C P and Sinfelt J H 1981
　　in "Nuclear and Electron Resonance Spectroscopies Applied
　　to Material Sciences" (Elsevier North-Holland)
Treglia G, Ducastelle F and Spanjaard D 1980a Phys. Rev. B 21 3729
Treglia G, Ducastelle F and Spanjaard D 1980b Phys. Rev. B 22 6472
Turner A M, Cheng Y J and Erskine J L 1982 Phys. Rev. Lett. 48 348
Turner A M and Erskine J L 1983 Phys. Rev. B 28 5628
Tyson J, Owens A H Walker J C and Bayreuther G 1981 J. Appl. Phys.
　　52 2487
Umrigar C and Wilkins J W 1985 Phys. Rev Lett. 54 1551
Upton T H and Goddard W A 1979 Phys. Rev. Lett. 42 472
Victora R H and Falicov L M 1985 Phys. Rev. B 31 7335
Wang C S and Callaway J 1977 Phys. Rev. B 15 298
Wang C S and Freeman A J 1980 Phys. Rev. B 21 4585
Wang C S and Freeman A J 1981 Phys. Rev. B 2 43644
Wang D-S, Freeman A J and Krakauer H 1981a Phys. Rev. B 24 1126
Wang D-S, Freeman A J and Krakauer H 1981b J. Appl. Phys. 52 2502
Wang D-S, Freeman A J and Krakauer H 1982 Phys. Rev. B 26 1340
Wang D-S, Freeman A J and Weinert M 1983 J. Mag. Magn. Matls.
　　31-34 891
Weinert M (unpublished)
Weinert M 1981 J. Math. Phys. 22 2433
Weinert M, Wimmer E and Freeman A J 1982 Phys. Rev. B 26 4571
Weinert M and Freeman A J 1983a J. Mag. Magn. Matls. 38 23
Weinert M and Freeman A J 1983b Phys. Rev. B28 6262
Weinert M and Davenport J W 1985 Phys. Rev. Lett. 54 1547
Weinert M and Davenport J W (unpublished)
Wendel H and Martin R M 1979 Phys. Rev. B 29 5251
Wilson R J and Miller Jr. A J 1980 Surf. Sci. 128 70
Wimmer E, Krakauer H, Weinert M and Freeman A J 1981 Phys. Rev.
　　B 24 864
Wimmer E 1984 J. Phys. C 17 L365
Wimmer E, Freeman A J and Krakauer H 1984 Phys. Rev. B 30 3113
Woodruff D P, Smith N V, Johnson P D and Roger W A 1982 Phys.
　　Rev. B 26 2943
Yokoyama G, Hirashita H, Oguchi T, Kambaya T and Gondaira K I
　　1981 J. Phys. F 11 1643
Yu I, Gibson A A V, Hunt E R and Halperin W P 1980b Phys. Rev.
　　Lett. 44 348
Yu I and Halperin W P 1981 J. Low. Temp. Phys. 45 189
Zajac G, Bader S D and Friddle R J 1985 Phys. Rev. B 31 4947
Zunger A and Freeman A J 1977 Phys. Rev. B 15 4716

Chapter 2

Ferromagnetism of Transition Metals at Finite Temperatures

H. Capellmann

Institut Laue-Langevin, Grenoble, France

and

Institut für Theoretische Physik, TH Aachen, Aachen, FRG

2.1 Introduction

In this article a brief overview of our present understanding of transition metal magnetism will be given. The main emphasis will be on finite temperature properties, a topic of which the knowledge is not yet as far developed as that of ground state properties. Nevertheless a qualitative understanding has evolved within the last decade of bulk properties and the basic theoretical ideas will be presented. The question of how the surface influences and changes the finite temperature bulk properties is less clear at this time from a theoretical point of view, more research on this topic is necessary to achieve an understanding.

In the transition metals Cr, Mn, Fe, Co, Ni the electrons responsible for the magnetic properties in the ferromagnetically (Fe, Co, Ni) or antiferromagnetically (Cr, Mn) ordered phases participate in the Fermi-surface. These electrons are itinerant, their wavefunctions are delocalized, therefore no truely localized magnetic moments exist. The delocalized nature of the wavefunctions is due to the possibility of the electrons to move around rather freely in the crystal (free hopping processes). It is this feature, which distinguishes itinerant magnetism form the magnetism due to localized moments, such as in the insulating transition metal oxides and in rare earth metals. In the latter systems well localized electronic wavefunctions not participating in the Fermi surface give rise to localized magnetic moments. The magnetism then can be discussed concentrating on the magnetic degrees of freedom alone, governed by some magnetic Hamiltonian (like a Heisenberg Hamiltonian).

This separation of the magnetic degrees of freedom from the translational degrees of freedom is not
possible in itinerant systems. The relevant d-electrons are delocalized: they can move freely through
the crystal. The kinetic energy associated with this
electron movement is characterized by the bandwidth W,
which is of the order of 5 eV. This energy is much
larger than the thermal energy $k_B T_c$ leading to a phase
transition from the magnetically ordered to the paramagnetic state. Therefore it is not possible to separate magnetic degrees of freedom from the translational degrees of freedom. For a discussion of the magnetic properties, one has to deal with the full many-
body problem. This is the reason why it took so long
until even a qualitative understanding of itinerant
magnetism evolved.

The ground state of the ferromagnetic metals Fe,
Co, Ni is characterized by a bandstructure, in which
the usual spin degeneracy of simple metals has been
lifted: Electrons described by the same wavevector
but different spin quantum numbers up or down have different single particle energies, the difference

$$\Delta_k = \varepsilon_{k\uparrow} - \varepsilon_{k\downarrow} \tag{1}$$

being the exchange splitting. Current state of the art
bandstructure calculations, Wang and Callaway (1977), are able

to account for ground state properties quite accurately, the average exchange splittings obtained for Fe, Co and Ni are of the order of 1.5 eV, 1 eV, and 0.6 eV, respectively. The methods, based on spin density functional theory, can also be applied sucessfully to describe the antiferromagnetic ground states of Cr and Mn, although for computational reasons a number of simplifications are imposed: For Cr a commensurate spin structure and for Mn a simplified crystal structure is used in the bandstructure calculations, Kübler (1980).

Although the ground state of the elements on the right of the 3d series is magnetically ordered with, taking iron as an example an average magnetic moment of 2.2 μ_B per Fe atom, no localized magnetic moments exist. Let us illustrate this important difference using a simple one band model with an assumed ferromagnetic ground state: The number of occupied spin up levels $N_\uparrow$ is larger than the corresponding number of occupied spin down levels $N_\downarrow$:

$$N_s = \sum_k F(\varepsilon_{ks}),$$ \hfill (2)

f is the T = 0 Fermi function.

Let us now make a very fast Gedanken experiment measuring the spin density at some arbitrary lattice site. The meaning of "fast experiment" is that the measuring time Δt is small compared to hW^{-1} where h is the Planck's constant and W the bandwidth

$$\Delta t < {}^h/_W.$$ \hfill (3)

This means that the energy transfer in the experiment must be large compared to the bandwidth. The result of such an experiment is characterized by the probabilities

$$w_s = \frac{1}{N} \sum_k f(\varepsilon_{ks}) \qquad (4)$$

where N is the total number of sites.

The probability to measure a spin up, exactly one electron with spin up occupying the site, is $w_\uparrow (1-w_\downarrow)$, the probability to find exactly one electron with spin down being $w_\downarrow (1-w_\uparrow)$. The probability of finding a total spin zero is given by the sum of the probabilities for the event of finding two electrons, which is $w_\uparrow w_\downarrow$ and for the event of finding no electrons at all, which is $(1-w_\uparrow)(1-w_\downarrow)$. Electron-electron correlation effects will acutally suppress the probability of finding two electrons simultaneously, but this does not change the conclusions of the discussion and shall be neglected here.

The average over the probabilities for these four different events leads to the average moment $<\vec{\mu}_i>$ per atom in the ground state. One has to keep in mind, however, that the amplitude of the magnetic moment is not constant in the ground state but has very fast quantum fluctuations. This is due to the fact that the electrons are itinerant and have wavefunctions which are phase coherent over large distances, therefore the electron density ρ_i and as a consequence the spin density $\vec{s}_i$ as well are not described by sharp quantum numbers even in the ground state, because the operators ρ_i and $\vec{s}_i$ do not commute with the Hamiltonian H

$$[\rho_i, H] \neq 0 \; ; \quad [\vec{s}_i^2, H] \neq 0. \qquad (5)$$

The expression "fast quantum fluctuations" is used to indicate that these fluctuations will show up directly in fast experiment with a short time constant (and hence large energy transfer). This typical "quantum fluctuation time"

t_q is of order 10^{-15} for the 3d transition metals

$$t_q \approx h/W \approx 10^{-15} s; \qquad (6)$$

t_q is the typical time for a 3d electron to move from one lattice site to another. The existence of this typical time constant is characteristic for itinerant systems and has no equivalence in localized magnetism. This will turn out to be of considerable importance for the behaviour at finite temperatures and the unusual properties of the paramagnetic phase.

In systems having well localized magnetic moment the amplitude of the moments at sites i are constants of motion

$$[\vec{S}_i^2 , H_{loc}] = 0 \qquad (7)$$

and the resulting spin dynamics are quite different from that of itinerant systems.

In the latter a "slow" measurement of the spin density with a typical time constant larger than t_q (which means typical energy transfers much smaller than W) will average over the quantum fluctuations and will directly measure the average moment per atom.

The spin flip excitation spectrum at low temperatures consists of single particle excitations and collective excitations. The single particle excitations, which correspond to transitions from a spin up to a spin down band, form a continuum (the Stoner continuum). Excitations with small or vanishing energy transfer ω are in general possible for finite momentum transfer q only; single particle excitations with zero momentum transfer (q = 0) cost a finite amount of energy ($\omega \neq 0$), the exchange splitting Δ. The fact that no single particle excitations for small q and ω

exist, leads to well defined collective excitations in that region: spin waves. For the ferromagnetic transition metals the dispersion is quadratic at small q:

$$E_{sw}(q) = Cq^2 \tag{8}$$

C being the spin wave stiffness constant. For large q the spin waves enter the single particle continuum and become overdamped not existing as well defined excitations any more.

An average spin wave energy E_{sw} is of the order of 50 meV

$$E_{sw} \sim 50 \text{ meV}, \tag{9}$$

which is two orders of magnitude smaller than the bandwidth. The typical time constant t_{sw} associated with this energy

$$t_{sw} \approx \frac{h}{E_{sw}} \approx 10^{-13} \text{s}. \tag{10}$$

is much slower than the fast t_q. This clear separation of timescales makes possible a visualization of the spin waves in the following way.

Averaging over fast quantum fluctuation (10^{-15}s) leads to average moments per atom. These quantum averaged moments now themselves may fluctuate slowly in time if the system is in some excited configuration. The spin wave configuration corresponds to a slow wave like precession (10^{-13}s). In the following the expression "atomic moment" or average "moment per atom" will be used, and the quotation marks are to attract the readers' attention to the fact that these objects have a definite meaning only after averaging over the very fast quantum fluctuations.

2.2 Finite temperature properties; what drives the phase transition?

In itinerant systems the possibilities for a phase transition from a magnetically ordered state at low temperatures to a paramagnetic state at high temperatures are more diverse than in localized systems. Several different driving mechanisms for the phase transition have been proposed in the past.

A) The exchange splitting Δ might be strongly temperature dependent and vanish at T_c leading to a usual paramagnetic metal above T_c. This mechanism is the basic assupmption of Stoner theory, Wohlfarth (1953). It implies that the amplitude of the "atomic moments" decreases strongly with temperature and vanishes at T_c. This might be detected in an experiment with a typical time constant slightly faster than the (slow) magnetic fluctuation time t_{sw} (10^{-13}s for Fe) $\Delta t \lesssim t_{sw}$. This is much slower than the quantum fluctuation time (10^{-15})s) and the experiment would therefore average over those fluctuations, but fast enough to observe slow magnetic fluctuations on the scale t_{sw}. If the exchange splitting is strongly temperature dependent this intermediate timescale experiment should see the amplitude of the "atomic moment" decrease with T and vanish at T_c , "slow" magnetic fluctuations of significant intensity above T_c are absent according to Stoner theory. These predictions of Stoner theory are contrary to the true situation in transition metals, as can be seen most directly in paramagnetic neutron scattering experiments, Déportes et al (1981); Brown et al (1982); Brown et al (1982, 1983). The physical reason why this mechanism is not relevant in transition metals is the high cost in energy to make the exchange splitting Δ vanish. Typically at the magnetic transition temperature the thermal energy $k_B T_c$ is an order of magnitude smaller than Δ itself. To make Δ vanish, however, one needs thermal energies of the order of the exchange splitting. If the transition temperature is calculated using Stoner theory, the result obtained is typically too large by a factor of 5 to 10, which among other failures, indicates, that Stoner theory does not contain the essential ingredients for the phase transition.

B) A second mechanism, invoked as a possible driving force for the phase transition is based on amplitude fluctuations, Moriya and Kawabata (1973; Moriya (1979,1982); Murata and Doniach (1972. In this picure an average "atomic moment" still persists above T_c, but it is argued that the amplitude should show very strong fluctuations even on the long time scale of $t_{sw} \approx 10^{-13}$s, and that these slow amplitude fluctuations should destroy long range magnetic order above T_c.

To avoid confusion, I want to give the definition of what is meant by amplitude of the "atomic moment". It is important to realized that the "atomic moment" is an object already averaged over the very fast quantum fluctuations. Let us take a typical pure quantum state $|\kappa\rangle$ of the many body system, i.e. some vector in Hilbert space, $|\kappa\rangle$ is supposed to have a finite probability of being occupied at the temperature of interest i.e. it contributes with finite weight to the statistical averages. This will be the case if its energy per site is low enough. The amplitude $|\vec{m}_i^\kappa|$ of the "atomic moments" in the configuration $|\kappa\rangle$ (the word "configuration" will be used for a pure quantum state) is defined as the amplitude of the following quantum mechanical expectation value:

$$\vec{m}_i^\kappa = g\mu_B \langle\kappa| \vec{S}_i |\kappa\rangle \quad . \tag{11}$$

$\vec{S}_i$ is the total spin density operator for a lattice cell at site i. The quantum mechanical expectation value ensures the average over the fast quantum fluctuations in the configuration. The amplitudes $|\vec{m}_i^\kappa|$ might differ from one site to another, and for a given i they might differ from one $|\kappa>$ to another $|\kappa'>$. Remark that this definition of the amplitude is different from $(<\vec{S}_i^2>)^{1/2}$, which would be non zero even in a simple metal like Na, which is almost free electron like.

If amplitude fluctuations drive the phase transition an experiment on the intermediate time scale $\Delta t \lesssim t_{sw}$ should show finite "atomic moments" persisiting above T_c, but differing markedly in amplitude from one lattic point to another. This mechanism has mainly been invoked for "weak itinerant ferromagnets", Moriya and Kawabata (1973); Moriya (1979, 1982); Murata and Doniach (1972, i.e. systems with low T_c (<50 K) and a small moment at T = 0 (< $0.5\mu_B$).

C) A third mechanism - and the dominant one for the ferromagnetic transition metals - is based on transverse fluctuations of the magnetization (for example spin waves are transverse fluctuations): the "atomic moments" have almost constant amplitude, but do not point into the same direction at different lattice points. This type of fluctuation is the only one possible in localized magnetic systems. There, $\vec{S}_i^2$ corresponds to a sharp quantum number S(S+1) with S being an integer or half integer. The amplitude is a constant of motion here, only transverse fluctuations are allowed. In itinerant systems $\vec{S}_i^2$ does not correspond to a sharp quantum number due to the delocalized electronic wave functions leading to the fast quantum fluctuations. After averaging over these fast fluctuations, however, the amplitudes $|\vec{m}_i|$ might be almost the same for all i and all configurations |κ>which carry appreciable weight in the thermal averages.

This is advocated in the "fluctuating band theory" to be the case for the ferromagnetic transition metals, Capellmann (1979,1982), Korenman et al (1977,1979). Transverse fluctuations cost low energy, typical spin wave energies are of the order of 0.05 eV, and a macroscopic occupation of spin wave excitations may lead to a phase transition destroying long-range magnetic order. This picture is similar to what happens in a localized magnetic system, where only transverse fluctuations are possible. We shall see, however, that the itinerancy of the electrons in the transition metals leads to characteristic and drastic differences to localized systems in the magnetic properties of the paramagnetic phase.

The mechanisms B (amplitude fluctuations) and C(transverse fluctuations) might actually be mixed and cooperate to produce the phase transition, in particular in weak itinerant systems, Sokoloff (1973,1975).

2.3 The paramagnetic phase

The modern theories of itinerant magnetism ,Moriya and Kawabata (1973); Moriya (1979,1982); Murata and Doniach (1972); Capellmann(1979,1982); Korenman et al (1977); Korenman and Prange (1979); Sokoloff (1973, 1975); Lonzarich (1984);Hertz and Klenin (1974); Hasegawa (1979,1980); Hubbard (1979, 1981); Cyrot (1984);You et al (1980); Lin-Chung and Holden (1981); Shastry et al (1981), have one common feature: The dominant fluctuations leading to the phase transition to the paramagnetic state in the 3-d transition metals are slow transverse fluctuations (to some ex-

tent accompanied by longitudinal fluctuations in weak itinerant ferromagnets) resulting in a paramagnetic state preserving some sort of "atomic moments" above the transition
temperature T_c. There has been considerable controversy, however, concerning the detailed behaviour of the magnetic
correlations functions above T_c.

In the "fluctuating band theory"(FBT)Capellmann (1979,
1982) Korenman et al (1977, Korenman and Prange (1979),
Sokoloff (1973, 1975), the phase above T_c in its magnetic
properties is still dominated by the itinerancy of the 3-d
electrons resulting in "exchange split local bands" with
considerable short range magnetic order. These "local bands"
can be thought of as being exchange split not with respect
to a global direction but with respect to the direction of
the local magnetization, which fluctuates (slowly compared
to the time scale of electron hopping processes t_q) in space
and time, having no long range order above T_c,

The disordered local moment picture (DLM),Hasegawa
(1979,1980), Hubbard (1979,1981), Shastry et al (1981),
Staunton et al (1984), on the other hand starts from the
assumption that the magnetic properties above T_c are close
to the properties of systems having well localized magnetic
moments such as insulators and rare earth metals, being
well described by e.g. a Heisenberg Hamiltonian: Although
the 3d electrons are itinerant, within the DLM picture
they are thought to establish local magnetic moments which
can be disordered even on the scale of nearest neighbour
distances above T_c.

Within the last several years extensive neutron scattering experiments (probing the magnetic correlation functions directly) and photoemission experiments (both angle and spin resolved) have been carried out and to some extent have settled the controversy . The results obtained indicate a considerable degree of short range magnetic order (SRMO) above T_c, thus excluding the DLM picture and favouring the point of view taken in the fluctuating band theory. Before discussing the theoretical ideas and the appropriate experiments it is useful for the understanding to go back to the definition and the meaning of "atomic moments" in itinerant systems:

Equ. (11) defines the "atomic moments" in some typical pure quantum state$|\kappa>$ of the paramagnetic phase, as a quantum mechnical expectation value of $\vec{S}_i$. We shall discuss time dependences of the operators (say $\vec{S}_i$) is given by

$$\vec{S}_i(t) = e^{-iHt} \, \vec{S}_i \, e^{iHt} \qquad (12)$$

The expectation values $\vec{m}_i(t)$ will in general be time dependent, and in the paramagnetic phase an average of $\vec{m}_i(t)$ over a sufficiently long time will vansih (or alternatively the average of $\vec{m}_i^\kappa$ over the $|\kappa>$ contributing significantly at some temperature T> T_c). The common feature of all modern theories,Moriya and Kawabata (1973); Moriya (1979, 1982); Murata and Doniach(1972); Capellmann (1979, 1982);Korenman et al (1977); Korenman and Prange (1979); Sokoloff (1973, 1975); Lonzarich (1984); Hertz and Klenin (1974);Hasegawa (1979, 1980); Hubbard (1979, 1981); Cyrot (1984);You et al (1980); Lin-Chung and Holden (1981); Shastry et al (1981); Staunton et al (1984), is that at temperatures above T_c, the $\vec{m}_i$ have a finite amplitude for all $|\kappa>$ contributing significantly to the statistical average. Transverse fluctuations of the $\vec{m}_i$ are supposed to be responsible for the phase transition.

The controversy between the fluctuating band theory and the disordered local moment picture concerns the degree of short range magnetic order of the $\vec{m}_i$ persisting into the paramagnetic phase.

The expression "short range magnetic order" at this point is still quite vague and needs further specification, which is best done discussing the spindensity-spindensity correlation function

$$C(R_i - R_j, t) \sim \ll \vec{S}_i(t) \quad \vec{S}_j(0) \gg \tag{13}$$

Where the symbol $\ll \quad \gg$ indicates a thermodynamic average

$$\ll A \gg = \frac{Tr\{e^{-\beta H}A\}}{Tr \, e^{-\beta H}} \tag{14}$$

The correlation function $C(R,t)$ will of course reflect the characteristic times and lengths of the system.

The basic point of the fluctuating band theory, Capellmann (1979, 1982), Korenman et al (1977), Sokoloff (1973, 1975), is that the itinerancy of the magnetic 3d electrons will influence the short range behaviour of the correlation function, Capellmann and Vieira (1982). This short range part of $C(R,t)$ depends on the specific short range forces in the system. A rough estimate of the characteristic length l_1 over which the itinerancy will dominate the form of $C(R,t)$ at temperature $T(\gtrsim T_c)$ is

$$l_1 \sim v_F \frac{h}{kT} \tag{15}$$

v_F is the Fermivelocity and l_1 is a typical length an itinerant electron can travel in a thermal fluctuation time h/kT. For the 3d transition metals l_1 will be large compared to typical nearest neighbour distances.

The FBT, Capellmann (1979, 1982); Korenman et al (1977); Sokoloff (1973,1975);Capellmann and Vieira (1982), argues that over distances short compared to l_1 the magnetic properties in itinerant systems closely resemble ground state properties due to the fact, that the itinerant d-electrons responsible for the magnetism are phase coherent of such distances. Thus significant disorder is possible only for distances larger than l_1. At these larger distances ($>l_1$) the "universal" behaviour (i.e. not depending on the specific short range forces) for the equal time correlation function $C(r)$

$$C(R) \sim \frac{1}{R} e^{-R/\xi} \tag{16}$$

should be recovered, where ξ is the correlation length.

To be more specific here, I shall call ξ the "thermodynamic correlation length". ξ is temperature dependent and diverges at T_c. Close to T_c where ξ is very large the critical properties show "universal behaviour", to a large extent they do not depend on the underlying microscopic picture and the short-range forces. One has to keep in mind, however, that the form $C(r)$ given in eq. (16) only describes the long-range behaviour, for distances beyond which the short range (system dependent) forces dominate the properties of the correlation function.

The differences between the "electronic coherence length" l_1 and the thermodynamic correlation length ξ shows also up in their different temperature dependences. Whereas the thermodynamic correlation length ξ scales with the relative temperature $(T-T_c) \cdot {}^1/T_c$, diverging with a universal exponent for $T \to T_c$, the (system dependent) electronic coherence length l_1 will show no anomalies around T_c , scaling with $^1/_T$ as argued in equ. (15). If we take iron as an example, T_c is of order 1000 K, and the melting temperature of order 1800 K, thus in the range above T_c, where the solid still exists, l_1 will not change drastically and we can speak of a typical electronic coherence length l_1. To understand the meaning of short range magnetic order it is also necessary to discuss the time dependence of the correlation function $C(R,t)$ (equ. 13), or alternatively its Fouriertransform

$$S(Q,\omega) \sim \int dt \sum_j e^{iQ(R_i - R_j)} e^{-i\omega t} \ll \vec{S}_i(t) \cdot \vec{S}_j(0) \gg \qquad (17)$$

It was argued above that for short distances (i.e. large Q) the electronic coherence length l_1 makes significant thermal disorder impossible, which means that $S(Q,\omega)$ should not contain Fouriercomponents of significant intensity in the large Q region for energies $\hbar\omega \lesssim k_B T_c$. This last specification concerning the frequency dependence of $S(Q,\omega)$ is of central importance: The low frequency (i.e. within thermal energy range) components of $S(Q,\omega)$ in the large Q region should have weak intensity according to the fluctuating band theory, whereas there should be considerable intensity spread over large energies of up to the band width (of order several eV for the transition metals):

When introducing the "quantum fluctuation time" t_q (of order $\hbar/W \sim 10^{-15}$s), the typical time for the itinerant 3d electrons to cover a nearest neighbour distance, it was implied that on this time scale also the local spin density should change drastically. This results in significant intensity for large Q (in the neighbourhood of Brillouin Zone B.Z. boundary) and high energy (up to the bandwidth) Fouriercomponents of $S(Q,\omega)$. These Fouriercomponents are a result of the quantum fluctuations in the system, they are present at all temperatures (also at $T = 0$). "Thermal disorder" is caused by fluctuations which have energies within the range kT, whereas the high energy (hω ≫ kT) fluctuations are a result of the underlying quantum mechanical properties of the itinerant electrons.

Short range magnetic order according to the FBT implies that for <u>large Q</u> the intensity of spin fluctuations in the energy range $h\omega \lesssim kT_c$ should be small. Considerable intensity, however, should be spread over the large energy region hω < W characteristic of the quantum fluctuations.

This implies a considerable difference to properties of magnetic insulators, rare earth metals or Heusler alloys, where truely localized magnetic moments determine the magnetic properties. These systems are usually well described by a Hamiltonian of the Heisenberg type:

$$H_{loc} = -\frac{1}{2} \Sigma J_{ij} \ \vec{S}_i \cdot \vec{S}_j. \tag{18}$$

All energies are then determined by the exchange constants J_{ij}, e.g. spinwave energies at low T as well as the transition temperature T_c , which is typically comparable to the maximum spin wave energy , Mook (1983). At temperatures above T_c <u>all</u> fluctuations are within thermal energy range also for large Q, leading to significant intensity of $S(Q,\omega)$ for $h\omega < kT_c$ for <u>all</u> Q, thus implying thermal disorder already over nearest neighbour distances.

Let us point out again that for the small Q part of the spectrum of itinerant systems and localized systems have qualitatively <u>similar</u> behaviour (universal behaviour independent of the short range forces). The characteristic differences due to the itinerancy of the magnetic electrons in the 3d transition metals show up at large Q (probing the short distances).

Within the disordered local moment picutre, Hasegawa
(1979,1980); Hubbard (1979, 1981); Shastry et al (1981);
Staunton et al (1984), on the other hand, it is argued
that although the electrons carrying the magnetism of
transition metals are itinerant, their magnetic properties
can nevertheless be described by an effective Hamiltonian
of the type H_{loc} (equ. 18) enabling disorder on a nearest
neighbour scale above T_c. The argument put forward is
roughly as follows:Imagine at some instant of time a magnetic configuration characterized by some specific distribution of spin up and spin down electrons (in equal amounts
above T_c) within the lattice. These electrons are itinerant
and are therefore able to move to different lattice sites
within a characteristic time t_q. The basic assumption of
the DLM picture is that over time intervals $t_{sw} \gg t_q$ the
electron hopping processes take place as if the lattice
sites carried labels "spin up" and spin down", such that
spin up electrons predominantly visit "spin up" sites and
spin down electrons predominantly visit the "spin down"
sites, thereby establishing "local moments". The labeling
itself is supposed to change on a characteristic time scale
$t_{sw} \gg t_q$, and above T_c to be largely uncorrelated over nearest neighbour distances ("disordered local moments").This
picture therefore implies considerable thermal disorder at
short distances leading to significant intensity of the
Fouriercomponents of $S(Q,\omega)$ for all Q (also large Q close
to the B.Z. boundary) within the thermal energy range
$h\omega \lesssim kT_c$.

The theoretical controversy (LBT versus DLM picture)
has been largely resolved in the last several years mainly
due to intensive neutron scattering an photoemission experiments. Neutron scattering is able to directly measure
the magnetic correlation function ($S(Q,\omega)$ is proportional

to the magnetic neutron acattering cross section for momentum transfer Q and energy transfer ω), which is the object
of the controversy. Due to the limited neutron energies
available, however, one is restricted to energy transfers
smaller than 0.1 to 0.2 eV (correspinding to thermal energies of 1000 to 2000 K), but again this is the controversial energy region. Photoemission on the other hand allows
the study of the underlying electronic structure (essentially the one electron Green's function) whichgives rise to
the magnetic properties. Obvious adavantages of photoemission are the high intensity light sources and wide range of
available energy transfers in particular when using synchrotron radiation. The electronic structure studied will reflect the properties of the magnetic correlation functions,
exactly how is not yet completely understood, however. A
number of different theories making these connections exist,
Korenman and Prange, (1980);Capellmann and Prange (1981);
Staunton et al (1984); Durham et al (1984),and to the extent that they predict significant differences when relying
either on the FBT, Korenman and Prange (1980);Capellmann
and Prange (1981), or the DLM picture, Staunton et al (1984);
Durham et al (1984),can serve to distinguish between the
two different pictures.

Results from both types of experiments seem to be incompatible with the disordered local moment picture and support the basic assumptions of the fluctuating band theory:
Recent neutron scattering experiments, Déportes et al
(1981); Brown et al (1982); Brown et al (1982,1983), using
polarized neutrons and polarization analysis were able to
determine the total $S(Q,\omega)$ integrated over all energy transfers smaller than typical thermal energies kT_c (of order
0.1 eV for Fe), but excluding the higher energy quantum
fluctuations. It was found that even well above T_c this

integrated intensity $S_{th}(Q)$ is strongly peaked around Q=0
dropping to very small values with increasing Q.

A typical Q_c over which this dropoff occurs is obtained from the observed peak in $Q^2S_{th}(Q)$ as a function of
Q(in localized systems $Q^2S_{th}(Q)$ monotonically increases
with Q) and taking iron as an example, Q_c turns out to be
order 0.4 $Å^{-1}$,considerably smaller than $Q_{B.Z.}$ (the Q values for the B.Z. boundary), which is of order 1.7 $Å^{-1}$.
Q_c depends only weakly on temperature above T_c. These findings directly confirm the basic ideas of the FBT.

How Q_c is used to construct a characteristic length
is largely a matter of definition: Q_c^{-1} would be of order
2.5 $Å$, whereas the characteristic wavelength $2\pi \cdot Q_c^{-1}$ would
be of order 16 $Å^{-1}$. The reader may choose his own defini-
tion, but should keep in mind that for most of Q-space
(Q_c <Q <$Q_{B.Z.}$) thermal magnetic fluctuations are strongly
suppressed in intensity when compared to a localized magnetic system. The volume of Q-space ($\sim Q_c^3$) in which very
strong intensity of thermal magnetic fluctuations exist
is only a few percent of the entire Brillouin zone, being
centered around Q = 0, i.e. long wavelengths.

Spin and angle resolved photoemission experiments,
Kisker et al (1985), are also incompatible with theories
based on the disordered lcoal moment picture , Staunton
et al (1984); Durham et al (1984), thus implying some short
range magnetic order. Recently these photoemission data
have been analized on the basis of a theory incorporating
magnetic clusters, Haines et al (1985). The conclusions
reached are that correlation lengths must exceed 4 $Å$. Although these photoemission experiments as a test of magnetic correlation functions are less direct than neutron

scattering, they provide valuable information about the underlying electronic structure, e.g. demonstrating the persistence of a "local exchange splitting" above T_c in iron, Kisker et al (1985). For nickel similar experiments are more difficult to interpret due to intrinsic (small exchange splitting; what is the "true " Δ at T = 0?) and experimental resolution (compared to Δ) difficulties.

Finally I shall briefly comment on another controversy (this time between different groups in neutron scattering), which has attracted considerable interest: "Spin waves above T_c".

More than 10 years ago it was claimed that for surprisingly long wavelength (i.e. small Q) several times the nearest neighbour distance spin waves with propagating character persisted into the paramagnetic phase in the itinerant systems Ni , Mook et al (1973), and Fe, Lynn(1975). This unsusual behaviour was taken as strong support for the LBT(short range order regions supporting propagating waves) These experiments, carried out at Oak Ridge, used the usual (unpolarized) neutron scattering technique. Later experiments using polarized neutrons carried out at Brookhaven, Steinsvoll et al (1983); Uemura et al (1983); Shirane et al (1984), disputed the propagating character of the magnetic fluctuations (as defined through a peak in constant Q scans of $S(Q,\omega)$ for finite ω), claiming diffusive behaviour (i.e. $S(Q,\omega)$ being peaked around $\omega = 0$). The resulting controversy involved experimental difficulties: It was claimed , Lynn (1983), that insufficient resolution used at Brookhaven, Steinsvoll et al (1983); Uemura et al(1983); Shirane et al (1984), lead to incorrect conclusions. Improved resolution experiments using polarized neutrons and polarization analysis at the ILL

Grenoble , Brown et al (1984), essentially confirmed the Brookhaven results for Fe (mainly diffusive behaviour with possibly a weak sidepeak at finite ω reflecting a small propagating component, the evidence for the latter being at the limit of significance due to experimental resolution and intensity problems). More recent work done at Oak Ridge, Mook and Lynn (1985), using furthermore improved resolution and intensity, however, again claims the existence of propagating spinwaves in Ni and Fe above T_c for quite long wavelengths (i.e. Q as small as 0.3 $\overset{\circ}{A}^{-1}$).

Although the question of propagating waves above T_c is of considerable interest, it is not decisive concerning the correctness of the LBT. Propagating waves or not is a question about the <u>functional form</u> of $S(Q,\omega)$ as a function of ω for fixed Q. The decisive condition for the LBT concerns the <u>total intensity</u> within the thermal energy range $\omega < kT_c$ of $S(Q,\omega)$, i.e. the absence of a significant short wave length, low energy intensity. This condition is indeed satisfied.

2.4 Concluding remarks

The general picture of transition metal magnetism at finite temperatures presented gives a summary of the qualitative physical ideas developed in the last decade which in connection with a number of crucial experiments have lead to a partial understanding. Although no theoretical details of mathematical nature were presented, hopefully the qualitative arguments were put into an accessible form and may serve as a guideline. The interested reader is referred to a collection of review articles, Capellmann (1986), on the theoretical as well as experimental developements for an extensive description of metallic magnetism.

References:

Brown, P.J., Déportes, J., Givord D., and Ziebeck K.R.A., J. Appl. Phys. 53 (1982) 1973.

Brown, P.J., Capellmann, H., Déportes J., Givord D., and Ziebeck K.R.A., J. Magn. Magn. Mat. 30 (1982) 243; 30 (1980) 335; 31-34 (1983) 295.

Brown, P.J. Capellmann H., Déportes, Givord D. and Ziebeck K.R.A., Sol. St. Comm. 52 (1984) 83.

Capellmann H., J.Phys. F4 (1979) 1466; Solid State Commun. 30 (1979) 7; Z. Phys. B 34 (1979) 29; J. Magn. Magn. Mat. 28 (1982) 250.

Capellmann H.and Prange R.E., Phys. Rev. B23 (1981) 4709.

Capellmann H., Vieira V., Sol. St. Comm. 43 (1982) 747.

Capellmann ed. Springer "Metallic Magnetism" (1986)

Cyrot, J. Magn. Magn. Mat. 45 (1984) 9.

Déportes J. Givord D., Ziebeck K.R.A., J.Appl. Phys. 52, (1981) 2074.

Durham P., Staunton J., Gyorffy B.L., J. Magn. Magn. Mat.45 (1984) 38.

Hasegawa H., J. Phys. Soc. Japan 46 (1979) 1504; 49 (1980) 178, 963.

Haines E., Clauberg R., Feder R., Phys. Rev. Let. 54 (1985) 932.

Hertz J.A., and Klenin M.A., Phys. Rev. B10 (1974) 1084.

Hubbard, Phys. Rev. B19 (1979) 2626; B20 (1979) 4584; B23 (1981) 5974.

Kisker E., Schröder K.,Campagna M., Gudat, W., Phys. Rev. Let. 52 (1984) 2285; Phys. Rev. B31 (1985) 329 and references therein.

Korenman V., Murray J.L., and Prange R.E., Phys. Rev. B16 (1977) 4032, 4048, 4058.

Korenman V. and Prange R.E., Phys. Rev. B19 (1979)4691,4698.

Korenman V. and Prange R.E.,Phys. Rev. Lett. 44 (1980)1291.

Kübler J., J. Magn. Magn. Mat. 20 (1980) 107, 277.

Lin-Chung P.J., and Holden A.J., Phys. Rev. B23 (1981) 3414.

Lynn J.W.,Phys.Rev. B11 (1975) 2624.

Lynn J.W.,Phys.Rev. B28 (1983) 6550.

Lonzarich G.G., J. Magn. Magn. Mat. 45 (1984) 43.

Mook H.A., J. Magn. Magn. Mat. 31-34, (1983) 250.

Mook H.A., Lynn J.W., Nicklow R.M., Phys. Rev. Let. 30, (1973) 556.

Mook H.A., Lynn J.W., J. Appl. Phys. 57 (1985) 3006.

Moriya T. and Kawabata A., J. Phys. Soc. Japan 34 (1973) 639; 35 (1973) 609.

Moriya T., J. Magn. Magn. Mat. 14 (1979) 1.

Moriya T., J. Phys. Soc. Japan 51 (1982) 420.

Murata K.K. and Doniach S., Phys. Rev. Lett 29 (1972) 285.

Shastry B.S., Edwards D.M. and Young A.P., J. Phys. C14 (1981) L665.

Shirane G., Steinsvoll O., Uemura Y.J., Wickstead J., J. Appl. Phys. 55 (1984) 1887.

Sokoloff J.B., Phys. Rev. Lett. 31 (1973) 1417; J.Phys.F5 (1975) 528, 1946.

Staunton J., Gyorffy B.L., Pindor A.J., Stocks G.M., Winter H., J. Magn. Magn. Mat. 45 (1984) 15.

Steinsvoll O., Majkrzak G.E., Shirane G., Wickstead J., Phys. Rev. Let. 51 (1983) 300.

Uemura Y.J., Shirane G., Steinsvoll O., Wickstead J., Phys. Rev. Let. 51 (1983) 2322.

Wang C. S. and Calloway J., Phys. Rev. B15 (1977) 298.

Wohlfarth E.P., Rev. Mod. Phys. 25 (1953) 211.

You M.V., Heine V., Holden A.J. and Lin-Chung P.J., Phys. Rev. Lett. 44 (1980) 1282.

Ziebeck K.R.A., J. Magn. Magn. Mat. 30 (1982) 243; 30 (1982) 335; 31-34 (1983) 295.

Chapter 3

Critical Behaviour at
Surfaces of Ferromagnets

K. Binder

Institut für Physik, Universität Mainz
D-6500 Mainz, Postfach 3980, FRG

3.1. Introduction

In a large but finite sample extensive thermodynamic quantities (such as the free energy F of the system) can be split into a "bulk" term(proportional to the volume V) and a surface correction. While in a magnet usually the magnetic order is usually affected by the surface over a distance of a few lattice spacings only, and then this surface correction is rather small, this clearly is no longer true near a second-order phase transition, where the correlation length ξ_b becomes large. Thus, we expect that surface corrections are larger near the critical temperature T_{cb} of the bulk and should exhibit interesting critical behaviour. Thus, if F is written as (A is the total surface area, T the temperature, H the magnetic field)

$$F/V = f_b \ (T,H) + (A/V)f_s \ (T,H), \quad V \to \infty \qquad (1)$$

the critical behaviour will show up in terms of characteristic power laws both for bulk properties and the surface free energy density f_s. E.g. if we consider bulk and surface specific heats C_b, C_s

$$C_b \ \propto \ \frac{\partial^2 f_b}{\partial T^2} \Bigg|_{H=0} \propto \big| \ 1-T/T_{cb} \ \big|^{-\alpha_b}, \ C_s \propto \frac{\partial^2 f_s}{\partial T^2} \Bigg|_{H=0} \propto \big| \ 1-T/T_{cb} \ \big|^{-\alpha_s} \ , \qquad (2)$$

we expect near T_{cb} a singular behaviour as described in Eq.(2) by the critical exponents α_b, α_s.

Now the surface modifies the magnetic order in many ways: translational invariance of the crystal is broken in the direction perpendicular to the surface, both by the fact that for the surface spins neighbours are missing, and because in general we expect that the magnetic interactions (exchange, anisotropy, etc.) near the surface may differ from the bulk. Thus clearly the local magnetization m_1 near the surface will differ from the bulk magnetization m_b, and hence there is again need to distinguish their critical exponents β_b, β_1:

$$m_b \equiv - (\frac{\partial f_b}{\partial H})_T \propto (1-T/T_{cb})^{\beta_b} \quad , \quad m_1 \equiv (\frac{\partial f_s}{\partial H_1})_{T,H} \propto (1-T/T_{cb})^{\beta_1} \quad ,$$

$$\{H=0, H_1=0\} . \tag{3}$$

For convenience of the theoretical formulation, we here have introduced a field H_1 acting on the surface spins only, and then m_1 in Eq.(3) is written as a derivative of f_s with respect to this field, similarly as m_b is the derivative of f_b with respect to the uniform field.

One should not think, however, that there is just one surface quantity corresponding to each bulk quantity: In analogy to Eq.(1), we can also split the total magnetization M into m_b and the "sur-face magnetization m_s",

$$M/V = m_b + \frac{A}{V} m_s, \quad m_s = - (\frac{\partial f_s}{\partial H})_{T,H_1} \propto (1-T/T_c)^{\beta_s} \{H=0, H_1=0\} . \tag{4}$$

The story is even more complicated than this when we consider spin correlation functions, because even in a cubic crystal directions parallel and perpendicular to the surface are no longer equivalent. Also, there is need to consider three types of response functions: The surface susceptibility χ_s as well as local susceptibilities χ_1, χ_{11} which have exponents γ_s, γ_1, γ_{11}, respectively, while there is only one susceptibility $\chi_b \equiv (\partial m_b/\partial H)_T \propto |1-T/T_{cb}|^{-\gamma_b}$ in the bulk,

$$\chi_s \equiv (\partial m_s/\partial H)_{T,H_1} \quad , \quad \chi_1 \equiv (\partial m_1/\partial H)_{T,H_1} \quad , \quad \chi_{11} \equiv (\partial m_1/\partial H_1)_{T,H} \quad , \tag{5}$$

$$\chi_s \propto | 1-T/T_{cb}|^{-\gamma_s} \quad , \quad \chi_1 \propto | 1-T/T_{cb}|^{-\gamma_1} \quad , \quad \chi_{11} \propto | 1-T/T_{cb}|^{-\gamma_{11}}. \tag{6}$$

Although H_1 is a field not available to the experimenter in his la-boratory, nevertheless all these quantities are accessible to experiment, at least in principle. For instance, the surface excess

quantities m_s, χ_s etc. can be measured by studying thin films of various thicknesses, and analyzing the data in terms of equations such
as Eq.(1), (4). The local quantities $\{ m_1, \chi_1, \chi_{11} \}$ are accessible
by local probes only, such as low energy electron diffraction (LEED),
Mössbauer effect, nuclear magnetic resonance (NMR), electron capture spectroscopy (ECS), totally reflected neutron or X-ray diffraction (Dietrich and Wagner 1984). For instance, in an antiferromagnet
one may use LEED to record the superstructure Bragg reflections, whose
intensities are related to the local staggered magnetization $\tilde{m}_1$
which is the appropriate order parameter there, while the intensity
of the diffuse scattering near the Bragg positions would be re-
lated to $\tilde{\chi}_{11}$, the staggered susceptibility corresponding to χ_{11}.
In a ferromagnet, these Bragg spots coincide with the spots due to
the crystallographic surface structure; but the magnetic contribution can be separated by using spin polarized electrons (SPLEED).
By this powerful technique, which is discussed in detail elsewhere in
this book, exponents such as β_1 [Eq.(3)] have in fact been estimated
rather reliably (see e.g. Alvarado et al. 1982).

So far we have tacitly assumed that the order at the surface
sets in at the same temperature as it does in the bulk. However, since
the interactions near the surface may be very different from the
bulk ones, this is not necessarily true. For instance, in a nearest-
neighbour Ising ferromagnet with exchange J in the bulk and exchange
$J_s = J(1+\Delta)$ in the surface plane there exists a critical value Δ_c,
where the surface starts to order at a temperature $T_{cs} > T_{cb}$ (Fig.1).
In the surface layer for $\Delta > \Delta_c$, one hence undergoes a "surface
transition", where the system has two-dimensional order which decays
exponentially as one moves into the bulk. Additional (weak) singularities occurring in f_s, m_1, etc. when one passes T_{cb} (the "extraordinary transition") will not be considered further here {see Ohno
and Okabe (1984) for recent work on this problem} . The "special
transition" occurring for $\Delta = \Delta_c$ is a multicritical point due to the
coincidence of two-dimensional criticality at the surface and three-

98

dimensional criticality in the bulk, while only the latter occurs for $\Delta \lessgtr \Delta_c$, and this then is termed the "ordinary transition".

A further complication is that the change of interactions at the surface may lead at the surface to an ordering phenomenon different from that of the bulk. A simple example of this is again provided by an Ising ferromagnet, where Δ is large but negative, so that the surface orders antiferromagnetically (Binder and Hohenberg 1974, Binder and Landau 1985). Conversely, ferromagnetic order might occur at the surface of antiferromagnets, etc., and various other types of "magnetic surface reconstruction" (Trullinger and Mills 1973) are conceivable as well.

This terminology already suggests an important analogy between surface effects on _magnetic_ phase transitions and surface effects on phase transitions in general: one just has to re-interpret m_b as the order parameter in the bulk and m_1 as the local order parameter at the surface for the phase transition in question - which may be a transition paraelectric-ferroelectric or any other type of structural transition, an order-disorder transition in a binary alloy, unmixing in binary mixtures, gas-liquid condensation of a fluid, etc. Less obvious is the fact that the "special transition" (Fig. 1) is related also to a conformational transition of a polymer adsorbed at a surface {the volume taken by the polymer coil changes from essentially three-dimensional to essentially two-dimensional, see Eisenriegler at al. (1982) }. This analogy arises because the problem of polymer configurational statistics is equivalent to magnetic transitions of the model of n-component spins in the limit $n \to 0$ (De Gennes 1979). A similar analogy arises for the "percolation problem" (Stauffer 1979), which can be viewed as the limit of the q-state Potts model (Wu 1982) for $q \to 1$, and consequently surface effects on magnetic transitions are analoguous to surface effects on percolation. We shall not exploit these analogies further and rather refer to a more detailed review (Binder 1983); we emphasize, however, that in some systems the quantity corresponding to the local field H_1

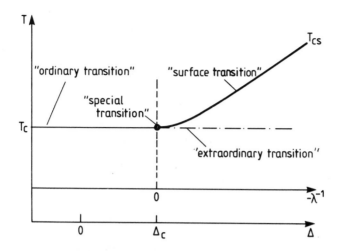

Fig. 1: Phase transitions occurring at the surface layer of a semi-
infinite magnetic system. In the case of nearest-neighbour
ferromagnetic interactions treated in the mean field approximation, the enhancement Δ is related to the inverse of the
extrapolation length λ [Δ_c corresponds to the case where $\lambda = \infty$].

at the surface is nonzero, unlike magnets: this happens at the surface
of fluid-gas systems or binary mixtures, where the energy may be lowered
when one species is enriched at the surface. This may lead to a "wetting
transition" {see Pandit at al (1982) for a review } where the system
forms a macroscopically thick domain of the preferred phase near the
surface, separated by an interface from the bulk. If one could generate
such local fields $H_1 \neq 0$ also in ferromagnets choosing them opposite
to the spontaneous magnetization in the bulk, this wetting transition
would just mean the formation of an oppositely oriented magnetic
domain close to the surface. While this phenomenon can hardly be
realized in magnets and hence will not be analyzed further, a related
phenomenon can occur when the transition from the ferromagnetic state

to the paramagnetic one is of first order in the bulk. Then the surface may have a disordering effect on the system, introducing a thin
layer which is nearly disordered near the surface. As one approaches
the bulk transition, the domain wall separating this disordered layer
from the ordered phase in the bulk gets depinned from the surface,
and via the growth of thickness of this "paramagnetic" layer m_1
vanishes continuously { as described by Eq.(3) } although m_b jumps to
zero at T_{cb} discontinuously (Lipowsky 1984 and references therein).
Again analoguous phenomena are predicted to occur possibly in a wide
variety of physical systems. As with phase transitions in the bulk,
magnets are likely to play the role of prototype examples also for all
these types of surface critical phenomena, and hence one can understand the large interest devoted to the pertinent theory. Hence, we
cannot attempt to review all the theoretical work here { for more
details and bibliography see Binder (1983) } , but rather try to summa-
rize the key points.

3.2. Mean Field Theory for the Semi-Infinite Ising Model

As for all phase transition problems, mean field theory may not be quantitatively accurate close to criticality, but nevertheless is useful to provide a qualitative understanding of the phenomena in question; this is also true here, where it helps to introduce the appropriate concepts.

If we consider a film of thickness L with two free surfaces but otherwise infinite in extent, the free energy per area can be written as an integral in terms of the coordinate z across the film (Binder and Hohenberg 1972, Lubinsky and Rubin 1975a)

$$\frac{F}{TA} = \int_0^L dz \{\tfrac{1}{2} Am^2(z) + \tfrac{1}{4} Bm^4(z) - Hm(z) + \tfrac{1}{2} C (\tfrac{\partial m}{\partial z})^2\} + 2F_s^{(bare)}/T; \quad (7)$$

here A, B, C are constants, with $A = A't$ changing sign at T_{cb} $\{t \equiv (T-T_{cb})/T_{cb}\}$, and $F_s^{(bare)}$ represents the local perturbation due to a surface, which we similarly expand in terms of $m_1 \equiv m(z=0)$ $\{=m(z=L)\}$,

$$F_s^{(bare)}/T = \tfrac{1}{2} C\lambda^{-1}m_1^2 - H_1 m_1, \quad (8)$$

where the constant λ has dimension of a length and is called extrapolation length (see Fig. 2 below). Requesting that m(z) yields a minimum of F leads to the Euler-Lagrange equation

$$Am(z) + Bm^3(z) - C(d^2m/dz^2) = H \quad (9)$$

with the boundary conditions

$$\frac{\partial m}{\partial z} - \frac{m}{\lambda} = -\frac{H_1}{C} \, , \ z=0 \, ; \quad \frac{\partial m}{\partial z} + \frac{m}{\lambda} = \frac{H_1}{C} \, , \ z = L \ . \quad (10)$$

The coefficients in Eq.(9) relate to the bulk quantities in the standard way as

$$m_b = \sqrt{-A/B} \propto (-t)^{1/2} \, , \ t \to 0, \ H=0 \, , \quad (11a)$$

$$\chi_b^+ = A^{-1} = A'^{-1}t^{-1}, \ t \to 0^+, \ H=0, \ \chi_b^- = (-2A)^{-1}, t \to 0^-, \ H=0; \quad (11b)$$

$$\xi_b^+ = \sqrt{C/A} \propto t^{-1/2}, \; t \to 0^+, \; H = 0, \quad \xi_b^- = \sqrt{-C/2A}, \; t = 0^-, \; H=0; \quad (11c)$$

Here superscripts $\pm$ in susceptibility and correlation length serve to distinguish temperatures above and below T_{cb}, and obviously Eq.(11) displays the "classical" mean-field exponents $\beta_b = 1/2$, $\nu_b = 1/2$ $\{\xi_b \propto |t|^{-\nu_b}\}$, $\gamma_b = 1$. We now for the moment make the system semi-infinite, $L \to \infty$, and then $m(z \to \infty) \to m_b$, $(\partial m/\partial z)_{z\to\infty} \to 0$; then Eqs.(9), (10) are easily integrated to yield an equation for m_1, namely

$$\frac{1}{2} A \, (m_b^2 - m_1^2) + \frac{1}{4}B \, (m_b^4 - m_1^4) + \frac{1}{2} C \, (\frac{m_1}{\lambda} - \frac{H_1}{C})^2 = H(m_b - m_1). \quad (12)$$

From this result it is easy to show that (for the moment we assume $\lambda > 0$)

$$m_1 = m_b \lambda / \xi_b^- \propto -t, \quad t \to 0^-, \; H=H_1=0, \; \text{i.e.} \; \beta_1 = 1, \quad (13a)$$

$$m_1 = \lambda H_1/C, \; \chi_{11} = \lambda/C, \quad t \to 0^+, \; H=0, \; H_1 \to 0, \; \text{i.e.} \; \gamma_{11} = 0, \quad (13b)$$

$$m_1 = \lambda H(A^{-1}/C)^{1/2}, \chi_1 = \lambda \, (A^{-1}/C)^{1/2} \propto t^{1/2}, \; H_1=0, H \to 0, \text{i.e.} \; \gamma_1 = \frac{1}{2}; \quad (13c)$$

in this way some of the exponents defined in the introduction follow straightforwardly. For obtaining the surface excess magnetization m_s and susceptibility χ_s, one needs to consider the full magnetization profile. Above T_c the term $Bm^3(z)$ in Eq.(9) may be neglected for small enough H, H_1, and hence Eqs.(9), (10) yield

$$m(z) = m_b - (m_b - m_1) \exp(-z/\xi_b^+), \quad (14a)$$

while below T_c the profile is described by

$$z/\xi_b^- = \int_{m_1/m_b}^{m(z)/m_b} d\psi \, \Big/ \sqrt{(1-\psi^2)^2/4 + m_b \chi_b^- H(\psi-1)}. \quad (14b)$$

It is hence seen that the scale for the order parameter variation is set by the correlation length ξ_b. Close to the surface we have an interplay between the length ξ_b and λ, see Eq.(13a) and Fig. 2, where also the case $\lambda < 0$ is included. Before we discuss this we remark that m_s can be obtained from the shaded area underneath the profile (Fig. 2a) as

$$m_s = \int_0^\infty dz \, [m_b - m(z)] = \xi_b^+ (m_b - m_1) \propto H \chi_b^+ \xi_b^+, \quad t \to 0, H \to 0, \quad (15)$$
$$(t > 0)$$

and hence it follows that $\gamma_s = \gamma_b + \nu_b = 3/2$; similarly one finds for $t < 0$ $\beta_s = \beta_b - \nu_b = 0$. For the surface specific heat exponent $\{Eq.(2)\}$ one finds $\alpha_s = \alpha_b + \nu_b = \frac{1}{2}$ ($\alpha_b = 0$).

Now we turn to the case $\lambda < 0$. Again we may use Eq.(12) to obtain χ_1 and χ_{11} for $t > 0$, but now we no longer take the limit $t \to 0^+$ as in Eq.(13b,c). Then we obtain instead

$$\chi_{11}^+ = (\lambda/C)/(1 + \lambda/\xi_b^+), \quad \chi_1^+ = (\lambda \chi_b^+/\xi_b^+)/1 + \lambda/\xi_b^+), \quad (16)$$

which shows that both χ_{11}^+ and χ_1^+ will diverge when $\xi_b^+ = -\lambda$, i.e. for

$$(T_{cs}/T_{cb} - 1) = (C/A') \lambda^{-2}. \quad (17)$$

Thus the surface starts to order before the bulk, as anticipated in Fig. 1. For small $t_s \equiv T/T_{cs} - 1$ one gets $\chi_1^+ \propto |t_s|^{-1}$, $\chi_{11}^+ \propto |t_s|^{-1}$, $m_1 \propto (-t_s)^{1/2}$; that means that the exponents β_1, γ_1, γ_{11} etc. at the surface transition are just the bulk (two-dimensional) exponents [of course, within Landau theory the bulk exponents do not depend on dimensionality at all]. While for $\lambda > 0$ $m(z)$ increases when one moves from the surface towards the bulk (Fig. 2a), for $\lambda < 0$ $m(z)$ is a decreasing function of z(Fig. 2c,d) and for $\chi^{-1} = 0$ the profile is completely flat (Fig. 2e). At this special transition in mean-field theory we have hence $\chi_1 = \chi_b$, $m_1 = m_b$, $\gamma_1^{sp} = \gamma_b = 1$, $\beta_1^{sp} = \beta_b = 1/2$. Note however, that $\chi_{11} \propto |t|^{-1/2}$, i.e. $\gamma_{11}^{sp} = 1/2$.

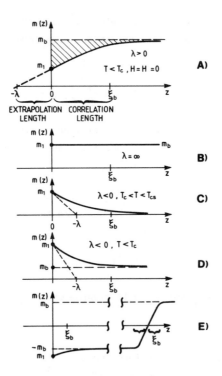

Fig. 2 Schematic magnetization profiles near a free surface, according
to meanfield theory. Various cases are shown: (a) Extrapolation
length λ positive. The transition from the disordered state to
this state is called the "ordinary transition". The shaded
area indicates the definition of the surface magnetization m_s.
(b) Extrapolation length λ infinite. The transition from the disordered state to this state is the "special transition".(c) Extrapolation length λ negative, temperature above the bulk but
below the surface critical temperature. Transition from this state
to the fully disordered one is the "surface transition".(d) Same
as (c) but $T < T_{cb}$. The transition between the states shown in
(c),(d) is called "extraordinary".(e) Surface field H_1 competing
with bulk order ($m_b > 0$, $H_1 < H_1^*$ such that $m_1 < -m_b$). In this case
a domain of oppositely oriented magnetization with macroscopic
thickness (wetting layer") would form at the surface.

Thus we see that the extrapolation length λ plays a crucial role in determining the surface critical behaviour, and it is natural to ask how this length is related to microscopic interaction parameters. Working out a layerwise molecular field approximation for a simple cubic Ising magnet with nearest-neighbour exchange J everywhere in the bulk but exchange J_s in the surface plane, one finds that the results reduce near $T_{cb} = 6 J/k_B$ to the above continuum theory, with $\lambda = a_o/(1-4 \Delta)$, $\Delta = J_s/J-1$, a_o being the lattice spacing. A distinction occurs, however, for the case considered in Fig. 2e: the transition from "non-wet" state to the "wet" state shown then is replaced by an infinite sequence of (first-order) transitions, where layer after layer jumps from a positive value m(z) to a negative one { these layering transitions "describe "multilayer adsorption", see Pandit at al. (1982) } . Another feature not yet anticipated in the continuum theory occurs if J_s is sufficiently negative: there may occur antiferromagnetic order at the surface of a ferromagnet. Fig. 3 shows the full molecular-field phase diagram. We shall return to this phenomenon of a surface magnetic order differing from the bulk ("magnetic surface reconstruction") in Sec. 4.3.

As a last topic of this section, we consider surface phenomena at a first-order transition within mean field theory (Lipowsky 1982,1984). We still start from an ansatz such as Eq.(7), but assume that the bulk free energy density f(m) is different: instead of the expression in curly brackets in Eq.(7) we now assume, for instance

$$f(m) + \frac{1}{2} C \left(\frac{\partial m}{\partial z}\right)^2 = \frac{1}{2} Am^2 + \frac{1}{2} Bm^4 + \frac{1}{6} Dm^6 - Hm + \frac{1}{2} C \left(\frac{\partial m}{\partial z}\right)^2, \quad (18)$$

with A as above but $B < 0$ $(D > 0)$. Then T_{cb} (where A changes sign) is not the phase transition temperature but rather the stability limit of the disordered phase; the magnetization m_b vanishes discontinuously at the higher temperature T^* given by $A'(T^*/T_c-1)=3B^2/(16D)$, where it jumps from $m_b^{crit} =\sqrt{-3B/(4D)}$ to zero. The order parameter m_1 at the surface exhibits a similar jump only for $C\lambda^{-1} < [A'(T^*/T_c-1)]^{1/2}$.

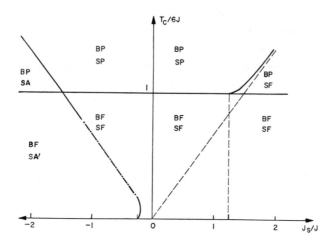

<u>Fig. 3:</u> Transition temperature for the surface of an Ising simple cubic
lattice with exchange J_s in the surface different from the
exchange J in the bulk, according to a layer-wise molecular
field approximation (Binder and Hohenberg 1974). Different phases are denoted as follows: BP (bulk paramagnetic), BF (bulk
ferromagnetic), SP (surface paramagnetic)

For $C \lambda^{-1}$ larger than this value , however, m_1 vanishes continuously,
$m_1 \propto | T-T^* |^{\beta_1}$, with $\beta_1 = \frac{1}{2}$. For $C \lambda^{-1} = [A'(T^*/T_c-1)]^{1/2}$ we en-
counter another multicritical point, with $\beta_1^m = 1/4$. This continuous
behaviour occurs because the profile m(z) develops an intrinsic
structure similar to Fig. 2 (e) [but with m(z) in the range between
0 and 1 instead between -1 and +1] : Thus m(z) is rather small over a
distance 1, and then rises steeply through an inflection point towards
m_b. The thickness 1 of this disordered layer diverges as $1 \propto | \ln |T^*-T||$
when one approaches the first-order transition. A detailed analysis
(Lipowsky 1982) reveals for the critical behaviour of the various sur-

faces quantities the exponents $\gamma_1 = 1/2$, $\gamma_{11} = 0$, $\alpha_s = 1$, $\beta_s = 0$, $\gamma_s = 1$; at the multicritical point, $\gamma_1^m = 3/4$, $\gamma_{11}^m = 1/2$, $\alpha_s^m = 1$, $\beta_s^m = 0$, $\gamma_s^m = 1$. The latter behaviour changes, however if one considers other forms of $f(m)$ [e.g. a term $\propto m^3(z)$ in Eq.(18)] or F_s^{bare}/T [involving higher powers than m_1^2 in Eq.(8)], see Lipowsky (1982,1984). In addition, if $H_1 > H_1^c$, one finds "surface induced order" when one approaches the first order transition from the disordered side, and again a layer, whose thickness grows logarithmically when one approaches the transition, intrudes from the surface. Defining $\Delta m_1 = m_b - m_1 \propto |T^* - T|^{\beta_1}$ one again finds $\beta_1 = \frac{1}{2}$. For $H_1 < H_1^c$ this transition is of first order (and also occurs for $H_1 = 0$ if λ is negative but $|\lambda| < |\lambda_{crit}|$).

3.3. Scaling Theory and Renormalization Group Results

We start by defining further critical exponents, considering the response to magnetic fields right at the transition, as well as the decay of correlation functions there. From the magnetization m_b, m_s, and m_1 one defines exponents δ_b, δ_s, δ_1, and δ_{11} as follows

$$m_b \propto H^{1/\delta_b}, \quad m_s \propto H^{1/\delta_s}, \quad m_1 \propto H^{1/\delta_1}, \quad T = T_{cb}, \quad H_1 = 0; \quad m_1 \propto H_1^{1/\delta_{11}}, \quad T = T_{cb}, H = 0 \quad (19)$$

With respect to the correlation function $\Gamma(\vec{r}, \vec{r}') = \langle S_{\vec{r}} S_{\vec{r}'} \rangle_T$ we have to distinguish between correlations in the bulk, which depend on relative distances only $\{\Gamma_b(\vec{r} - \vec{r}')\}$, perpendicular correlations $\Gamma_\perp(z,z')$ and parallel ones, $\Gamma_\parallel(\vec{\rho} - \vec{\rho}')$. Here the coordinates in the directions parallel to the surface are denoted by $\vec{\rho}$, and again translational invariance in these directions is implied. Off criticality the decay of all these correlation functions is exponential, the decay constant being the bulk conclation length ξ_b - with the exception of the vicinity of the surface transition, where $\Gamma_\parallel$ decays slower, with a two-dimensional correlation length. At bulk criticality, however, we have again power laws (d is the dimensionality):

$$\Gamma_b(r) \propto r^{-(d-2+\eta_b)}, \quad \Gamma_\parallel(\rho) \propto \rho^{-(d-2+\eta_\parallel)}, \quad \Gamma_\perp(z,z') \propto |z-z'|^{-(d-2+\eta_\perp)} \quad (20)$$

This serves to define the exponents η_b, $\eta_{\parallel}$ and $\eta_{\perp}$. Eq.(20) illustrates the anisotropy of critical correlations near the surface (this happens already in mean field theory, where $\eta_b=0$, $\eta_{\perp}=1$, $\eta_{\parallel}=2$ at the ordinary transition, while $\eta_b^m = \eta_{\perp}^m = \eta_{\parallel}^m = 0$ at the special transition). Finally, for a thin film of thickness L a "shift exponent" λ_s is defined by $T_c(L)/T_{cb}-1 \propto L^{-\lambda_s}$, $L \to \infty$, and the way in which T_{cs} merges at the special transition serves to define the "crossover exponent" φ, $T_{cs}(\Delta)/T_{cb}-1 \propto (\Delta/\Delta_c-1)^{1/\varphi}$ [in mean field theory $\varphi=1/2$].

Thus there is a wealth of critical exponents describing surface critical behaviour, as first pointed out by Binder and Hohenberg (1972). Fortunately, these exponents are not at all independent of each other, but rather finite size scaling (see Barber 1983 for a review) and surface scaling (Binder and Hohenberg 1972,1974,Barber 1973) implies that at the ordinary transition there is just one new independent exponent (in addition to the bulk ones), while at the special transition there are two. We consider scaling at the ordinary transition first and state that the singular part of the free energy of a thin film should be a generalized homogeneous function

$$f_{sing}(L,H,H_1,t) = L^{-(2-\alpha_b)/\nu_b}\,\tilde{F}(L^{(\beta_b+\gamma_b)/\nu_b}H,\ L^{\Delta_1/\nu_b}H_1,\ L^{1/\nu_b}t)\quad (21)$$

Note that the last argument in Eq.(21), $L^{1/\nu_b}t \propto (L/\xi_b)^{1/\nu_b}$, just says that the "length L scales with ξ_b" (finite size scaling). For $L \to \infty$, Eq.(21) must reduce to the standard scaling behaviour obeyed by the bulk, $f_{sing}^b (H,t)=t^{-(2-\alpha_b)}\,\tilde{f}_b (t^{-(\beta_b+\gamma_b)} H)$; this requirement has fixed both the prefactor $L^{-(2-\alpha)/\gamma_b}$ and the scaling power $L^{-(\beta_b+\gamma_b)/\gamma_b}$ of the bulk magnetic field, and hence only the scaling power of the surface field H_1 is left undetermined, and here hence enters the one additional exponent announced above (which we call Δ_1 by defining the scaling power of H_1 to be L^{Δ_1/ν_b}). All exponents {apart from the exponents referring to correlation functions in Eq.(20)} can now be obtained from Eq.(21), taking suitable derivatives and the limit $L \to \infty$. This leads to, after some simple algebra (Binder 1983),

$$\alpha_s=\alpha_b+\nu_b,\quad \beta_s=\beta_b-\nu_b,\quad \gamma_s=\gamma_b+\nu_b,\quad \beta_1+\gamma_1=\beta_1\delta_1 =\beta_b+\gamma_b,\quad (22a)$$

$$\gamma_1(\delta_{11}-1) = \gamma_{11}(\delta_1-1), \quad \beta_1(1+\delta_{11}) = \beta_s(1+\delta_s), \quad \gamma_{11}+\beta_1 = \beta_1 \delta_{11} = \Delta_1, \quad (22b)$$

$$2\gamma_1 - \gamma_{11} = \gamma_s, \quad 2-\alpha_s = 2\beta_1 + \gamma_{11}, \quad \lambda_s = 1/\nu_b. \quad (22c)$$

Noting that the susceptibilities can always be expressed as sums of corre-
lations via fluctuation relations $\{$ e.g. $k_B T \chi_b = \sum_{\vec{r}} \Gamma_b(\vec{r}), \ k_B T \chi_{11} = \sum_{\vec{\rho}} \Gamma_{\parallel}(\vec{\rho}) \}$ scaling assumptions for the correlation functions$\{$ e.g.
$\Gamma_{\parallel}(\rho) = \rho^{-(d-2+\eta_{\parallel})} \tilde{\Gamma}(\rho/\xi_b) \}$ lead to scaling laws relating the corre-
lation exponents to thermodynamic ones,

$$\gamma_1 = \nu_b(2-\eta_{\perp}), \quad \gamma_{11} = \nu_b(1-\eta_{\parallel}), \quad \eta_{\parallel} = 2\eta_{\perp} - \eta_b, \quad \beta_1 = \frac{\nu_b}{2}(d-2+\eta_{\parallel}). \quad (23)$$

Scaling theory also implies that the order parameter profile near the
surface is not always linear as in mean field theory (Fig. 2a), but
rather follows another power law,

$$m(z) \propto (-t)^\beta \tilde{m}(\frac{z}{\xi_b}) \propto (-t)^{\beta_1} z^{(\beta_1-\beta_b)/\nu_b}, \quad H=H_1=0, \quad z \ll \xi_b^-. \quad (24)$$

Finally we write down the scaling assumption for the special transition.
While Eq.(21) for the ordinary transition leads to

$$f_s(H,H_1,t) = |t|^{-2-\alpha_b-\nu_b} \tilde{f}_s(t^{-(\beta_b+\gamma_b)}H, t^{-\Delta_1}H_1), \quad (25a)$$

near the special transition there is one more important variable
$g \equiv \Delta - \Delta_c$ (cf. Fig.1) and hence

$$f_s^m(H,H_1,t,g) = |t|^{-2-\alpha_b-\nu_b} \tilde{f}_s^m(t^{-(\beta_b+\gamma_b)}H, t^{-\Delta_1^m}H_1, t^{-\varphi}g). \quad (25b)$$

Thus the exponent Δ_1^m adopts a different (multicritical) value, and in
addition the crossover exponent φ enters, which hence implies that two
independent exponents need to be determined.

We have obtained these exponents from mean field theory in the
previous section already. Now it is well known, of course, that mean

field theory does not provide an accurate description of the critical
behaviour of real systems (it would do so for d$>$ 4, see e.g. Fisher
1974). Thus there is need for other approaches to estimate these
exponents. The most powerful approach for the study of critical phenomena in general is the renormalization group approach (e.g. Fisher 1974).
It has been applied to surface critical behaviour rather early (Lubensky and Rubin 1975b), but encountered severe technical difficulties
which only have been overcome in more recent work (Diehl and Dietrich
1980 , 1981a-c, 1983; Diehl and Eisenriegler 1982, 1984; Diehl 1982;
Diehl etal. 1983, 1985; Dietrich and Diehl 1981; Reeve and Guttmann 1980,
1981; Guttmann and Reeve 1980; Reeve 1981). For further recent work
see also Rudnick and Jasnow (1982), Cardy (1983, 1984, 1985), Goldschmidt
(1983), Goldschmidt and Jasnow (1984), Nemirovsky and Freed 1985, Ohno
and Okabe (1983, 1984), and Burkhardt and Eisenriegler (1985).

An exposition of the renormalization group approach is beyond the
scope of this review, however (see Diehl 1982 for a recent brief review).
We shall restrict ourselves in quoting the main results, pointing out
the strengths as well as the weaknesses of this approach. Thereby the
need for alternative approaches for complementing the renormalization
group results (and for checking its accuracy) will become evident. Such
alternative approaches are systematic high temperature series and their
extrapolation, applied to surface criticality by Binder and Hohenberg
(1972, 1974),by Whittington et al. (1979), Guttmann et al. (1980), and
Ohno et al. (1984), and the Monte Carlo computer simulation method
(see Sec.4). Also we shall not discuss any work using the position space
renormalization group {see e.g. Burkhardt and Eisenriegler (1981) and
references therein }.

One particular merit of the field-theoretic renormalization group is
that it is not restriced to the Ising model of Sec. 2, where the magnetization m(z) is a scalar and which describes very anisotropic uniaxial
magnetic systems only; rather one considers a n-component vector field
$\vec{m}$ $(\vec{r})$ which hence encompasses the Heisenberg model (n=3), the XY-model
(n=2), the Ising model (n=1), polymer problems (n=0), and the spherical

model limit ($n \rightarrow \infty$). So instead of Eqs.(7), (8) the starting free energy function (the "Ginzburg-Landau-Wilson-Hamiltonian") is [assuming full isotropy in the bulk and a semi-infinite system with one free surface at z=0]

$$\frac{\mathcal{H}}{T} = \int d\vec{r} \; \{ \frac{1}{2} A \vec{m}^2(\vec{r}) + \frac{1}{4} B [\vec{m}^2(\vec{r})]^2 - \vec{H} \vec{m}(\vec{r}) + \frac{1}{2} C [\nabla \vec{m}(\vec{r})]^2 \} + \mathcal{H}_s^{(bare)}/T, \quad (26)$$

with

$$\mathcal{H}_s^{(bare)}/T = \int d\vec{r} \; \delta(z=0) [\frac{1}{2} C \lambda^{-1} \vec{m}^2(\vec{r}) - \vec{H}_1 \vec{m}(\vec{r}) + \frac{1}{2} C \lambda_1^{-1} m_1^2(\vec{r})] . \quad (27)$$

Eqs.(26), (27) are the obvious generalization of Eq.(7) to an n-component system which is fully isotropic in the bulk, while a surface anisotropy (which is assumed of uniaxial type in Eq.(27), favouring the 1-direction if $\lambda_1^{-1} < 0$) is included. Renormalization group theory shows that this surface anisotropy is an "irrelevant perturbation", as far as the ordinary transition is concerned. This means, the asymptotic critical exponents β_1, γ_1, γ_{11}, etc. remain unaltered. They are the same as the exponents for the fully isotropic situation, irrespective of the strength of this anisotropy [Note, however, that the width of the asymptotic critical region where these exponents actually are observed shrinks when the anisotropy increases, as will be discussed below] . On the other hand, surface anisotropy is relevant for the surface transition (in the case of uniaxial anisotropy the surface transition asymptotically is Ising-like, irrespetive of the number n of spin components) as well as for the special transition, where it leads to new kind of multicritical behaviour, the so-called "anisotropic special transition" (Diehl and Eisenriegler 1985, Ohne and Okabe 1984).

For the ordinary transition, renormalization group yields the exponents β_1, γ_1, γ_{11} etc. as systematic expansion in either the parameter $\epsilon = 4-d$ where d is the dimensionality [since the mean-field exponents become exact for d > 4] or in the parameter $\frac{1}{n}$ [since the limit n→∞ also is exactly soluble (Bray and Moore 1977)], as done by Ohno and Okabe (1983). E.g., the exponents η , β_1 following from the (second-order) ϵ-expansion are (Diehl and Dietrich 1981)

$$\eta_{\parallel} = 2 - \frac{n+2}{n+8}\,\epsilon - \frac{(n+2)\,(17n+76)}{2\,(n+8)^3}\,\epsilon^2, \tag{28}$$

$$\beta_1 = 1 - \frac{3}{2(n+8)}\,\epsilon - \frac{3(n+2)\,(12-n)}{8(n+8)^3}\,\epsilon^2, \tag{29}$$

while the $\frac{1}{n}$-expansion has been worked out to first order only (Ohno and Okabe 1983a), e.g.

$$\eta_{\parallel} = d - 2 + \frac{S_d}{n}\,\frac{8(d-2)\,(2d-5)}{d}\,\frac{\left[\Gamma(\frac{d}{2}-1)\right]^3\,\Gamma(3-\frac{d}{2})\,\Gamma(2d-6)}{\left[\Gamma(d-2)\right]^2\,\Gamma(d-3)} \tag{30}$$

Expansions for all the other exponents are readily obtained from the surface scaling laws listed above [Eqs.(22,23))] ; a particular merit of the renormalization group approach is that it fully confirms the homogeneity assumptions, Eqs.(21), (25), from which all the surface scaling laws are derived, to all orders in ϵ .

Eqs.(28), (29) are not strictly valid at the "marginal" dimensionality d=4, in the sense that logarithmic corrections appear to the leading power law, e.g.

$$m_1 \propto (-t)^1\,|\ln(-t)|^{3/(n+8)} \tag{31}$$

Eq.(31) was first proposed for an experimentally relevant "marginal" case, namely for uniaxial ferroelectrics and dipolar ferromagnets (Kretschmer and Binder 1979), such as $LiTbF_4$, the critical bulk behaviour to leading order of which corresponds to that of a short-range Ising model at d=4.

Another important success of the renormalization group approach is that it not only yields an expansion for the leading critical exponents but also identifies the physical sources which give rise to correction terms (and hence limit the critical region), and estimate the associate exponents (Diehl and Eisenriegler 1982). For m_1 one obtains

$$m_1 \propto (-t)^{\beta_1}\,\{1 + C_1\,(-t)^{\nu_b y_\lambda} + C_2(-t)^{\nu_b y_A} + C_3(-t)^{\nu_b w_b}\} \tag{32}$$

Here C_1, C_2 and C_3 are non-universal amplitudes; w_b is the bulk correction-to-scaling exponent; $y_\lambda \cong 1$ is the exponent associated with corrections due to the finiteness of λ^{-1} in the starting hamiltonian, Eq.(27); finally y_A arises from surface anisotropies, i.e. the term $C \lambda_1^{-1} m_1^2$ in Eq.(27)

$$y_A = 1 - \frac{n}{n+8} \varepsilon - [\frac{5n+22}{(n+8)^3} \frac{5}{3} n] \varepsilon^2 \approx 0.59 \ (n=3, d=3). \qquad (33)$$

Since surface anisotropies thus lead to correction terms $c_2 |t|^{0.4}$, even relatively small anisotropies may lead to measurable deviations from the asymptotic behaviour for $t \approx 0.2$.

Next we mention the renormalization group results for the isotropic special transition $(\lambda_1^{-1} \equiv 0)$, where now the crossover exponent Φ in addition to $\eta_{\parallel}^m$ must be calculated (Diehl and Dietrich 1983)

$$\eta_{\parallel}^m = -\frac{n+2}{n+8} \varepsilon - \frac{5(n+2)(n-4)}{2(n+8)^3} \varepsilon^2, \qquad (34a)$$

$$\varphi = \frac{1}{2} - \frac{n+2}{n+8} \frac{1}{4} \varepsilon + \frac{(n+2)}{8(n+8)^3} [8 \pi^2 (n+8) - (n^2+35n+156)] \varepsilon^2. \qquad (34b)$$

Ohno and Okabe (1983) obtain $\eta_{\parallel}^m$ in the 1/n-expansion as

$$\eta_{\parallel}^m = d-4 + \frac{1}{n} \frac{4(4-d)}{\Gamma(d-3)} [\frac{(6-d)\ \Gamma(2d-6)}{d\ \Gamma(d-3)} + \frac{1}{\Gamma(5-d)}], \ 3<d<4. \qquad (35)$$

Note that the surface transition for d=3, n>2 no longer exists, and hence the isotropic special transition also does not exist. For n=2 one expects that the surface transition is of the Kosterlitz-Thouless type, leading to a state with algebraically decaying correlations instead of true order (for a qualitative discussion, see Binder 1983).

However, in the presence of surface anisotropy the surface transition is predicted to be of Ising type (n≈1) for all n (Diehl and Eisenriegler 1984, Ohno et al.1985), and the line of surface transitions merges with the ordinary transition {for which surface anisotropy is "irrelevant" and gives rise to a correction term only, see Eq.(32) } at the "aniso-tropic special transition".

There the correlation involving spin components in the "easy" (e) direction decay differently from those involving the (n-1) hard (h) components; so one must distinguish between $\eta_{\|,e}$ and $\eta_{\|,h}$ (Diehl and Eisenriegler 1984).

$$\eta_{\|,e} = \eta_{\|}^m + 2(n-1)\left[\frac{\epsilon}{(n+8)} + \frac{4n+2}{(n+8)^3}\epsilon^2\right], \tag{36a}$$

$$\eta_{\|,h} = \eta_{\|} + 2\left[\frac{1}{n+8}\epsilon + \frac{10n+50}{(n+8)^3}\epsilon^2\right], \tag{36b}$$

and one must also distinguish between two crossover exponents [in a scaling relation similar to Eq. (25b), two variables relating to λ^{-1} and λ_1^{-1} instead of the one variable g are needed]. The corresponding 1/n expansion results (Ohno et al. 1985) are now valid for 2<d<4 - the anisotropic special transition does exist for $n \gtrsim 2$ and d = 3.

It must be emphasized, however, that all the expansions quoted for the exponents are asymptotic expansions only, and using equations such as Eq.(29) for d=3, i.e. ϵ =1 (or Eq.(30) for n=1,2,3) yields results the accuracy of which is rather uncertain. Diehl (1982) suggests that for the ordinary transition and n=1 $0.78 \lesssim \beta_1 \lesssim 0.82$, while for n=3 $0.81 \lesssim \beta_1 \lesssim 0.88$. These values result from either direct use of Eq.(29) or use of Eqs.(28), (23) together with the appropriate value of ν_b. For the n=1 special transition, $\Phi \approx 0.68$ and $0.19 \lesssim \beta_1^m \lesssim 0.25$. Of course, the true accuracy of such extrapolations can only be asserted by other methods. Thus it is gratifying that the high temperature series (Ohno et al. 1984) yields $\beta_1 = 0.77 \pm 0.02$ (n=1) and $\beta_1 = 0.81 \pm 0.04$ (n=3), for the ordinary transition, in fair overlap with the renormalization group results. These results for n=3 are also in reasonable agreement with the SPLEED data for m_1 in Ni due to Alvarado et al. (1982).

Finally we stress that the renormalization group theory not only yields exponents but also makes predictions for other universal quantities, such as "scaling functions" { f_s in Eq.(25a) or suitable derivatives, or $\tilde{m}$ (ζ) in Eq. (24)} or "critical amplitude ratios". By critical amplitudes one means prefactors in power laws such as Eqs.(2-6).

Thus, writing $\xi_b = \xi_b^{\pm} |t|^{-\nu_b}$, $\hat{m}_b = m_b(-t)^{\beta_b}$, $m_s = \hat{m}_s(-t)^{\beta_s}$, $\chi_b = \hat{\chi}_b^{\pm} |t|^{-\gamma_b}$, $\chi_s = \chi_s^{\pm} |t|^{-\gamma_s}$, etc., we define critical amplitudes $\xi_b^{\pm}$, $\hat{m}_b$, $\hat{m}_s$, etc. Now the rule is that from quantities the exponents of which combine into a scaling relation one also may form an amplitude combination which is a dimensionless universal constant (Okabe and Ohno 1984, Diehl et al. 1985). These can again be calculated in an ε-expansion. E.g., some universal amplitude ratios corresponding to scaling relations Eq.(22a) are, to first order in ε (Diehl et al.1985)

$$\hat{m}_s/(\hat{m}_b\,\xi_b^-) = -2\,\ln 2\,(1-0.05\varepsilon), \quad \hat{m}_s^m/(\hat{m}_b\xi_b^-) = \pi\varepsilon/6, \tag{37a}$$

$$\hat{\chi}_s^+/(\hat{\chi}_b^+\,\xi_b^+) = -\left[1 - \frac{n+2}{n+8}\,\varepsilon\pi\left(\frac{1}{\sqrt{3}} - \frac{1}{2}\right)\right], \quad \hat{\chi}_s^-/(\hat{\chi}_b^-\,\xi_b^-) = \frac{3}{2}\,(1+0.94\varepsilon). \tag{37b}$$

However, a comparison with the available Monte Carlo results reveals that these results for $\varepsilon=1$ are no longer quantitatively reliable. But again the renormalization group approach is valuable as it justifies three-scale-factor universality for the surface free energy at the ordinary transition-fixing scales of H, H_1 and t in Eq.(25a), which involves three material dependent factors, one is left with one universal function f_s. At the (isotropic) special transition, there is another scale-factor for g [Eq. (25b)], and hence one has four-scale-factor universality. These statements are the obvious generalizations of the well-known two-scale-factor universality in the bulk (Stauffer et al.1972).

3.4. Monte Carlo Simulations

Early work on the Ising model with free surfaces (Binder 1974, Binder and Hohenberg 1974) yielded evidence in favour of the scaling relation for the shift exponent, $\lambda_s = 1/\nu_b$ {Eq.(22c)} . The estimate for β_1 obtained for the simple cubic nearest neighbour Ising ferromagnet, exchange J in the bulk and exchange $J_s = J$ in the surface plane, was $\beta_1 \approx 0.66 \pm 0.1$ but it is now clear that this value is too small because of crossover towards the special transition. Landau (1976b) studied the surface excess quantities C_s, m_s, χ_s, also for a square lattice with free (one- dimensional) boundaries (Landau 1976a), and confirmed the corresponding scaling relations {Eq.(22a) }. He also estimated critical amplitudes and found the critical amplitude ratio $\hat{\chi}_s^+/\hat{\chi}_s^- \approx 3.14(d=2)$ and $\hat{\chi}_s^+/\hat{\chi}_s^- \approx 0.78$ (d=3), while the corresponding ε-expansion is (Diehl et al.1985) $\hat{\chi}_s^+/\hat{\chi}_s^- = -\frac{4\sqrt{2}}{3}$ (1-0.15ε) and hence disagrees with the Monte Carlo data for ε =1,2 dramatically.

More complete studies of the ordinary and special transition were recently performed by Binder and Landau (1984,1986) and by Kikuchi and Okabe (1985); Fig. 4 shows the surface layer magnetization for a variety of values of J_s/J and t. It is seen that for $J_s/J \lhd 1$ the slope on the log-log plot is independent of J_s/J and thus can be interpreted as $\beta_1 \approx 0.78^+_-0.02$. This conclusion is corroborated by additional but less extensive data of Kikuchi and Okabe (1985); These authors also study χ_1 and the variation of m_1 at T_c with H and H_1, and hence check some of the additional scaling relations compiled in Eq.(22). For $J_s/J \gtrsim 1$ the slope in Fig. 4 is steadily decreasing with increasing J_s/J and hence can only be interpreted as an "effective exponent"; this behaviour must be interpreted as a crossover towards the special transition, which is estimated to occur for $J_{sc}/J \approx 1.50^+_-0.03$. This implies for the multicritical value $\beta_1^m \approx$ 0.175$^+_-$0.025, somewhat smaller than the ε-expansion estimate quoted above. In fact, the data of Fig. 4 are consistent with the crossover scaling hypothesis, Eq.(25b), see Fig. 5b: all the curves of Fig. 4 collapse on two branches - one for the surface transition (upper branch) and one for the ordinary transition (lower branch.). The crossover exponent is esti- mated as $\varphi \approx 0.56^+_-0.04$, again distictly smaller than the ε-expansion estimate quoted above; but the present estimate is also corroborated from a study of the surface transition temperature $[T_{cs}/T_{cb}-1 \propto (J_s/J_{sc}-1)^{1/\varphi}]$, see Fig. 5a. Finally Fig. 6 shows the scaling of the mag- netization profile, thus confirming Eq.(24). In fact, these data are in rather good agreement with corresponding ε-expansions (Diehl et al. 1985) for the scaling function $\tilde{m}$.

Much less definitive results exist so far for the Heisenberg model with free surfaces. Early work(Binder and Hohenberg 1974) studied a lattice as small as 16^3, with one free 16x16 surface; at the opposite surface an effective field (simulating bulk behaviour) was acting. Fig. 7 shows

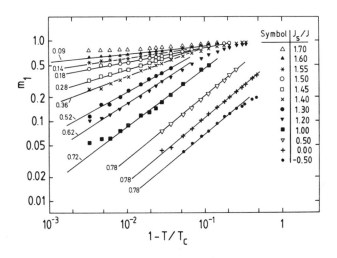

Fig. 4: Log-log plot of surface-layer magnetization vs reduced temperature, for various ratios of the exchange J_s in the surface planes
to the exchange J in the bulk. Slopes of the straight lines, as indicated, yield effective exponents β_1^{eff}. Data are from Monte
Carlo simulation of 50 x 50 x 40 lattices with two free 50 x 50
surfaces (and otherwise periodic boundary conditions) using preferential surface sites sampling. From Binder and Landau (1984).

that the magnetization near the surface is rather strongly perturbed,
$m_b-m(z) \propto 1/z$ rather than the exponential variation $m_b-m(z) \propto \exp(-z/\xi_b)$
found in the Ising model {see e.g. Binder and Landau (1984)} . This implies
that in the Heisenberg case for H=0 the surface magnetization is infinite,
Eq.(4) makes sense for H≠0 only. This long-range perturbing effect of a
surface can already be understood by spin-wave theory; in fact, at low temperatures the data are in reasonable quantitative accord with spin wave predictions (Fig.8).

Only rather crude data are available for the critical region, however;
Binder and Hohenberg (1974) suggested from their data $\beta_1 \approx 0.75 \pm 0.10$, which
is not too far from the ε-expansion results but less accurate; Müller-

118

Fig.5: a) Log-log plot of the
difference between sur-
face critical temperatu-
re T_{cs} and bulk critical
temperature T_c vs. $J_s - J_{sc}$.
Monte Carlo estimates
are from the analysis of
m_1 for $J_s/J=1.6$ and 1.7;
series expansion estima-
tes are based on series
for χ_{11} at $J_s/J=1.8$
and 2.0 (Binder and Ho-
henberg 1974).
b) Scaled surface layer
magnetization $m_1(T_c/T-1)^{-\beta_1^m}$
plotted vs. the scaling
variable $x=(T_c/T-1)^{-\varphi}$
$|J_s/J_{sc}-1|$, for $J_{sc}/J=1.5$,
$\beta_1^m=0.175$ and $\varphi=0.56$. Arrow
denotes data for m_1 at J_{sc},
broken straight lines show
asymptotic slopes of the
scaling function for large x.
From Binder and Landau
(1984).

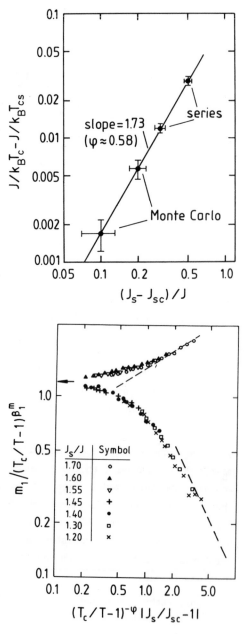

Krumbhaar (1976) studying m_1 at T_{cb} as function of H found $\delta_1=2.3\pm0.1$.
Again these estimates are consistent with the surface scaling laws, but
a better accuracy clearly would be desirable.

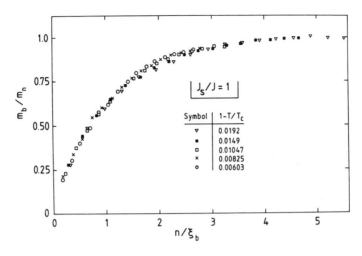

Fig. 6: Scaling plot for the magnetization profile of a semi-infinite
Ising magnet at $J_s/J=1$. From Binder and Landau (1984)

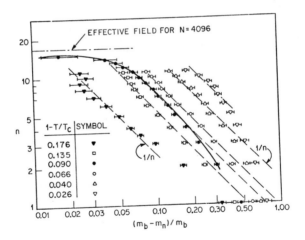

Fig. 7: Magnetization m_n in the n'th layer of a simple cubic Heisenberg
ferromagnet with a free surface at n=1 (and a self-consistent
effective field acting at n=16) at various temperatures. Applying this effective field at all boundaries m_b was estimated
independently. From Binder and Hohenberg (1974)

Finally we turn to studies of Ising ferromagnets with antiferromagnetic
surface exchange (Binder and Landau 1985). This models "magnetic surface reconstruction", as noted above. The resulting phase diagram
(Fig. 9) is qualitatively similar to the mean-field result (Fig. 3).
The transition line of the antiferromagnetic surface ordering is hardly affected by the onset of order in the bulk; it always stays rather
close to the asymptote which it reaches for $J_s/J \to -\infty$. This behaviour
means that the surface layer basically does not notice much the coupling to the underlying ferromagnetic order in the bulk.Even at the multicritical point (which occurs near $J_s/J \approx -2.0$), where simultaneously

Fig.8: Normalized magnetization m_n at the
n'th layer of a clas-
sical Heisenberg ferro-
magnet with a free
surface at layer 1
for low temperatures.
Arrows show lowest-
order spin wave pre-
dictions.From Binder
and Hohenberg (1974)

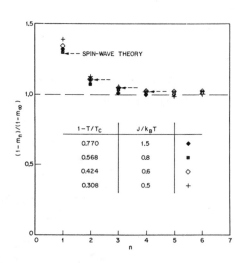

$1-T/T_C$	J/k_BT	
0.770	1.5	◆
0.568	0.8	■
0.424	0.6	◇
0.308	0.5	+

bulk ferromagnetic critical behaviour and antiferromagnetic surface
critical behaviour occur, the two-dimensional critical behaviour of the
antiferromagnetic surface (exponent $\beta=1/8$) and the ordinary transition
in the second layer (exponent $\beta_1 \tilde{=} 0.78$) do not influence each other (Fig.
10). More interesting behaviour is expected again for isotropic magnets
with antiferromagnetic surface exchange, but no numerical results are
available as yet.

Fig. 9: Monte Carlo results
for the transition
temperature of a
simple cubic Ising
lattice with exchange
J_s in the surface
different from the
bulk. Broken straight
lines show asymptotic
behaviour of $T_N(J_s, J)$
for $J_s/J \to -\infty$, and the
bulk transition tempera-
ture T_{cb}. Phases shown
are denoted as in Fig.3.
From Binder and Lan-
dau (1985).

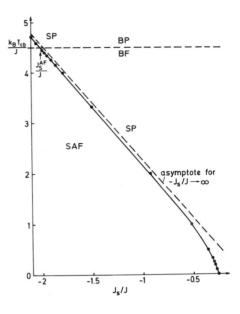

3.5. Summary

There is now a wealth of predictions from renormalization group
theory for the static critical behaviour of surfaces of ferromagnets,
including second-order ε -expansion results for the critical exponents, both at the ordinary and the isotropic as well as anisotropic
special transitions; also some results for corrections to scaling,
various scaling functions and amplitude ratios are available - the
latter only to first order in ε , however. The general framework of
the surface scaling theory and the classification of the various
transitions at the surface now seems well understood.

Unfortunately, only a part of these predictions so far has been
tested by Monte Carlo computer simulations, and even less results on
surface critical behaviour from experiment on real materials are available. {So far, evidence for critical behaviour at the ordinary
transition in Ni has been presented [Alvarado et al.1982] , and evi-

dence for $T_{cs} > T_{cb}$ in Gd (0001) has been found(Weller et al.1985)} .
This is clearly a field where, in spite of the enormous technical
difficulty, much more work is needed. Very likely, some surprises
in this area may be forthcoming, and stimulate further theoretical
developments.

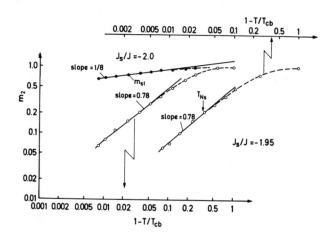

Fig. 10: Log-log plots of the staggered magnetization in the first
layer, m_{st}, vs. the temperature distance from the antiferromagnetic ordering temperature T_{NS} at $J_s/J = -2.0$ and of the
second layer magnetization m_2 vs. the temperature distance
from T_{cb} at $J_s/J = -2.0$ and -1.95. Note the shifts in coordinate scales. Straight lines indicate the respective
exponents as quoted. From Binder and Landau (1985).

3.6. References

Alvarado SF, Campagna M and Hopster H 1982 Phys. Rev.Lett. $\underline{4851}$

Barber MN 1973 Phys.Rev.$\underline{B8}$ 407

Barber MN 1983 in Phase Transitions and Critical Phenomena Vol 8 ed
    C Domb and JL Lebowitz (New York: Academic Press)

Binder K 1983 in Phase Transitions and Critical Phenomena Vol 8 ed
    C Domb and JL Lebowitz (New York: Academic Press)

Binder K and Hohenberg PC 1972 Phys.Rev.$\underline{B6}$ 3461

—————— 1974 Phys.Rev.$\underline{B9}$ 2194

Binder K and Landau DP 1984 Phys.Rev.Lett.$\underline{52}$ 318

—————— 1985 Surf.Sci. $\underline{151}$ 409

—————— 1986 in preparation

Bray AJ and Moore MA 1977 Phys.Rev.Lett. $\underline{38}$ 735

Burkhardt TW and Eisenriegler E 1981 Phys.Rev. $\underline{B24}$ 1236

—————— 1985 preprint

Cardy J 1983 J.Phys.$\underline{A16}$ 3617

—————— 1984 J.Phys.$\underline{A17}$ L385

—————— 1985 preprint

De Gennes PG 1979 Scaling Concepts in Polymer Physics (Ithaca: Cornell
    University Press)

Diehl HW 1982 J.Appl.Phys. $\underline{53}$ 7914

Diehl HW and Dietrich S 1980 Phys.Lett.$\underline{80A}$ 408

—————— 1981a Z.Phys.$\underline{B42}$ 65

—————— 1981b Phys.Rev.$\underline{B24}$ 2878

—————— 1981c Z.Phys.$\underline{B43}$ 281

—————— 1983 Z.Phys.$\underline{B50}$ 117

Diehl HW and Eisenriegler E 1982 Phys.Rev.Lett. $\underline{48}$ 1767

—————— 1984 Phys.Rev.$\underline{B30}$ 300

Diehl HW, Dietrich S and Eisenriegler E 1983 Phys.Rev.$\underline{B27}$ 2937

Diehl HW, Gompper G and Speth W 1985 Phys.Rev.$\underline{B31}$ 5841

Dietrich S and Diehl HW 1981 Z.Phys.$\underline{B43}$ 315

Dietrich S and Wagner H 1984 Z.Phys.$\underline{B56}$ 207

Eisenriegler E, Kremer K and Binder K 1982 J.Chem.Phys. $\underline{77}$ 6296

Fisher ME 1974 Rev.Mod.Phys. $\underline{46}$ 597

Goldschmidt YY 1983 Phys.Rev.$\underline{B28}$ 4052

Goldschmidt YY and Jasnow D 1984 Phys.Rev.B29 3990

Guttmann AJ and Reeve JS 1980 J.Phys.A13 2495

Guttmann AJ, Torrie GM and Whittington SG 1980 J.Magn.Mag.Mat.15-18 1091

Kikuchi M and Okabe Y 1985 Progr.Theor.Phys.73 32

Kretschmer R and Binder K 1979 Phys.Rev.B20 1065

Landau DP 1976a Phys.Rev.B13 2997

—— 1976b Phys.Rev.B14 255

Lipowsky R 1982 Phys.Rev.Lett. 49 1575

—— 1984 J.Appl.Phys.55 2485

Lubensky TC and Rubin MH 1975a Phys.Rev.B12 3885

—— 1975b Phys.Rev.B11 4533

Müller-Krumbhaar H 1976 J.Phys.C9 345

Nemirovsky AM and Freed KF 1985 Phys.Rev.B31 3161

Okabe Y and Ohno K 1984 Phys.Rev.B30 6573

Ohno K and Okabe Y 1983 Progr.Theor.Phys.70 1226

—— 1984 Progr.Theor.Phys.72 736

Ohno K, Okabe Y and Morita A 1984 Progr.Theor.Phys.71 714

Pandit R, Schick M and Wortis M 1982 Phys.Rev.B26 5112

Reeve JS 1981 Phys.Lett.81A 237

Reeve JS and Guttmann AJ 1980 Phys.Rev.Lett. 45 1581

Reeve JS and Guttmann AJ 1981 J.Phys.A14 3357

Rudnick J and Jasnow D 1982 Phys.Rev.Lett. 48 1059

Stauffer D 1979 Phys.Repts. 54 1

Stauffer D, Ferer M and Wortis M 1972 Phys.Rev.Lett.29 345

Trullinger SE and Mills DL 1973 Solid State Commun.12 819

Weller D, Alvarado SF, Gudat W, Schröder K and Campagna M 1985 Phys.Rev.
    Lett. 54 1555

Whittington SG,Torrie GM and Guttmann AJ 1979 J.Phys.A12 2449

Wu FY 1982 Rev.Mod.Phys. 54 235

Chapter 4

Principles and Theory
of
Electron Scattering and Photoemission

R. Feder

Theoretische Festkörperphysik, FB 10, Universität Duisburg GH,
D-4100 Duisburg, FRG

4.1 Introduction

The previous Chapters dealt with bound polarized electrons,
which are part of the surface system (solid/vacuum interface region
with or without adsorbed "foreign" atoms) and determine its magnetic
structure both in the zero-temperature ground state and at elevated
temperatures, in particular near the ferromagnetic to paramagnetic
phase transition temperature. This Chapter reviews the conceptual
foundation and the theoretical treatment of a variety of scattering
methods, in which free polarized electrons, incident on the surface or
emitted from it, are employed to obtain information on physical properties (geometrical, electronic and magnetic) of the surface system.

We recall (for a more detailed introduction cf. Kessler 1976/85)
that the spin polarization of a "beam" of free polarized electrons is
characterized (in the non-relativistic limit) by a polarization vector

$$\underset{\sim}{P} = tr[\underset{\sim}{\sigma}\rho]/tr[\rho] , \tag{1}$$

where ρ is the statistical operator (density matrix) describing the
beam, and the vector $\underset{\sim}{\sigma}$ comprises the (2x2) Pauli spin matrices

$$\sigma_1 = \begin{pmatrix} 0 & 1 \\ 1 & 0 \end{pmatrix}, \quad \sigma_2 = \begin{pmatrix} 0 & -i \\ i & 0 \end{pmatrix}, \quad \sigma_3 = \begin{pmatrix} 1 & 0 \\ 0 & -1 \end{pmatrix} \tag{2}$$

$P = |\underset{\sim}{P}|$ is referred to as the degree of polarization, with $P = 1$ corresponding to complete (100 %) polarization. With respect to a given pre-
ferential direction $\underset{\sim}{e}$, the polarization is $\underset{\sim}{P}\cdot\underset{\sim}{e} = (N^+-N^-)/(N^++N^-)$, where
$N^+(N^-)$ is the number of electrons with spin parallel (antiparallel)
to $\underset{\sim}{e}$.

As schematically indicated in Fig. 1, the relevant scattering
methods can be classified according to their input and output channels
as follows. Given an incident electron beam of kinetic energy E, momen-
tum $\underset{\sim}{k}$ and polarization vector $\underset{\sim}{P}$, there are three distinct possibilities:
(a) detection (measurement) of elastically scattered electrons with

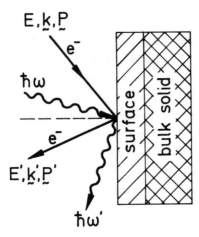

Fig. 1:

Schematic indication of
scattering methods invol-
ving free polarized elec-
trons (e⁻) and linearly
or circularly polarized
photons ($\hbar\omega$).

$E' = E$, $\underset{\sim}{k}'$ and $\underset{\sim}{P}'$ (elastic electron scattering or, more specially for
crystals, elastic low-energy electron diffraction (LEED)); (b) detec-
tion of outgoing electrons with $E' < E$, $\underset{\sim}{k}'$ and $\underset{\sim}{P}'$; this comprises inelastic electron scattering (energy loss spectroscopy), Auger emission
and true secondary electron emission (SEE); (c) detection of outgoing
photons with $\hbar\omega' \leq E - E_F$ (bremsstrahlung, appearance potential spectroscopy). For incident electromagnetic radiation (photons $\hbar\omega$), the
most interesting polarized-electron emission process is photoemission.
(Processes with photons in both the input and output channel, like
Compton or Raman scattering, are not considered here). While the energies involved will be mentioned later when discussing the various methods, it should already be noted here that electrons with energies
below about 200 eV have mean free paths in solids of the order of a
few nearest-neighbour distances and are therefore particularly suited
for studying the surface region, which typically comprises a few mon-
atomic layers parallel to the surface. The observable quantities, in
which the information on the surface system is coded, are the cross

section (intensity) and polarization of the outgoing electrons or photons as functions of the polarization of the incident projectiles, of the energies and of the scattering geometry. To extract the maximal physical information from measurements of these quantities, numerical calculations on the basis of a quantitative theory of the respective technique are required.

The Chapter is organized as follows. A prerequisite common to all the above methods, the interaction of an ingoing or outgoing electron with the surface system, is discussed in Section 4.2 and represented by an effective one-electron Dirac Hamiltonian, which contains a magnetic field in cases with magnetic ordering. The subsequent Sections deal with the individual polarized-electron methods, explaining the physical principles, outlining formal theoretical treatment and presenting some general theoretical results. (More specific theoretical results will be shown in Part II of the book in conjunction with experimental data). Section 4.3 addresses the theoretically simplest method, elastic scattering, with an emphasis on spin-polarized low-energy electron diffraction (SPLEED) from systems with perfect two-dimensional lattice periodicity parallel to the surface. Inelastic electron scattering and true secondary electron emission are treated in Section 4.4. Photoemission mainly in the (vacuum) ultraviolet regime ($\hbar\omega$ below about 70 eV) is introduced in Section 4.5, with special emphasis on ferromagnets at low and near-T_c temperatures, and on spin-orbit coupling effects (break-down of nonrelativistic dipole selection rules and spin polarization of the photocurrent) for non-magnetic materials. Section 4.6 briefly deals with bremsstrahlung ("inverse photoemission") and its dependence on the spin polarization of the incident electron beam. The final Section 4.7 summarizes the state of the art of the theory of spin-dependent electron scattering and emission, and concludes with a remark on the relation between theory and experiment.

4.2 One-Electron Picture

The electron spectroscopic methods, which will be introduced in the following Sections, all have in common that an electron is added to or removed from a surface system (most generally a ferromagnetic solid with atoms of a different species adsorbed at its surface), i.e. we are dealing with single particle excitations of an inhomogeneous many-electron system. To make this complicated many-body problem theoretically tractable, it is necessary to reduce it to an effective one-electron problem with several simplifying assumptions.

4.2.1 Relativistic Electron-Ferromagnet Hamiltonian

The ground state of a ferromagnetic surface system is successfully described by spin-density functional formalism, with its one-electron equations involving effective electrostatic and magnetic fields (cf. Chapter 1, monographs by Lundqvist and March (1983), Dahl and Avery (1984), Callaway and March (1984), and references therein). To account quantitatively for spin-orbit coupling and further relativistic effects (cf. also Malli (1983)), especially when dealing with large-Z materials, a relativistic version is required (for details cf. Ramana and Rajagopal (1983)). For the present problem of describing one-electron excitations of the surface system it would thus seem adequate to employ an excited-state version of relativistic density functional theory (as proposed nonrelativistically by Sham and Kohn (1966)), i.e. a Dirac equation involving as an "optical potential" a complex, non-local and energy-dependent self-energy matrix, which is a functional of the charge and spin density of the ground state. Since non-locality implies an integro-differential equation, the solution of which is beyond present-day means, one has to resort to a local approximation for the self-energy operator, which is then expressed in terms of an effective electrostatic potential $\tilde{V}(E,\underset{\sim}{r})$ and an effective magnetic vector potential $A(E,r)$, where E is the one-electron energy. Retaining, as in the ground state formalism (cf. Chapter 1), only the magnetic dipole part of the coupling to $\underset{\sim}{A}$ by means of an effective magnetic field $\tilde{B}(E,\underset{\sim}{r})$, one has the effective Dirac Hamiltonian

$$H = c\underset{\sim}{\alpha}\underset{\sim}{p} + \beta mc^2 + V(E,\underset{\sim}{r}) - \beta\bar{\underset{\sim}{\sigma}}\underset{\sim}{B}(E,\underset{\sim}{r}) , \qquad (3)$$

where c is the velocity of light, m the electron rest mass. $V(E,r) = e\widetilde{V}(e,\underset{\sim}{r})$ and $B(E,\underset{\sim}{r}) = e\hbar/(2mc)\widetilde{\underset{\sim}{B}}(E,r)$, with the electron charge $e = -|e|$. The (4x4) matrices $\underset{\sim}{\alpha} = (\alpha_1,\alpha_2,\alpha_3)^T$, β and $\bar{\underset{\sim}{\sigma}} = (\bar{\sigma}_1,\bar{\sigma}_2,\bar{\sigma}_3)$ can be expressed in terms of the (2x2) Pauli matrices $\underset{\sim}{\sigma} = \underset{\sim}{\rho}$ (cf. eq.(2)) as the tensor products

$$\alpha_k = \rho_1\otimes\sigma_k, \quad \beta = \rho_3\otimes 1_2, \quad \bar{\sigma}_k = 1_2\otimes\sigma_k; \quad k = 1,2,3 . \qquad (4)$$

Eq.(3) is distinct from the formally identical one-electron Hamiltonian of ground state theory, since V and $\underset{\sim}{B}$ are energy-dependent and complex, with the imaginary parts accounting for spin-dependent lifetimes.

In the nonrelativistic limit, with terms up to the order $1/c^2$, the Dirac Hamiltonian eq.(3) is approximated by the (2x2) form (cf. textbooks on quantum mechanics)

$$H = \frac{p^2}{2m} + V(E,\underset{\sim}{r}) - \underset{\sim}{\sigma}\cdot\underset{\sim}{B}(E,\underset{\sim}{r}) -$$

$$- \frac{p^4}{8m^3c^2} + \frac{\hbar^2}{8m^2c^2} \Delta V(E,\underset{\sim}{r}) + \frac{\hbar}{4m^2c^2} \underset{\sim}{\sigma}\cdot[\nabla V(E,\underset{\sim}{r})\times\underset{\sim}{p}] . \qquad (5)$$

The first line is the standard Pauli Hamiltonian; the second line gives the relativistic mass correction, the Darwin term and spin-orbit coupling, which reduces to the familiar $\underset{\sim}{\sigma}\cdot\underset{\sim}{L}$ form in the case of a central field. Rigorously, an approximation to the order $1/c^2$ has some further terms involving $\underset{\sim}{\sigma}$ and $\underset{\sim}{B}$. Practically, these terms are however negligible, as has been explicitly demonstrated by electron scattering calculations for 3d ferromagnets and for Gd (Ackermann and Feder 1984). This is also plausible, since there is already the lower-order Pauli term in eq.(5), which determines magnetic spin effects (like exchange splitting of electronic energy bands and exchange-induced scattering asymmetries). If spin-orbit coupling is neglected and if the effective magnetic field $\underset{\sim}{B}(E,\underset{\sim}{r})$ has a uniform direction throughout the system, chosen as the z-axis, the (2x2) matrix eq.(5) is diagonal, consisting of two "scalar relativistic" (or - without the mass correction and Darwin terms - Schrödinger) Hamiltonians with a spin-up potential $V^+(E,\underset{\sim}{r})$ and a spin-down potential $V^-(E,\underset{\sim}{r})$, respectively, given as

$$V^{\pm}(E,\underset{\sim}{r}) = V(E,\underset{\sim}{r}) \mp B(E,\underset{\sim}{r}) \quad , \tag{6}$$

where $B = \underset{\sim}{B} \cdot \hat{\underset{\sim}{z}}$. Whether spin-orbit coupling, which is always present in reality, may be neglected or not, depends on the physical properties and mechanisms one wishes to study, and on the atomic number Z of the material (recalling that spin-orbit interaction increases with Z). If one is, for example, interested in spectral fine structure, it is indispensable already for hydrogen. In studying surface magnetism with the aid of polarized free electrons, the possibility of neglecting spin-orbit coupling will be discussed later in conjunction with electron scattering (Section 4.3.7) and photoemission (Section 4.5.7). If it has to be included, calculations should be based on the Dirac Hamiltonian eq.(3) rather than on its two-component approximation eq.(5). This is - as a consequence of the Coulombic potential near the nuclei - necessary for quantitative accuracy especially for large-Z elements (cf. Meister and Weiss 1968). With regard to computing time and storage, eq.(3) (implying four first-order differential equations) is no more demanding than eq.(5) (two second-order differential equations).

The above framework is of course a fortiori valid for nonmagnetic systems (for which $\underset{\sim}{B}$ in eq.(3) is identically zero). For ferromagnetic systems it corresponds to the "itinerant" or band theory ("Stoner model") of ferromagnetism (cf. Chapter 2 of this book, reviews by Jelitto and Ziegler (1983), Moriya (1984) and Wohlfarth (1984), and references therein) if the effective magnetic field $\underset{\sim}{B}(E,\underset{\sim}{r})$ is parallel to a global quantization axis and has the same (position-dependent) values in all the atomic cells containing the same chemical species. This "mean-field" version is appropriate only at temperatures, which are low enough for the effects of thermal spin fluctuations (on the physical quantity under consideration) to be negligible. For higher temperatures, especially in the vicinity of the ferromagnetic phase transition temperature, (Curie temperature), $\underset{\sim}{B}(E,\underset{\sim}{r})$ in eq.(3) or (5) varies from unit cell to unit cell in both direction and magnitude, with some short-range correlation. This corresponds to the modern "fluctuating-local-moment" picture of transition metal ferromagnetism (cf. Chapter 2 and above-cited reviews), and will be pursued somewhat further in conjunction with photoemission theory (Section 4.5.7). We now turn to a specification of the effective fields.

4.2.2 Effective One-Electron Potentials

To know the "optical potentials" V and $\underset{\sim}{B}$, which embody the eletron-solid interaction in eq.(5), is firstly interesting in its own right and secondly indispensable for realistic computational applications of a variety of electron spectroscopy theories (to be presented below). According to eq.(6), V and B can be expressed in terms of spin-dependent effective potentials $V^s(E,\underset{\sim}{r})$ with s = +/- for spin up/down electrons:

$$V = (V^+ + V^-)/2 , \quad B = -(V^+ - V^-)/2 . \tag{7}$$

This nonrelativistic procedure appears reasonable also beyond the non-relativistic limit on the following grounds: relativistic effects originate mainly from very close to the nuclei, whilest the "magnetic" d and f electrons (together with some sp hybridization contribution) have their highest charge density further away. Near the nuclei one thus has $V \approx V^+ \approx V^-$ (and $B \approx 0$), i.e. an almost nonmagnetic situation, whilest the neglect of spin-orbit coupling and further relativistic corrections, which underlies a non-zero spin splitting ($V^+ \neq V^-$, $B \neq 0$) in eq.(6), is tolerable in the region of d or f electron concentration.

The problem is - in the ferromagnetic case - thus reduced to constructing the spin-dependent, energy-dependent and complex potentials $V^S(E,\underline{r})$, which are local approximations to actually non-local self-energies. They are therefore - except at the Fermi energy - fundamentally distinct from the energy-independent real one-electron potentials of ground-state density functional theory, especially when the latter are obtained in the popular local-density approximation to the exchange-correlation energy. Discussions of this issue and further references may be found in reviews by Williams and Barth (1983) and Callaway and March (1984). We shall also briefly return to it in the context of photoemission theory (Section 4.5.1).

The effective one-electron potentials V^S consist of an electrostatic (Hartree) part V_H and a self-energy part Σ^S, which in turn is made up of an exchange part Σ^S_X and a Coulomb correlation part Σ^S_C:

$$V^S(E,\underline{r}) = V_H(\underline{r}) + \Sigma^S(E,\underline{r}), \text{ with } \Sigma^S = \Sigma^S_X + \Sigma^S_C . \tag{8}$$

Spin and energy dependence thus enter via exchange and correlation. Further, V_H and Σ^S_X are real, whilest Σ^S_C is generally complex, with its imaginary part accounting for the finite life-time of quasiparticles, or, in a scattering situation, for the loss of flux out of the elastic channel into the inelastic scattering channels. Deferring a detailed treatment of inelastic processes (e.g. excitation of electron-hole pairs, plasmons and magnons) to a later Section (4.4.4), we are dealing at present only with the sum of their effects on a one-electron state in terms of an imaginary potential contribution $\text{Im}\Sigma^S_C = : V^S_i$. $\text{Re}\Sigma^S_C$ is connected with $\text{Im}\Sigma^S_C$ by a Kramers-Kronig relation (cf. e.g. Hedin and Lundqvist 1969). Inelastic processes therefore also manifest themselves in the real part of V^S, and contribute to its difference from the (real and energy-independent) one-electron potential of ground state density functional formalism.

For an inhomogeneous system with slowly varying ground state spin densities $\rho^S(\underline{r})$, a local-density approximation to Σ^S has been proposed (by Sham and Kohn (1966) and, in its spin-dependent form, by Gunnarsson and Lundqvist (1976)) (using atomic units $\hbar=1$, $m=1$, $|e|=1$):

$$\Sigma^s(E,\underset{\sim}{r}) = \Sigma_h^s(k^s(\underset{\sim}{r}), \rho^+(\underset{\sim}{r}), \rho^-(\underset{\sim}{r})), \text{ with } s = +/- , \qquad (9)$$

where Σ_h^s is the self-energy of a quasiparticle of momentum $k^s(\underset{\sim}{r})$ and spin s in a homogeneous electron gas with spin densities $\rho^+(\underset{\sim}{r})$ and $\rho^-(\underset{\sim}{r})$; the local momentum $k^s(\underset{\sim}{r})$ is defined by

$$(k^s(\underset{\sim}{r}))^2/2 = E - V_H(\underset{\sim}{r}) - \Sigma_h^s(k^s(\underset{\sim}{r}), \rho^+(\underset{\sim}{r}), \rho^-(\underset{\sim}{r})), \qquad (10)$$

or a more approximate form obtained from eq.(10) by replacing $V_H(\underset{\sim}{r})$ by $E_F - E_F^h(\rho^+(\underset{\sim}{r}), \rho^-(\underset{\sim}{r}))$. For a discussion of this approximation see e.g. Hedin and Lundqvist (1971). At the Fermi energy, i.e. for $E=E_F$, $V^s(E,\underset{\sim}{r})$ coincides with the ground state effective potential $V^s(\underset{\sim}{r})$ as obtained in the local-spin-density approximation. Above and below E_F, however, it departs from $V^s(\underset{\sim}{r})$. This is explicitly seen, if Σ_h^s is calculated in the Hartree-Fock approximation, i.e. if exchange but not correlation is taken into account. One then has

$$V^s(E,\underset{\sim}{r}) = V_H(\underset{\sim}{r}) - \alpha^s(E,\underset{\sim}{r}) \, 3 \, (6\rho^s(\underset{\sim}{r})/8\pi)^{1/3} , \qquad (11)$$

$$\alpha^s(E,\underset{\sim}{r}) = \frac{4}{3} [0.5 + \frac{(1-\eta^2)}{4\eta} \ell n \, |\frac{1+\eta}{1-\eta}|] , \qquad (12)$$

with $\eta = k^s/k_F^s$ and $k_F^s = (6\pi^2\rho^s(\underset{\sim}{r}))^{1/3}$. The exchange term in eq.(11) has the well-known "$X\alpha$" form, but with an energy-dependent α, which has the Dirac (1930) / Gaspar (1954) / Kohn-Sham (1965) value $2/3$ at E_F, and decreases (increases) with E above (below) E_F. A slight modification of the exchange term, proposed by Slater, Wilson and Wood (1969), consists in defining k_F^s for eq.(12) along with k^s by eq.(10) (with $E = E_F$). Potentials of the type of eq.(11) have been found superior to (energy-independent) ground-state potentials in scattering situations involving electrons with kinetic energies ranging from about 20 eV to several hundred eV: elastic electron-atom scattering (Awe et al. 1983, Kemper et al. 1983) elastic low-energy-electron diffraction (LEED) (Ford et al. 1982) and spin-polarized LEED (Feder 1981, Feder et a. 1983a and refs. therein). Calculation of Σ_h^S (eq.(9)) in approximations beyond Hartree-Fock, i.e. inclusion of correlation effects, leads to a complex $V^S(E,r)$. The real part of such an effective potential, obtained from random-phase-approximation results of Hedin and Lundqvist (1971), has been successfully employed in LEED calculations by Echenique and Titterington (1977) and Neve et al. (1983). This is compatible with the aforementioned adequacy of an exchange-only real part, since addition of a real correlation part appears to have a comparatively minor effect on scattering results (as was explicitly found in electron scattering calculations from Xe atoms by Awe et al. (1983)).

The real potential thus obtained is shown schematically in Fig. 2 for the case of a non-magnetic semi-infinite crystal $(V_r^+ = V_r^- = V_r)$.

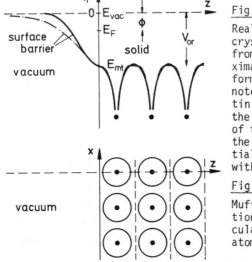

Fig. 2:

Real part V_r of effective electron-crystal potential, as obtained e.g. from eq.(9), without shape approximation (——) and in muffin-tin form (····). E_{vac}, E_F and E_{mt} denote the vacuum, Fermi and muffin tin levels, respectively. V_{or} is the (energy-dependent) real part of the "inner potential", and ϕ the work function. Surface potential barrier without (——) and with (—·—·) image asymptotics.

Fig. 3:

Muffin-tin potential approximation sketched in plane perpendicular to surface (containing the atomic nuclei (•)).

For easier tractability, the shape of $V_r(E,r)$ is commonly approximated by the "muffin-tin" form (cf. Fig. 3 and dotted line in Fig. 2), which is spherically symmetric inside (usually touching) spheres around the nuclei ("crystal atoms") and spatially uniform (with the real "inner potential" value $V_{or}(E)$) in the interstitial region. The exchange-correlation-induced energy dependence appears both in the spherical potential parts and in $V_{or}(E)$, with the latter typically (e.g. for Ni and Fe) decreasing from about 14 eV for $E \approx E_{vac}$ to about 11 eV at 100 eV (cf. e.g. Jennings and Thurgate 1981, Lindgren et al. 1984, Tamura et al. 1985, and refs. therein). For electron scattering purposes, the muffin-tin approximation is generally adequate inside the solid, but less so in the surface region for "open" surfaces (e.g. clean fcc(110) surfaces or low-coverage adsorbate system). The surface potential barrier (in Fig. 2 for z<0) has in the "near region" also lattice-periodic variations parallel to the surface ((x,y)plane). How safely these can be ignored, depends on the type of study and will be discussed below (in particular Section 4.3.5). For $z \to -\infty$, surface potential barriers obtained in a local-density approximation obviously reach zero together with the ground-state electron density. The asymptotically correct image potential form $(\propto (4z)^{-1})$, which is due to the response of the electron density to an external electron, requires a non-local-density exchange-correlation approximation (cf. e.g. Lundqvist and March 1983, Almbladh and Barth 1985 and refs. therein). The same applies to the asymptotic dipole potential form in electron-atom scattering (which is important at very small angles).

As for the imaginary potential $V_i^S = Im\Sigma_C^S$, the local-density approximation eq.(9) appears to be useful for electron kinetic energies well above 100 eV (cf. calculations of the extended X-ray absorption fine structure (EXAFS) by Lee and Beni (1977) and by Tran Thoai and Ekardt (1981)), but to fail below about 100 eV (cf. electron-atom scattering calculations by Woolfson et al. (1982)). The latter is not surprising on physical grounds: employing Σ_h of a homogeneous electron gas of the entire density $\rho(\underline{r})$ not only implies the doubtful concept of local plasmon excitation, but furthermore involves the core electrons in such excitations, which is particularly grave at energies below core excitation thresholds. By contrast, a spatially uniform V_{oi} of typically about 4 eV, as obtained for a homogeneous electron gas (cf. Quinn and Ferrel (1958) and Lundqvist (1969) corresponding to the average valence electron density, has been very successful in a vast number of LEED and photoemission calculations (cf. e.g. Feder 1981, Jona et al. 1982, Neve et al. 1983, Marcus and Jona 1984, Feder et al. 1984, and refs. therein). We recall that $V_{oi}(E)$ determines the electron life-time τ and the (intensity) mean free path λ according to

$$\tau(E) = (2|V_{oi}(E)|)^{-1} \text{ [hart}^{-1}\text{] and } \lambda(E) = \tau\sqrt{2(E+V_{or})} \text{ [bohr] },(13)$$

where E is the kinetic energy (in hartree; atomic units: $\hbar = m_e = 1$) (relative to the vacuum zero). The energy dependence of V_{oi} and the corresponding λ is illustrated in Fig. 4 for a typical metal, as deter-mined empirically via comparison with experimental (crystal current) data (Tamura et al. 1985 and refs. therein). Starting from zero

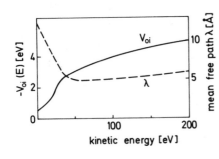

Fig. 4:

Uniform (and spin-averaged) imagi-nary part V_{oi} of the effective potential for Fe (cf. Tamura et al. 1985) and corresponding in-tensity attenuation coefficient (mean free path) λ.

at the Fermi energy (not shown; about 5 eV below the vacuum level), V_{oi} at energies below about 20 eV is mainly due to electron-hole pair excitations; its subsequent steep rise is associated with plasmon-like collective excitations, and its form above about 30 eV is given by $0.85 \ E^{1/3}$. The corresponding mean free path is of the order of a few lattice constants, which is an important contributing reason for the surface sensitivity of electron spectroscopy techniques.

The spatial variation of V_i has so far received comparatively little attention. In addition to the above-cited local-density appro-ximations, model calculations of its spatial inhomogeneity due to lo-calized 3d and deeper core electron excitations were performed by Ing and Pendry (1975) and Beni et al. (1976). Observable effects were identified by elastic spin-polarized LEED (cf. Section 4.3) calculations (by Feder and Kirschner (1981) for 4f excitation on W, and by Tamura and Feder (1982) for 3d excitation on Fe). These calculations employed, in addition to a uniform V_{oi}, a local spherically symmetric model

$$V_i(E,r) = w \, \rho_c(r)/(E-V_r(E,r))^2 \quad , \tag{14}$$

where $\rho_c(r)$ is the charge density of the relevant localized electrons (4f or 3d in the quoted work) and w is an adjustable strength parameter. Spatial inhomogeneity of V_i was further treated - in LEED calculations by Lindgren et al. (1984) - by a two-parameter model allocating different homogeneous V_i values to alternating sub-atomic layers (ion-core planes and interstitial planes) parallel to the surface. It seems desirable to transcend such adjustable-parameter models by realistic ab-initio calculations of the underlying inelastic scattering processes (cf. Section 4.4). Two further types of spatial inhomogeneity of V_i are associated with the surface itself: firstly, the very existence of the surface with its strongly decaying valence charge density gives rise to anisotropic terms (obtained via density gradient corrections to Σ_h by Rasolt and Davis (1979), which can be significant for the topmost atomic layer (cf. Rasolt and Davis 1980); secondly, surface plasmon excitation enhances V_i in the surface region. For ferromagnets V_i^+ is generally different from V_i^- due to the spin-dependence of inelastic scattering (as will be discussed in Section 4.4).

In the energy range from several eV below the Fermi level to several eV above, which has become experimentally accessible in great detail by photoemission (cf. 4.5 and Chapters 10-12) and bremsstrahlung (cf. 4.6 and Chapter 12), the effective potential $V^S(E,r)$ (cf. equations (8) to (11)) - which determines the quasi-particle energy band structure of solids and their surfaces - is at present still a subject of intense research interest. While there is general consensus that

- except for $E=E_F$ and cases of fortuitous agreement - the ground state
one-electron potential is not appropriate, there is a range of opini-
ons as to what one should and can actually use. A pragmatic approach
adopts a real potential of the form of eq.(11) but takes α as a constant or function of energy not given by eq.(12) but to be adjusted
to fit experimental data. This "Xα method", which actually effectively
includes not only exchange but also some correlation, has, for example,
been successfully used by Eckardt et al. (1984) in calculating noble
metal band structures in very good agreement with experimental data.
If inelastic processes are important (as in the case of transition metals with partially filled d bands), addition of an imaginary potential
part of the form $\beta(E-E_F)$, with β as an adjustable constant, appears
adequate (cf. Feder et al. 1983, and Ackermann and Feder 1985b). The
linear behaviour in $(E-E_F)$ of this V_i, rather than the $(E-E_F)^2$ form
known for the homogeneous electron gas very near E_F, appears plausible
from d-band phase space arguments and is supported by experiment. A
first-principles approach using the local-density form eq.(9) with a
random-phase-approximation to Σ_h was recently carried out by Sacchetti
(1982, 1983), and Petrillo and Sacchetti (1984) for several non-magnetic 3d metals. The resulting energy-dependent shifts and broadenings of
the quasi-particle bands relative to the ground-state density functional band structure appear reasonable, but a quantitative check by comparing with experimental photoemission data has not yet been made. For
the ferromagnetic transition metals Fe, Co and Ni, which have been a
particular challenge to theory, it is important to note that calculations of the ground state spin densities are improved by including
intra-atomic correlation effects beyond those contained in the usual
local-density approximation (cf. e.g. Oles and Stollhoff (1984) and
refs. therein) (For shortcomings of the local-density approximation in
calculating the ground state of ferromagnetic Fe, cf. also Wang et al.
1985). Features observed in photoemission like narrowing of the d bands
- compared to local-density ground-state results -, reduction and anisotropy of the exchange splitting for spin densities of e_g and t_{2g}
symmetry, are then already partially accounted for in the ground state,

and consequently require less severe "self-energy-corrections". It still remains to be investigated, whether a density-functional approach to the quasi-particle potentials $V^S(E, \underline{n})$ on the basis of such "highly correlated" ground state spin densities is viable. Such a first-principles procedure would seem more appealing than self-energy calculations within a Hubbard model involving adjustable parameters (cf. e.g. Liebsch 1981, Treglia et al. 1982, and refs. therein). If the above-mentioned e_g/t_{2g} anisotropy is important (like in the case of Ni), the effective potentials have to account for it, and can consequently not be approximated by the muffin-tin form (cf. Figs. 2 and 3). For semi-conductors and insulators, for which groundstate density functional theory notoriously underestimates the gap between valence and conduction bands by typically 20-50 %, there has recently been substantial progress by calculations of the quasi-particle self-energy, especially also in the local-density approximation (equations (9) and (10)) (cf. Pickett and Wang 1984 and refs. therein).

With this - necessarily brief - digression on self-energy approximations in the vicinity of the Fermi energy, we conclude the Section on the effective quasi-particle picture, which is the foundation common to the theoretical treatment of the various electron spectroscopic methods, to which we shall now turn.

4.3   Elastic Low-Energy Electron Scattering

Amongst the various surface spectroscopy techniques involving low-energy (up to about 200 eV) polarized free electrons (cf. Fig. 1), elastic scattering is the simplest from the theoretical point of view.

4.3.1 Basic Concepts

While there has recently also been some interest in scattering from amorphous materials ("metallic glasses") and crystal surfaces with disordered or incommensurate adsorbates, by far the most studies have been concerned with the coherent scattering from crystalline surfaces,

i.e. low-energy electron diffraction (LEED) or spin-polarized LEED
(SPLEED). As indicated in Fig. 5, a monoenergetic beam of electrons

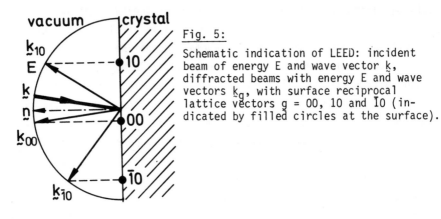

Fig. 5:

Schematic indication of LEED: incident
beam of energy E and wave vector $\underset{\sim}{k}$,
diffracted beams with energy E and wave
vectors $\underset{\sim}{k}_g$, with surface reciprocal
lattice vectors g = 00, 10 and $\bar{1}0$ (indicated by filled circles at the surface).

of kinetic energy E and wave vector $\underset{\sim}{k}$ (with components $\underset{\sim}{k}''$ parallel to
the surface and $\underset{\sim}{k}^z$ normal to it) (corresponding to a de Broglie wave
length of the order of the crystal lattice spacing) is diffracted at
the surface into beams with energy E and wave vectors $\underset{\sim}{k}_g$ such that
their components are given as

$$\underset{\sim}{k}_g'' = \underset{\sim}{k}'' + \underset{\sim}{g} \quad \text{and} \quad k_g^z = (2E - \underset{\sim}{k}_g''^2)^{1/2} \quad , \tag{15}$$

where $\underset{\sim}{g}$ denotes the (two-dimensional) surface reciprocal lattice vectors. Momentum parallel to the surface is thus conserved modulo a sur-
face reciprocal lattice vector $\underset{\sim}{g}$. Beams with real k_g^z in eq.(14) actually emerge from the surface (cf. also Ewald circle construction in
Fig. 5) and can be detected ("propagating" beams), e.g. as spots on a
fluorescent screen. Obviously, such "spot pattern" gives the size and
shape of the surface unit mesh.

    In conventional (unpolarized or spin-averaged) LEED, maximal in-
formation on the surface is obtained by firstly measuring the intensities I_g of the diffracted beams as functions of E and $\underset{\sim}{k}$ (which is
usually characterized by polar angle of incidence θ with respect to
the surface normal $\underset{\sim}{n}$ and by azimuthal angle φ) and secondly comparing

them to their theoretical counterparts calculated for assumed surface structural models (involving e.g. reconstruction of the topmost atomic layers of clean surfaces or positions of adsorbed atoms relative to the "substrate"). The "true" surface structure is identified by optimal agreement according to visual or automated criteria for a sufficiently large "data base" (e.g. intensity versus energy profiles of several beams for several polar and azimuthal angles of incidence). This LEED intensity analysis is firmly established as a powerful method of surface structure determination. (Details, results and references to the vast number of original articles may be found in the following monographs and review articles: Pendry 1974, Duke 1974, Van Hove and Tong 1979, Heinz and Müller 1982, Jona et al. 1982, Marcus and Jona 1982, Marcus and Jona 1984).

In spin-polarized LEED (SPLEED) one extends the conventional set-up either by a spin-polarized primary beam (generated by a "source" of polarized electrons, cf. Chapter 5), by spin analysis of the diffracted beams (with the aid of a "spin detector", cf. Chapter 5), or, more ambitiously, by both. In general one thus has the basic observable quantities $I_g(E,\theta,\phi;\underline{P})$ and $P_g(E,\theta,\phi;\underline{P})$. If the surface is ferromagnetically ordered with magnetization $\underline{M}(\underline{r})$ - i.e. an effective magnetic field $\underline{B}(E,r)$ (cf. eq.(3)) - parallel to some unit vector $\underline{m}$, these quantities depend, in addition, on $\underline{M}$. Since spin effects become manifest by reversal of $\underline{P}$ and/or $\underline{M}$, it is convenient to distinguish, for each beam, four intensities $I_{g\sigma}^{\mu}$, where $\sigma=+(-)$ refers to the primary beam polarization $\underline{P}$ parallel (antiparallel) to some unit vector p, and $\mu=+(-)$ to $\underline{M}$ (and thereby $\underline{B}$) antiparallel (parallel) to unit vector $\underline{m}$. (The latter definition implies that $\mu=+$ is associated with the majority spin direction of the ferromagnet parallel to $\underline{m}$). The four intensities I_{σ}^{μ}, (dropping the beam index g in the following), which are observable upon separate reversal of $\underline{P}$ and $\underline{B}$, are equivalent to their sum I and three scattering asymmetries defined as

$$A_{so} = (I_+^+ + I_+^- - I_-^+ - I_-^-)/I \quad , \tag{16a}$$

$$A_{ex} = (I^+_+ + I^-_- - I^+_- - I^-_+)/I \quad , \tag{16b}$$

$$A_u = (I^+_+ + I^+_- - I^-_+ - I^-_-)/I \quad . \tag{16c}$$

The physical meaning of these asymmetries is indicated by the subscripts "so" (for spin-orbit), "ex" (for exchange) and "u" (for unpolarized). For non-magnetic materials, the only relevant spin-dependent mechanism is spin-orbit coupling (cf. eq.(5), or $\sigma \cdot L$ for the central fields of the muffin-tin model); $\underline{B} = 0$ implies $I^+_\sigma = I^-_\sigma =: I_\sigma$, and consequently $A_{ex} = 0 = A_u$, leaving only

$$A_{so} = (I_+ - I_-)/(I_+ + I_-) \quad ; \tag{17}$$

this is the spin-orbit-induced "up/down" asymmetry, which is plausible from the fact that spin-up ($\sigma=+$) and spin-down ($\sigma=-$) electrons experience different effective scattering potentials due to different sign of the spin-orbit coupling term (for details cf. e.g. Kessler 1976/1985). For ferromagnets, the exchange term σB (eq.(5)) reverses sign with $\underline{B}$, which implies $I^+_\sigma = I^-_\sigma$; since A_{so} (eq.(16a)) contains, for each spin direction σ, the sum $I^+_\sigma + I^-_\sigma$, magnetic exchange effects (approximately) cancel and still leave spin-orbit coupling as the dominant origin of A_{so}. However, A_{ex} and A_u (eq.(16b,c)) are now in general non-zero. If spin-orbit coupling is ignored, we see from eq.(5) that $I^\mu_\sigma = I^{-\mu}_{-\sigma}$, and therefore $A_{so} = 0$ and $A_u = 0$, whilest A_{ex} reduces to a purely exchange-induced asymmetry $A^{(0)}_{ex}$. As for A_u, the above implies that it can be non-zero only if both spin-orbit coupling and magnetic exchange interaction are present. Its physical meaning is apparent from its definition eq.(16c): for each $\mu=\text{sign}(\underline{B} \cdot \underline{m})$, the sum $I^\mu_+ + I^\mu_-$ corresponds to the scattered intensity due to a primary beam, which is an incoherent superposition of two oppositely ($\sigma=+/-$) polarized beams (of equal intensity), i.e. equivalent to an unpolarized beam. A_u is therefore an asymmetry obtained from an _unpolarized_ incident beam upon reversal of the magnetization direction. Semi-classically, its origin may be visualized as follows: the unpolarized electrons get (partially) polarized by spin-orbit coupling (in the vicinity·of nuclei) and subsequently scattered by the "magnetic" electrons (mainly 3d electrons in the case of Fe, Co, Ni) with a strength depending on the direction of $\underline{M}$ relative to the spin-orbit-induced spin polarization vector. Obviously, the above definitions, introduced for crystalline surfaces, are applicable to targets of arbitrary geometry (i.e. amorphous materials) by assigning the I_σ^μ to general scattering directions (rather than the discrete beam directions $\hat{k}_g$).

Retrieval of the information on surface properties (especially magnetic ones), which is "coded" in experimental data of the above spin-dependent quantities, requires - as will be discussed in Section 4.3.7 - quantitatively realistic theoretical calculations. At the outset of a theoretical treatment, it is convenient to express the observable quantities in terms of a scattering matrix, which relates the plane-wave four-spinor v_g (the solution of the Dirac equation in vacuum, which characterizes the diffracted beam associated with surface reciprocal lattice vector g) to the incident-beam spinor u. Since a transformation between plane-wave four spinors is uniquely specified by transforming the two "large components (cf. Rose 1961), it suffices to have the (2x2) scattering matrices S_g:

$$\tilde{v}_g = S_g \tilde{u} , \qquad (18)$$

where $\tilde{v}_g$ and $\tilde{u}$ are the Pauli spinors consisting of the two large components of v_g respectively. If the incident beam is not in a pure spin state $\tilde{u}$, but polarized with $|P|<1$, it has to be described by the statistical operator (spin density matrix)

$$\rho = (1 + \underline{P}\underline{\sigma})/2 . \qquad (19)$$

To obtain the ensuing statistical operator ρ_g, we recall that a partially polarized beam is a statistical mixture of two beams fully polarized parallel and antiparallel to $\underset{\sim}{P}$. From eqs. (18,19) we then have

$$\rho_g(\underset{\sim}{P}) = S_g \frac{1}{2} (1 + \underset{\sim}{P}\underset{\sim}{\sigma}) S_g^+ .\tag{20}$$

For a ferromagnet, S_g and therefore ρ_g depend on the effective magnetic field $\underset{\sim}{B}$. The intensity I_g (relative to the primary-beam intensity) and the surface-normal current reflection coefficient j_g of the gth beam are obtained as

$$I_g(\underset{\sim}{P},\underset{\sim}{B}) = tr[\rho_g(\underset{\sim}{P},\underset{\sim}{B})]/tr[\rho(\underset{\sim}{P})] ; j_g = (k_g^z/k^z)I_g \tag{21}$$

i.e. we have - with a reversal of $\underset{\sim}{P}$ and $\underset{\sim}{B}$ - the above four basic intensities I_σ^μ and thence the three scattering asymmetries (eq.(16)). The polarization vector $\underset{\sim}{P}_g$ is also determined by ρ_g(eq.(20)), according to eq.(1). Calculation of the spin-dependent observable quantities amounts thus to the calculation of the scattering matrices S_g (eq.(18).

Before outlining a formalism, which is appropriate for such calculation, it is interesting and important to present some general features, which follow from time-reversal and spatial symmetries without recourse to specific model assumptions and numerical calculations.

4.3.2 Time Reversal and Spatial Symmetries

The key to symmetry-induced properties of the spin-dependent observable quantities in LEED (as was shown by Feder (1980) and by Dunlap (1980), where more details may be found) lies in the behaviour of the (2x2) scattering matrices S_g (cf. eq.(18)) (which correspond to the T matrix in general scattering theory, cf. e.g. Roman 1965) under the respective symmetry transformations, if the S_g are expressed in the form

$$S_g(\underset{\sim}{B}) = a_g(\underset{\sim}{B}) + \underset{\sim}{b}_g(\underset{\sim}{B}) \cdot \underset{\sim}{\sigma} ,\tag{22}$$

where the scalar a_g and the three-vector $\underset{\sim}{b}_g$ are four unknown complex numbers. Substituting eq.(22) into eq.(20), and the resulting ρ_g together with ρ from eq.(19) into eq.(21), one obtains for the intensity

$$I_g(\underset{\sim}{P},\underset{\sim}{B}) = |a_g(\underset{\sim}{B})|^2 + |b_g(\underset{\sim}{B})|^2 + c_g(\underset{\sim}{B})\cdot\underset{\sim}{P} \quad , \tag{23a}$$

$$c_g(\underset{\sim}{B}) = 2Re[a_g(\underset{\sim}{B})b_g^*] + i(b_g^*(\underset{\sim}{B}) \times b_g(\underset{\sim}{B})) \quad . \tag{23b}$$

For two incident beams with polarization $\underset{\sim}{P}$ and $-\underset{\sim}{P}$, this implies a scattering asymmetry

$$A_g(\underset{\sim}{P},\underset{\sim}{B}) = c_g(\underset{\sim}{B})\cdot\underset{\sim}{P}/[|a_g(\underset{\sim}{B})|^2 + |b_g(\underset{\sim}{B})|^2] =: A_g(\underset{\sim}{B})\cdot\underset{\sim}{P} \quad , \tag{24}$$

where A_g is an "asymmetry vector". In the nonmagnetic case ($\underset{\sim}{B}=0$), this asymmetry becomes the spin-orbit-induced asymmetry A_{so} (eq.(17)); for a ferromagnet without spin-orbit coupling it is equivalent to the exchange-induced A_{ex} (eq.(16b)). Another quantity of particular interest, the polarization vector $\underset{\sim}{P}_g$ occurring for an unpolarized incident beam ($P=0$) is similarly obtained (as $tr_\sigma\rho_g$):

$$P_g(\underset{\sim}{B}) = d_g(\underset{\sim}{B})/[|a_g(\underset{\sim}{B})|^2 + |b_g(\underset{\sim}{B})|^2] \quad , \tag{25a}$$

$$d_g(\underset{\sim}{B}) = 2Re[a_g(\underset{\sim}{B})b_g^*(\underset{\sim}{B})] - i(b_g^*(\underset{\sim}{B})\times b_g(\underset{\sim}{B})) \quad . \tag{25b}$$

P_g is seen to differ from the asymmetry vector A_g (eq.(24)) only by the sign of the second term in d_g (eq.(25b)) and c_g (eq.(23b)). While therefore in general $P_g \neq A_g$, $c_g \cdot c_g = d_g \cdot d_g$ entails that the lengths of the two vectors are always equal. In the absence of spin-orbit coupling, S_g (eq.(22)) is diagonal (with quantization axis parallel to $\underset{\sim}{B}$), i.e. there is only one non-zero component of b_g (parallel to $\underset{\sim}{B}$), and consequently $d_g=c_g$. Purely exchange-induced P_g and A_g are thus equal and parallel to the magnetization axis.

Symmetry effects are readily obtained from the above expressions. Transformation of eq.(22) by time reversal, which in particular reverses the magnetic field $\underset{\sim}{B}$, implies that $a_g(\underset{\sim}{B})\rightarrow a_g(-\underset{\sim}{B})$ and $b_g(\underset{\sim}{B})\rightarrow -b_g(-\underset{\sim}{B})$.

For non-magnetic systems, eqs.(23-25) then reveal

$$I_g(\underset{\sim}{P}=0) \to I_g(P=0),\ A_g \to A_g,\ P_g \to P_g,\ \underset{\sim}{P}_g \to -\underset{\sim}{A}_g,\ A_g \to -\underset{\sim}{P}_g \quad . \qquad (26)$$

For the intensity and the degrees of asymmetry and polarization, one thus finds the "reciprocity theorem" (cf. textbooks on quantum mechanics): these quantities are not changed by interchanging electron gun and detector in LEED. For ferromagnets, $I_g(P=0,\underset{\sim}{B}) \to I_g(P=0,-\underset{\sim}{B})$, or $I_+^\mu + I_-^\mu \to I_+^{-\mu} + I_-^{-\mu}$; the reciprocity theorem applies therefore to $I = I_+^+ + I_-^+ + I_+^- + I_-^+$ and (from eq. 16c), with a change of sign, to the scattering asymmetry A_u for an unpolarized primary beam.

Further general results arise from spatial symmetries of the diffraction set-up (crystal plus directions of the incident beam and the outgoing beam under consideration) (or more formally: Hamiltonian (eqs.(3,5) plus boundary conditions). Especially interesting and practically important are geometries in which the scattering plane (defined by incident and scattered beam directions $\hat{\underset{\sim}{k}}$ and $\tilde{\underset{\sim}{k}}_g$) is parallel to a mirror plane of the crystal (cf. Fig. 6, hatched plane). We

Fig. 6:

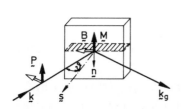

Special SPLEED geometries with scattering plane (with normal $\underset{\sim}{n}$) parallel to a mirror plane (hatched) of the surface system (with normal $\underset{\sim}{s}$). Magnetic field $\underset{\sim}{B}$ and magnetization $\underset{\sim}{M}$ normal (filled arrow) or parallel (empty arrow) to mirror plane.

recall that reflection at a plane transforms axial vectors (like $\underset{\sim}{\sigma}$, $\underset{\sim}{P}$, $\underset{\sim}{B}$) such that the component perpendicular to the plane remains unchanged, while the components parallel to the plane change sign. For non-magnetic systems, invariance of the T matrix S_g (eq.(22)) under the mirror plane reflection thus dictates that the components parallel to the plane, b_g'', vanish. Eqs.(24) and (25) then yield

$$P_g'' = A_g'' = 0 , \qquad P_g^\perp = A_g^\perp , \qquad (27)$$

i.e. the polarization vector (due to unpolarized incident electrons) and the scattering asymmetry vector are equal and normal to the scattering plane. Time reversal (cf. eq.(26)) therefore reverses each of them (relative to a fixed direction, but not to the actual normal to the scattering plane, which itself gets reversed). For the specular beam at normal incidence ($\theta=0$), this implies $\underset{\sim}{P}_{00}=\underset{\sim}{A}_{00}$. These results are identical to those generally valid in electron-atom scattering (cf. Kessler 1979/1985). They have been verified, within the limits of experimental uncertainty, by SPLEED experiments (for details and references cf. review by Feder (1981), in particular Figs. 6 and 8 therein). If the surface normal has a three-fold rotation axis and 3 mirror planes, specular beam "rotation diagrams" of $\underset{\sim}{P}$ and $\underset{\sim}{A}$ versus azimuthal angle ϕ exhibit three-fold rotation plus reflection symmetry, but $|\underset{\sim}{P}|$ and $|\underset{\sim}{A}|$ become six-fold symmetric by virtue of time reversal (for more details cf. Feder (1980, 1981) and Dunlap (1980)). Remarkable effects can also arise from an absence of symmetry: for non-centrosymmetric crystals (e.g. ZnS, GaAs), an unpolarized beam incident at $\theta=0$ generally produces a specularly reflected beam with non-zero spin polarization. (This case will be considered further in the context of photoemission in Section 4.5).

For ferromagnetic systems two mirror plane configurations are particularly suitable (cf. Fig. 6). In both, the magnetization $\underset{\sim}{M}$, i.e. $\underset{\sim}{B}$, is aligned parallel to the surface for experimental reasons: to avoid uncontrollable deflection of the incident electron beam as well as rotation of its $\underset{\sim}{P}$ by a magnetic field outside the crystal. In the first type of geometry, $\underset{\sim}{P}$ and $\underset{\sim}{B}$ are in the scattering plane. Invariance of S_g (eq.(22)) under the mirror operation implies $a_g(\underset{\sim}{B})=a_g(-\underset{\sim}{B})$, $\underset{\sim}{b}_g''(\underset{\sim}{B}) = -\underset{\sim}{b}_g''(-\underset{\sim}{B})$ and $\underset{\sim}{b}_g^{\perp}(\underset{\sim}{B}) = \underset{\sim}{b}_g^{\perp}(-\underset{\sim}{B})$, and thence from eq.(23) $I_g(\underset{\sim}{P},\underset{\sim}{B}) = I_g(-\underset{\sim}{P},-\underset{\sim}{B})$, i.e. in the simplified notation preceding eq.(16): $I_\sigma^\mu = I_{-\sigma}^{-\mu}$. This is also obvious from applying the mirror operation to the complete diffraction set-up (including $\underset{\sim}{P}$ and $\underset{\sim}{B}$). According to eq.(16), the scattering asymmetries A_{so} and A_u then vanish, whereas A_{ex} is in general non-zero. It is, however, not identical with the exchange-induced scattering asymmetry $A_{ex}^{(o)}$ obtained in the absence of spin-orbit coupling, since the latter rotates the spin in the scattering plane and thereby changes the effective magnetic interaction $\underset{\sim}{g} \cdot \underset{\sim}{B}$. The difference is particularly striking for $\underset{\sim}{P} \perp \underset{\sim}{B}$: while $A_{ex}^{(0)}=0$, A_{ex} may assume sizeable values (cf. Ackermann and Feder 1984). Time reversal transforms I_{σ}^{μ} into $I_{-\sigma}^{-\mu}$ and consequently does not change A_{ex}. The asymmetry and polarization vectors $\underset{\sim}{A}_g$ and $\underset{\sim}{P}_g$ have components parallel to the mirror plane, which are mainly determined by the exchange interaction, and a normal component, which mainly stems from spin-orbit coupling. In the second type of special geometry for ferromagnets (cf. Fig. 6), $\underset{\sim}{P}$ and $\underset{\sim}{B}$ are perpendicular to the scattering plane (mirror plane). Reflection symmetry of eq.(22) then requires $b_g''(\underset{\sim}{B}) = -b_g''(\underset{\sim}{B})=0$, whilest $a_g(\underset{\sim}{B})$ and $b_g^{\perp}(\underset{\sim}{B})$ remain unchanged, as in the non-magnetic case. Therefore, again, $A_g''(\underset{\sim}{B}) = P_{-g}''(\underset{\sim}{B})=0$, and $A_g^{\perp}(\underset{\sim}{B}) = P_g^{\perp}(\underset{\sim}{B})$ non zero. The three special asymmetries A_{so}, A_{ex} and A_u are also generally non-zero. Time reversal transforms $A_g^{\perp}(\underset{\sim}{B})$ into $A_g^{\perp}(-\underset{\sim}{B})$, changes the sign of A_{so} and leaves A_{ex} unchanged (relative to a fixed direction normal to the scattering plane) (cf. non-magnetic "limit", discussed around eq.(27), in which A_{so} coincides with $A_g^{\perp}(\underset{\sim}{B}=0)$). For the specular beam at $\theta=0$ we therefore have again $A_{so}=0$ but A_{ex} need not vanish; accordingly, polarization parallel to $\underset{\sim}{B}$ occurs for an unpolarized primary beam. If there is an additional mirror plane normal to the scattering plane and parallel to the surface normal, the unpolarized-beam asymmetry A_u (eq.(16c)) can - rather than via a reversal of $\underset{\sim}{B}$ - be expressed as a left-right scattering asymmetry for fixed $\underset{\sim}{B}$. (Details on this as well as symmetry results for hcp(0001) surfaces, relevant for Co and Gd, have been given by Tamura et al. (1984)).

The above symmetry arguments and results have been formulated for systems with perfect two-dimensional periodicity parallel to the surface. They are, however, easily extended to surfaces with disorder and in particular of amorphous systems (like metallic glasses). Firstly, for all quantities the reciprocal-lattice label g, relating to discrete diffracted beams, is replaced by a general-direction momentum $\underset{\sim}{k}'$ (with $|\underset{\sim}{k}'|=\sqrt{2E}$) after scattering. Since time reversal is independent of geometry, the respective results remain unchanged. As for spatial symmetries, these are obviously absent in amorphous materials. They are, however, effectively retrieved with regard to electron scattering, since deviations arising from non-symmetric local atomic environments cancel (approximately) in the summation over local T matrices (within the coherence area of the beam) and over scattered beam density matrices (from the coherent "domains" of the surface area illuminated by the incident beam). These effective symmetries should be those of a single "crystal atom". In particular, any scattering plane normal to the surface thus is an effective mirror plane.

Time reversal and spatial symmetries thus lead, via the T matrix parts S_g (eq.(22)), to general features of the observable scattering asymmetries and polarizations. Such features are not only interesting in their own right, but also valuable as a check on experimental data and on numerically calculated results. Some further consequences of symmetry will be addressed below: for experimentally more convenient alternative scattering asymmetries (in Section 4.3.7), and for intermediate quantities in theoretical formalism (section 4.3.5).

4.3.3 Single-Site Scattering

We now present, with an emphasis on general structure rather than on technical details, the "state-of-the-art" theory of elastic low-energy electron scattering, which has proven to be adequate for quantitative realistic calculations. While historically formalisms including either spin-orbit coupling or magnetic exchange interaction were developed separately (cf. Feder (1981) and references therein), it is more satisfactory to consider a "unified" theory, based on the Dirac Hamiltonian for a ferromagnet (eq.(3)), and thence to proceed to specializations when possible. Decomposition of the effective potentials $V^s(\underline{r})$ (cf. eq.(7)) into lattice sums $\sum_i V_a^s(\underline{r}-\underline{R}_i)$ over "crystal atom" potentials V_a^s, where $\underline{R}_i$ denotes the nuclear coordinates (lattice sites) and indices referring to chemically or environmentally inequivalent potentials V_a^s have been suppressed, naturally decomposes the scattering problem into two parts: (a) scattering by a single site (crystal atom) and (b) multiple scattering between the sites. This does not yet presuppose crystal periodicity, but also applies to alloys and amorphous systems.

Scattering by a single site is most practicably treated within the muffin-tin approximation (cf. Fig. 3), i.e. for spherically symmetric $V_a^S(r)$. The Dirac Hamiltonian eq.(3) thus has spherically symmetric V and B inside spheres of radius R, but is itself not spherically symmetric because of $\tilde{\sigma} \cdot B$. The usual relativistic partial wave analysis (cf. Rose 1961) is therefore not applicable. A suitable generalization (due to Feder et al. (1983b) and Strange et al. (1984)) proceeds as follows. Taking the interstitial constant potential as energy zero level, V vanishes for r>R. Further, B=0 is assumed for r>R, which is a reasonable approximation, as the "magnetic" electrons are mainly inside the spheres. Choosing, without loss of generality, the magnetic field in eq.(3) parallel to the z-axis, the magnetic coupling term becomes $-\beta\tilde{\sigma}_3 B$. Inside a sphere, the four-spinors $\phi(r)$ of the time-independent Dirac equation ($H\phi=E\phi$, with H from eq.(3) as just specialized) are expanded in spherical waves with radial functions f and g

$$\phi(\underset{\sim}{r}) = \underset{\kappa\mu}{\Sigma} \frac{1}{r} \begin{pmatrix} f_\kappa^\mu(r) & \chi_\kappa^\mu(\hat{\underset{\sim}{r}}) \\ ig_\kappa^\mu(r) & \chi_{-\kappa}^\mu(\hat{\underset{\sim}{r}}) \end{pmatrix} \quad ; \tag{28}$$

χ are the two-component spin angular functions

$$\chi_\kappa^\mu(\hat{\underset{\sim}{r}}) = \underset{\sigma=\pm 1/2}{\Sigma} C(\ell\tfrac{1}{2}j; \mu-\sigma,\sigma) \ Y_\ell^{\mu-\sigma}(\hat{\underset{\sim}{r}}) \ \chi^\sigma \ , \tag{29}$$

where C are the Clebsch-Gordan coefficients, Y the spherical harmonics and χ the two Pauli basis spinors (cf. Rose 1961 or textbooks on quantum mechanics). The quantum numbers κ (with $\kappa =\mp 1, \mp 2, ...$) incorporate the orbital and total angular momentum numbers ℓ and j:

$$j = |\kappa| - \frac{1}{2}, \ \ell = j + S_\kappa/2, \text{ with } S_\kappa = \kappa/|\kappa| \quad ; \tag{30}$$

i.e. $\ell=-\ell-1$ for $j=\ell+\frac{1}{2}$ and $\kappa=\ell$ for $j=\ell-\frac{1}{2}$, for each κ and thereby j, the magnetic quantum numbers are $\mu=-j$, $-j+1$, ..., $j-1$, j. Substitution of eq.(28) into the Dirac equation leads to the following system of coupled differential equations for the radial functions:

$$\hbar c \frac{d}{dr} f_\kappa^\mu = -\hbar c \frac{\kappa}{r} f_\kappa^\mu + (E-V+mc^2) g_\kappa^\mu +$$

$$+ B\langle \chi_{-\kappa}^\mu | \sigma_3 | \chi_{-\kappa}^\mu \rangle g_\kappa^\mu + B\langle \chi_{-\kappa}^\mu | \sigma_3 | \chi_{\kappa-1}^\mu \rangle g_{-\kappa+1}^\mu \; ; \qquad (31a)$$

$$\hbar c \frac{d}{dr} g_\kappa^\mu = \hbar c \frac{\kappa}{r} g_\kappa^\mu + (V-E+mc^2) f_\kappa^\mu +$$

$$+ B\langle \chi_\kappa^\mu | \sigma_3 | \chi_\kappa^\mu \rangle f_\kappa^\mu + B\langle \chi_\kappa^\mu | \sigma_3 | \chi_{-\kappa-1}^\mu \rangle f_{-\kappa-1}^\mu \; . \qquad (31b)$$

For $B=0$, these equations are seen to reduce to the standard central field Dirac equations (cf. Rose 1961). The magnetic field introduces couplings, which depend on μ via the matrix elements of σ_3, corresponding to the chosen magnetic axis. The last term in eq.(31b) couples (except for $\kappa=-1$, i.e. $j=\frac{1}{2}$ and $\ell=0$) to states with different j for the same ℓ, i.e. destroys j as good quantum number. The last term in eq.(31a) couples to states with $\ell\pm 2$. For each μ, eq.(31) thus represents two infinite sets of coupled equations, one for even and one for odd values of ℓ, i.e. for states of even and of odd parity. The physical origin of the two coupling terms is elucidated by a two-component approximation to the Dirac Hamiltonian. The j coupling (between $j=\ell+\frac{1}{2}$ and $\ell-\frac{1}{2}$) arises from the simultaneous presence of magnetic ($\sigma \cdot B$) and spin-orbit ($\sigma \cdot L$) interaction in the $1/c^2$ approximation eq. (5). Since H from eq.(5) commutes with L^2, there is no mixing between different ℓ. Such is brought about by a higher term involving B and σ. As discussed after eq.(5), this term is practically negligible. Eq.(31a,b) therefore practically consists of four coupled equations for each μ.

We now outline the solution of the relativistic scattering problem by a single sphere (crystal atom) (for details see Feder et al.

(1983b), Strange et al. (1984) and Ackermann (1985) in the ferromagnetic case, and Rose (1961) and quantum mechanics textbooks in the non-magnetic case). In the field-free region outside the sphere (for $r>R$), eq.(31a,b) is easily solved to yield, with eq.(28), incoming spherical waves with amplitudes A_κ^μ and outgoing ones with amplitudes B_κ^μ (with $\hbar=m=1$):

$$
A_\kappa^\mu \begin{pmatrix} j_\ell(kr)\chi_\kappa^\mu(\hat{r}) \\ i\varepsilon S_\kappa j_{\bar{\ell}}(kr)\chi_{-\kappa}^\mu \end{pmatrix}, \qquad B_\kappa^\mu \begin{pmatrix} h_\ell^{(1)}(kr)\,\chi_\kappa^\mu(\hat{r}) \\ i\varepsilon S_\kappa h_{\bar{\ell}}^{(1)}(kr)\,\chi_{-\kappa}^\mu(\hat{r}) \end{pmatrix}, \tag{32}
$$

where $\varepsilon = c/(E+c^2)$, $E = (c^4+k^2c^2)^{1/2}$, $\bar{\ell}=\ell-S_\kappa$ (cf. eq.(30)), and j_ℓ ($h_\ell^{(1)}$) are spherical Bessel (Hankel) functions; the χ_κ^μ are given in eq.(29). In this $(\kappa\mu)$ representation, the atomic t-matrix $\underset{\approx}{t}$ is defined by

$$
\underset{\sim}{B}^{(s)} = \underset{\approx}{t}\underset{\sim}{A} \quad \text{with } \underset{\sim}{B}^{(s)} = (\ldots, B_\kappa^\mu, \ldots) \text{ and } \underset{\sim}{A} = (\ldots, A_\kappa^\mu,) . \tag{33}
$$

It is calculated by first numerically solving eq.(31 a,b) inside the sphere and then matching at R the general solution with a linear combination of outside solutions (eq.(32)) with specified A_κ^μ to determine the scattered B_κ^μ. Since, for $B||\hat{z}$, there is no μ coupling in eq. (31a,b), $\underset{\approx}{t}$ is diagonal in μ. With respect to κ, there are off-diagonal elements corresponding to the dominant coupling between $j=\ell+\frac{1}{2}$ and $j=\ell-\frac{1}{2}$, and the weak $\Delta\ell=\pm 2$ coupling. For non-magnetic systems, for which these couplings are absent, $\underset{\approx}{t}$ is also diagonal in κ and its elements t_κ are $(\exp(2i\delta_\kappa)-1)$, where δ_κ are the usual relativistic scattering phase shifts (commonly denoted by δ_ℓ^+ and δ_ℓ^- relating to $j=\ell+\frac{1}{2}$ and $j=\ell-\frac{1}{2}$). If there is a magnetic field with arbitrary direction rather than $||\hat{z}$, $\underset{\approx}{t}$ is no longer diagonal in μ. It can be obtained from the μ-diagonal one by applying the appropriate rotation representation in orbital angular momentum and spin space (cf. Tamura et al. 1984 and Ackermann 1985).

Using the above atomic t-matrix (eq.(33)), the important case of an incident plane wave u·exp(ikr), where u is an arbitrary four-spinor, is readily described. Denoting by v(k,k') the spinor amplitude of the asymptotic scattered wave exp(ikr)/r and recalling that free four-spinors are completely specified by their two "large" components (cf. Rose 1961), we look for a (2x2) matrix $\underset{\sim}{T}$, which transforms the two large components of u into those of v. $\underset{\sim}{T}$ is found by expanding u·exp (ikr) into spherical waves, which specifies the A_κ^μ (cf. eq.(32)), substituting $\underset{\approx}{t}A$ for the outgoing spherical wave amplitudes (eq.(32)) and taking the asymptotic form:

$$T_{\sigma'\sigma}(\underset{\sim}{k'},\underset{\sim}{k}) = -(4\pi i/k) \sum_{\kappa\mu\kappa'\mu'} i^{(\ell-\ell')} t_{\kappa\mu}^{\kappa'\mu'} C\ (\ell,\tfrac{1}{2};\mu-\sigma,\sigma)\times$$

$$\times C(\ell',\tfrac{1}{2},j';\ \mu'-\sigma',\sigma') \times (Y_\ell^{\mu-\sigma}\ (\hat{\underset{\sim}{k}}))^* \ Y_{\ell'}^{\mu'-\sigma'}(\underset{\sim}{k'})\ , \qquad (34)$$

where κ and κ' determine j,ℓ and j', ℓ' according to eq.(30), and C are again Clebsch-Gordan coefficients. If the incident (plane-wave) electrons form a spin mixture described by a density matrix ρ_0, the density matrix after scattering is $\rho = T\rho_0 T^+$, and differential cross section and spin polarization vector are given by $tr[\rho]/tr[\rho_0]$ and $tr[\underset{\sim}{\sigma}\rho]/tr[\rho]$.

For non-magnetic systems, for which $\underset{\approx}{t}$ is diagonal in μ and κ, eq.(34) reduces to the form (cf. Rose 1961)

$$\underset{\sim}{T}(\underset{\sim}{k'},\underset{\sim}{k}) = f(\Theta) + g(\Theta)\ \hat{\underset{\sim}{n}}\cdot\underset{\sim}{\sigma}, \text{ with } \hat{\underset{\sim}{n}} = (\underset{\sim}{k}\times\underset{\sim}{k'})/|\underset{\sim}{k}\times\underset{\sim}{k'}|\ ; \qquad (35)$$

$\Theta = \sphericalangle(\underset{\sim}{k'},\underset{\sim}{k})$ is the scattering angle, and the "direct" scattering amplitude $f(\Theta)$ and the "spin flip" amplitude $g(\Theta)$ are given in terms of the phase shifts $\delta_\ell^\pm$ (for $j=\ell\pm\tfrac{1}{2}$) and Legendre polynomials P_ℓ and P_ℓ^1 as

$$f(\Theta) = \frac{1}{2ik} \sum_\ell \{(\ell+1)[\exp(2i\delta_\ell^+)-1] + \ell[\exp(2i\delta_\ell^-)-1]\}P_\ell(\cos\Theta) \qquad (36a)$$

$$g(\Theta) = \frac{1}{2ik} \sum_\ell [-\exp(2i\delta_\ell^+) + \exp(2i\delta_\ell^-)]P_\ell^1(\cos\Theta)\ . \qquad (36b)$$

From eq.(35), cross section I, spin polarization P (directed along the scattering-plane normal $\hat{n}$) and asymmetry A_{so} (for primary beams fully polarized parallel and antiparallel) (cf. symmetry considerations in Section 4.3.2) are

$$I(\Theta) = |f(\Theta)|^2 + |g(\Theta)|^2, \ P(\Theta) = A(\Theta) = -2 \ \text{Im}[f(\Theta)g^*(\Theta)]/I(\Theta) \ . \quad (37)$$

The number of ℓ required for convergence (in eqs.(34) and (36a,b)) increases with energy and with the atomic number. It typically ranges between 4 to 7 for kinetic energies below 100 eV. If spin-orbit coupling is neglected, $\delta_\ell^+ = \delta_\ell^-$ and consequently g = 0 (in eq.(36b)) and P = A = 0 (in eq.(37)); f reduces to the standard nonrelativistic scattering amplitude. For a ferromagnetic crystal atom without spin-orbit coupling, T (eq.(34)) becomes diagonal (for spin quantization axis parallel to $\underset{\sim}{B}$), with elements $f^+(\Theta)$ and $f^-(\Theta)$ obtained by nonrelativistic partial wave analysis from potentials $V^+(r)$ and $V^-(r)$ (cf. eq.(6)), respectively.

In the above, the "crystal atom" potential was assumed as spherically symmetric. It is, however, straightforward to incorporate non-spherical potential contributions, which arise, for example, from the anisotropy of the valence d shell or of covalent bonds. Expansion ot the potential $V(E,\underset{\sim}{r})$ and, for ferromagnets, also of the effective magnetic field $\underset{\sim}{B}(\underset{\sim}{r})$, in spherical harmonics leads to radial equations, which contain - compared to eqs.(31a,b) - additional (κ,μ) coupling terms. These give rise to corresponding off-diagonal elements of the atomic t-matrix (eq.(33)). The elastic plane-wave-scattering matrix $\underset{\sim}{T}(\underset{\sim}{k}',\underset{\sim}{k})$ retains is general form (eq.(34)), but not longer reduces to the simpler form of eq.(35) in the non-magnetic case.

4.3.4 General Many-Site Scattering

Having solved the elastic scattering problem by the potential in a cell around site $\underset{\sim}{R}_i$ in terms of an atomic t-matrix t_i (cf. eq.(33), with wiggly underlines omitted and label i added to allow for chemically and environmentally inequivalent sites), scattering by many-site systems like molecules, clusters, liquids, amorphous solids and crystals can be treated by adding up single-site events. We first outline the generel idea and then specialize to systems with perfect lattice periodicity parallel to a planar surface.

For an electron beam (described by $\psi^{(0)}$), incident on a system with N atomic sites, the total wave field ψ_i incident on site i consists of a "direct" part $\psi_i^{(0)}$ plus the sum of the wave fields scattered from the other sites j. To incorporate all possible multiple scattering processes, this incident field has to satisfy the self-consistency condition

$$\psi_i = \psi_i^{(0)} + \sum_{\substack{j \neq i}}^{N} G_{ij} t_j \psi_j \; ; \; i = 1,..,N; \tag{38}$$

where the one-electron Green function G_{ij} describes the propagation (in the uniform complex inner potential) from site j to site i. After dropping in eq.(38) the indices, i.e. regarding ψ and $\psi^{(0)}$ as vectors, t as a site-diagonal matrix and G as a general matrix, we solve for ψ and thence obtain the total scattered wave field ψ^{SC} as

$$\psi^{SC} = t\psi = t(1-Gt)^{-1} \psi^{(0)} =: \tau\psi^{(0)} \; , \tag{39}$$

i.e. related to the incident field $\psi^{(0)}$ by the newly defined t-matrix τ. Its site-elements τ_{ij} can be interpreted as "scattering path operators" (cf. e.g. Gyorffy and Stocks 1977): for a wave $\psi_j^{(0)}$ incident on site j, they describe the contributions to the total wave field emanating from site i, which are due to all multiple scattering paths starting at j and ending at i. More details and references on the above general multiple scattering theory may be found in reviews by Jona et al.(1982) and by Gyorffy and Stocks (1977). Multiple scattering equations for τ_{ij} in the $(\kappa\mu)$- and the (ℓms) angular momentum respresentations have recently been written by Strange et al. (1984). A more explicit treatment for systems with two-dimensional periodicity is given below.

Returning briefly to eq.(39), we note that the (2x2) t-matrices

$T(\underline{k}',\underline{k})$ relating incident and outgoing plane-wave spinors for the
N-atom system are easily obtained in terms of τ. If multiple scattering between the sites is ignored, i.e. Gt in eq.(39) omitted, one re-
trieves the "single-scattering" or "pseudo-kinematic" approximation (in
which atomic scattering is still treated exactly rather than in the
first Born approximation as in "kinematic" theory). If all sites are
identical (but not necessarily periodically arranged), the summation
over the site index i (implicit in eq.(39)) yields in this approxima-
tion $T(\underline{k}',\underline{k})$ as a product of the single-site matrix and a "structure
factor". The latter is spin-independent, if the inelastic mean free
path is spin-independent, which is certainly the case for nonmagnetic
systems. Consequently, the spin polarization after scattering is the
same as for an isolated "crystal atom". As regards the applicability
of the single-scattering approximation to elastic low-energy electron
scattering, it is clearly inadequate for crystalline solids (cf. e.g.
Feder 1981, Jona et al. 1982 and refs. therein). For liquids or amorphous solids (e.g. metallic glasses), it has some qualitative use, but
multiple scattering has to be taken into account for quantitative purposes (cf. Schilling and Webb 1970, Unguris et al. 1984, Chapter 7 of
this book).

4.3.5 Diffraction by Two-Dimensionally Periodic Systems

For crystalline materials with perfect lattice periodicity in
planes parallel to the surface, scattering of slow electrons, i.e. LEED,
is efficiently treated in two successive stages: (a) within a layer of
atoms (with one or several atoms per unit cell) (intralayer multiple
scattering) and (b) between layers (interlayer multiple scattering). In
the following, the essentials of spin-polarized LEED theory are presented. Details and references on relativistic LEED from non-magnetic
systems may be found in a review by Feder (1981). The extension to
ferromagnets is due to Feder et al. (1983b) and Tamura et al. (1984).

Intralayer multiple scattering is, for potentials of the muffin-
tin form (cf. Fig. 3), treated by means of a "layer-KKR" method, i.e.

a two-dimensional version of the Green function band structure method of Korringa, Kohn and Rostoker (cf. e.g. Callaway 1976). The wavefield incident from both sides on a single (sub-surface) layer (chosen with the nuclei in the (x,y) plane at z=0) (see Fig. 7) is represented

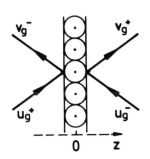

Fig. 7:

Scattering of plane-wave field $u_g^{\pm}$ incident on a single layer of identical crystal atoms (cf. eq.(40)).

in terms of plane waves with spinor amplitudes u_g^{σ}:

$$\sum_{g\sigma} u_g^{\sigma} \exp(i\underset{\sim}{k}_g^{\sigma} \cdot \underset{\sim}{r}) \text{ with } \underset{\sim}{k}_g^{\sigma} = (\underset{\sim}{k}_g^{\parallel}, \sigma[2(E+V_0)+k_g^{\parallel 2}]^{1/2}) , \qquad (40)$$

where $\sigma = +(-)$ for z<0(>0) and g enumerates the surface reciprocal lattice vectors; the "internal" wave vectors $\underset{\sim}{k}_g^{\sigma}$ differ from those in vacuum (cf. eq.(15)) via the complex inner potential V_0 (cf. Fig. 2 and 3). The relativistic multiple scattering treatment is somewhat simplified by the fact that in the interstitial region between the spheres the two small components of a four-spinor are of the order $(E+V)/(mc^2)$ and may therefore be neglected for low-energy electrons. (For details on this approximation cf. Koelling 1969). The spinors u_g^{σ} in eq.(40) can thus be restricted to the two large components u_{gs}^{σ} with s = +(-) corresponding to spin up (down). Intralayer scattering is treated in the total angular momentum representation used above for single-site scattering. Expansion of eq.(40) with respect to $j_{\ell}(kr)\, \chi_{\kappa}^{\mu}(\underset{\sim}{r})$, where χ_{κ}^{μ} are the spin angular functions (eq.(29)) and j_{ℓ} the spherical Bessel functions, yields the "direct" incident amplitudes (onto each crystal atom) as

$$A_{\kappa\mu}^{(0)} = 4\pi i^{\ell} \sum_{g\sigma s} u_{gs}^{\sigma} \; C(\ell\tfrac{1}{2}j; \; \mu-s,s)\,[Y_{\ell}^{\mu-s}(\hat{\underset{\sim}{k}}_g^{\sigma})]^* \qquad (41)$$

Denoting by $A_{\kappa\mu}$ the total amplitude of the partial wave $(\kappa\mu)$ incident on site R_j, the wave field scattered in the direction of a specified site R_o is (cf. eqs.(32,33)):

$$\sum_{\kappa'\mu'} t_{\kappa\mu\kappa'\mu'} \; A_{\kappa'\mu'} \; h_\ell^{(1)}(kr_j) \; \chi_\kappa^\mu(r_j), \quad \text{with } r_j = R_o - R_j \tag{42}$$

Re-expansion about R_o and summation over $R_j \neq R_o$ gives the scattered wave field incident on R_o as

$$A^{(s)} = \sum_{\kappa'\mu'} \; A_{\kappa'\mu'} \; X_{\kappa'\mu'\kappa\mu} \quad , \tag{43}$$

with the multiple scattering matrix elements

$$X_{\kappa\mu\kappa'\mu'} = \sum_s \sum_{\kappa''\mu''} t_{\kappa''\mu''\kappa\mu} C(\ell''\tfrac{1}{2}j''; \mu'',s) \; C(\ell'\tfrac{1}{2}j';\mu',s) \; \times$$

$$\times \sum_j \exp(ik_o'' r_j) \; G_{\ell'',\mu''-s; \; \ell',\mu'-s} (r_j) \quad , \tag{44a}$$

where C are Clebsch-Gordan coefficients and the non-relativistic matrix G (cf. Pendry 1974) has elements

$$G_{\ell m\ell'm'}(r_j) = \sum_{\ell''m''} 4\pi(-1)^{(\ell-\ell'-\ell'')/2}(-1)^{m'+m''} h_{\ell''}(kr_j) \; \times$$

$$\times \; Y_{\ell''}^{-m''}(r_j) \int Y_\ell^m(\hat{r}) \; Y_{\ell'}^{-m'}(r) \; Y_{\ell''}^{m''}(r)dr \quad . \tag{44b}$$

The total incident amplitude $A_{\kappa\mu}$ is now obtained in terms of $A_{\kappa\mu}^{(o)}$ from the self-consistent scattering condition $A_{\kappa\mu} = A_{\kappa\mu}^{(o)} + A_{\kappa\mu}^{(s)}$ (which is a special case of eq.(38)) by substituting eq.(43):

$$A_{\kappa\mu} = \sum_{\kappa'\mu'} \; A_{\kappa'\mu'}^{(o)} \; (1-X)^{-1}_{\kappa'\mu'\kappa\mu} \quad . \tag{45}$$

The total wave field scattered by the layer is given by eq.(42) summed over R_j. We re-expand it in terms of outgoing plane waves

$$\sum_{g\sigma s} v_{gs}^{\sigma} \chi^{s} \exp(i\underset{\sim}{k}_g\underset{\sim}{r}), \text{ with } \sigma = -(+) \text{ for } z<0(>0) \quad , \tag{46}$$

and substitute $A_{\kappa\mu}$ from eq.(45) wit $A_{\kappa\mu}^{(0)}$ from eq.(41). The scattered-wave two-spinors $\tilde{v}_g^{\sigma}$ are thus expressed by the incident-wave ones $\tilde{u}_g^{\sigma}$ (cf. also Fig. 7), i.e. we have the layer T-matrix (a special case of τ in eq.(39)), and, by adding the unscattered plane waves, the S-matrix, which transforms the ingoing into the total outgoing field (cf. also Fig. 7):

$$v_{gs}^{\sigma} = \sum_{\sigma'=\pm} \sum_{g'} \sum_{s'} M_{gg'ss'}^{\sigma\sigma'} u_{g's'}^{\sigma'} \quad , \tag{47}$$

with the S-matrix elements given as

$$M_{gg'ss'}^{\sigma\sigma'} = \delta_{gg'ss'}^{\sigma\sigma'} + (8\pi^2)/(kAk_g^+) \sum_{\kappa\mu} \sum_{\kappa''\mu''} t_{\kappa''\mu''\kappa\mu} \times$$

$$\times C(\ell''\tfrac{1}{2}j''; \mu'',s) i^{-\ell''} Y_{\ell''}^{\mu''-s}(\hat{\underset{\sim}{k}}_g^{\sigma}) \times$$

$$\times \sum_{\kappa'\mu'} i^{\ell'} C(\ell'\tfrac{1}{2}j'; \mu's') [Y_{\ell'}^{\mu'-s'}(\hat{\underset{\sim}{k}}_g^{\sigma'})]^{*} (1-\underset{\sim}{X})_{\kappa'\mu'\kappa\mu}^{-1} \tag{48}$$

where $\delta_{gg'ss'}^{\sigma\sigma'}$ is a Kronecker symbol (=1 if $\sigma = \sigma'$, $g = g'$ and $s = s'$; = 0 otherwise), A the area of the two-dimensional unit cell, and the elements of $\underset{\sim}{X}$ are given by eq.(44a).

If N denotes the number of two-dim. reciprocal lattice vectors $\underset{\sim}{g}$ (i.e. the number of "beams") required for convergence, the layer S-matrix $\underset{\sim}{M}$ (eq.(48)) has dimensions (4N×4N), rather than (2N 2N) in the spinless formalism (cf. e.g. Jona et al. 1982, Pendry 1974). Eq.(48) is valid for relativistic scattering from ferromagnets with arbitrary uniform direction of the effective magnetic field $\underset{\sim}{B}(r)$. For $\underset{\sim}{B}$ normal to the layer, the atomic T-matrix (with elements $t_{\kappa\mu\kappa''\mu''}$) is diagonal in μ, but has off-diagonal elements in κ)especially coupling between $j=\ell\pm\tfrac{1}{2}$ (as discussed in Sections 4.3.3). For non-magnetic systems, it is furthermore diagonal in κ with elements $[\exp(2i\delta_\kappa)-1]/2$, where δ_κ are the relativistic phase shifts. The general symmetry properties, which were discussed above (Section 4.3.2), also apply to $\underset{\approx}{M}$ and its (2x2) spin-submatrices $M_{\underset{\sim}{gg'}}^{\sigma\sigma'}$. In numerical work, they provide a test for results and permit a simplification of the calculations. For a single atomic layer (non-magnetic, or ferromagnetic with $\underline{B}$ perpendicular to the layer) with one atom per unit cell, there is an additional symmetry element: the mirror operation at the internuclear plane. It then suffices to compute $M_{gg'ss'}^{\sigma\sigma'}$ (eq.(48)) for propagation direction indices $(\sigma,\sigma') = (+,+)$ and $(+,-)$.

For dealing with interlayer scattering, it is useful to transform - by simple matrix algebra - the single-layer S-matrix $\underset{\approx}{M}$ (eq.(48)) into the (4N×4N) "transfer matrix" $\underset{\approx}{Q}$, which relates the plane wave spinor amplitudes (u_g^-, v_g^+) on the z>0 side of the layer to (u_g^+, v_g^-) on the z<0 side. By virtue of Bloch's theorem, diagonalization of $\underset{\approx}{Q}$ yields for any given E and $\underline{k}^{||}$ a set of values k^z, i.e. the spin-dependent bulk band structure $E_s(\underline{k}^{||},k^z)$. The corresponding Bloch waves provide an accurate means of treating diffraction by a semi-infinite stack of identical layers (i.e. a truncated bulk crystal): denoting by u_g^+ and v_g^- the incident and reflected plane-wave spinor amplitudes, respectively, matching of the plane-wave field outside to a linear combination of Bloch waves inside yields the (2N×2N) "bulk reflection matrix" $\underset{\sim}{R}^b$:

$$v_{gs}^- = \underset{g's'}{\Sigma} \; R_{gg'ss'}^b \; u_{g's'}^+ \quad . \tag{49}$$

Since in LEED only those Bloch waves are relevant, which carry current into the crystal (for details see Pendry 1974), it is computationally more efficient to avoid calculating simultaneously (like in the above $\underset{\approx}{Q}$ matrix diagonalization method) those carrying current out of the crystal. This is achieved in an elegant scheme (developed by Jepsen (1980) for spinless LEED and generalized to SPLEED by Feder et al. (1981); see also Feder 1981), in which $\underset{\approx}{Q}$ is brought to a block trian-gular form rather than diagonalized. This new Bloch wave method shares with the conventional one the advantage of reliable applicability also

when the absorptive potential V_i is very small (i.e. at very low energies). (For some discussion and references on finite-crystal and perturbation schemes cf. Feder 1981).

The transition region between vacuum and bulk, the "selvedge",
usually involves, in addition to the surface potential barrier (see
Fig. 2), atomic layers which differ from the bulk layers chemically
(adsorbates), geometrically (reconstruction with a larger unit cell,
relaxation of interlayer spacings for one or several layers (see e.g.
Davis and Noonan 1983, Gauthier et al. 1984)) or magnetically (with
effective fields $\underset{\sim}{B}$ different from that in the bulk, cf. Chapters 1 and
7 of this book). The transfer matrices $\underset{\sim}{Q}_n$ of these atomic layers (enu-
merated by n=1, ..., L) are obtained from the corresponding scattering
matrices $\underset{\sim}{M}_n$, which are calculated according to eq.(48). For ferromagnets with near-surface fields $\underset{\sim}{B}_n$, which differ from the bulk field $\underset{\sim}{B}$
by a factor $b_n = B_n/B$, the matrices $\underset{\sim}{Q}_n$ need not be calculated ab initio.
As shown by Feder and Pleyer (1982) and by Tamura et al. (1984), they can
in very good approximation be obtained from the bulk layer matrix $\underset{\sim}{Q}_b$
(calculated for $+\underset{\sim}{B}$ and $-\underset{\sim}{B}$) by first-order Taylor expansion around $\underset{\sim}{B}$:

$$\underset{\sim}{Q}_n(\pm\underset{\sim}{B}_n) = [\underset{\sim}{Q}_b(\underset{\sim}{B}) + \underset{\sim}{Q}_b(-\underset{\sim}{B})]/2 \pm b_n[\underset{\sim}{Q}_b(\underset{\sim}{B}) - \underset{\sim}{Q}_b(-\underset{\sim}{B})]/2 \quad . \tag{50}$$

The surface barrier has the two-dimensional periodicity of the
topmost atomic layer and can therefore be Fourier-expanded as
$\sum_g V_g(z)\exp(i\underset{\sim}{g}\underset{\sim}{r})$. Substituting this and $\sum_g \psi_g(z)\exp(i\underset{\sim}{k}_g''\underset{\sim}{r})$ into the Dirac
or Schrödinger equation, one obtaines a set of coupled ordinary differ-
ential equations for the $\psi_g(z)$, numerical integration of which yields
the barrier matrices $\underset{\sim}{M}_0$ and $\underset{\sim}{Q}_0$ (cf. Tamura and Feder 1983 and 1985).
Since spin-orbit coupling is negligible in barrier scattering at low
energies, the Schrödinger equation (possibly with a different potential
for spin-up and spin-down in the case of ferromagnets) suffices. If the
"corrugation" of the barrier is neglected, i.e. only $V_{00}(z) =: V(z)$ re-
tained, the differential equations are decoupled, and $\underset{\sim}{M}_0$ and $\underset{\sim}{Q}_0$ are
diagonal in g. Whether barrier corrugation may actually be disregarded

or not, depends, for barrier-sensitive observable features, on whether these are significantly influenced by the near-surface part of the barrier. For example, LEED fine structure features (associated with beam emergence thresholds; Rydberg resonances) are determined by the image asymptotic behaviour and can therefore be described by one-dimensional surface barrier models (cf. e.g. Jones et al. 1984, Baribeau et al. 1985 and refs. therein). On the other hand, spin-polarized LEED features from W(001) were found to depend strongly on corrugation (Tamura and Feder 1983 and 1985).

Scattering by the entire selvedge (surface barrier plus special near-surface atomic layers) is described by the product matrix $\underset{\sim}{Q} = \underset{\sim}{Q}_{\sigma\sigma'} \times \underset{\sim}{Q}_1 \times \ldots \times \underset{\sim}{Q}_l$. Partitioning $\underset{\sim}{Q}$ into four (2N×2N) submatrices $\underset{\sim}{Q}^{\sigma\sigma'}$, where (σ,σ') are the propagation direction indices (cf. eqs.(40) and (47)), and combining these via simple matrix algebra with the (2N×2N) bulk reflection matrix $\underset{\sim}{R}^b$ (eq.(49)), one obtains the reflection matrix $\underset{\sim}{R}$ for the complete system. Its elements $R_{gg'ss'}$ with g'=1 relating to the incident beam (g'=(0,0)) form for each g a (2x2) spin matrix. These are just the (2x2) scattering matrices S_g as defined in eq.(18), from which the statistical operator for the gth LEED beam and thence its spin polarization and spin dependent intensities are obtained according to eqs.(20), (1) and (21).

4.3.6 Disorder and Lattice Vibration Effects

If there are deviations from perfect lattice periodicity, elastic scattering is no longer confined to the discrete set $\{|\underset{\sim}{k}_g>\}$ of outgoing plane waves (i.e. sharp spots of the LEED pattern, but also non-zero for $\underset{\sim}{k}' \neq \underset{\sim}{k}_g$ (incoherent or diffuse scattering), appearing as a "background" intensity between the LEED spots. An important class of such structural disorder comprises systems with some disorder restricted to the surface region, e.g. low-coverage adsorbate atoms on an ordered substrate. (For a recent theoretical treatment and references see Saldin et al. 1985).

Another category, which is omnipresent experimentally, is associated with thermal lattice vibrations. This involves, strictly speaking, inelastic processes (electron-phonon scattering) (cf. e.g. Duke 1974, and Section 4.4 of this book). The effect on the elastic channel (zero-phonon scattering) may, however, also be obtained by means of the Born-Oppenheimer approximation, i.e. by treating elastic scattering from "frozen" atomic displacement configurations and averaging the T-matrices over the configurations (cf. e.g. Pendry 1974). Assuming a Debye spectrum of lattice modes and neglecting correlations between the atomic vibrations, the result amounts to a renormalization of the single-site T-matrix $\underset{\sim}{t}$ (cf. eq.(33)) (essentially by a Debye-Waller factor) and subsequent use of the renormalized $\underset{\sim}{t}$ in the multiple scattering calculations. (For details see Tamura et al. 1984). If $\underset{\sim}{t}$ is diagonal and expressed in terms of scattering phase shifts $\delta_\ell^\pm$, its renormalization is equivalent to replacing the $\delta_\ell^\pm$ (which are real for real $V^S(r)$ and real energy) by effective complex ones $\tilde{\delta}_\ell^\pm$. The additional imaginary part accounts for the removal of flux out of the elastic channel. (For details see e.g. Feder 1981, Duke 1974, Pendry 1974 and refs. therein). As is well known, the attenuation of the LEED beam intensities deviates from the simple Debye-Waller form as a consequence of multiple scattering, which is modified with increasing temperature firstly by thermal lattice expansion and secondly by the above attenuation of single-site scattering. For the spin polarization $\underset{\sim}{P}_g$ in LEED from non-magnetic systems, this leads to a significant temperature dependence (decrease or increase), whereas without multiple scattering (i.e. in the atomic-like kinematic approximation) $\underset{\sim}{P}_g$ would not change with temperature (cf. Kirschner and Feder 1981, Feder 1981, Chapter 6 of this book, and refs. therein). In the case of ferromagnets, the temperature dependence of the exchange-induced scattering asymmetry A_{ex} (cf. eq.(16b)) is of course mainly governed by the decrease of the magnetization (cf. following Section).

4.3.7 Determination of Magnetic and Other Surface Properties

Geometrical structure (arrangement of atomic sites), electronic charge density and spin density (magnetization, magnetic structure)

of the surface region manifest themselves, via the structure of the
effective fields $V(E,\underset{\sim}{r})$ and $\underset{\sim}{B}(E,\underset{\sim}{r})$ (cf. eq.(3)) or the equivalent
spin-dependent potentials $V^S(E,\underset{\sim}{r})$ (cf. eq.(7)), in the observable
(SP)LEED beam intensities, spin polarizations and scattering asymmetries taken as functions of the energy E and the surface-parallel mo-
mentum $\underset{\sim}{k}_{\parallel}$ of the primary beam (or the polar angle θ with respect to
the surface normal and an azimuthal angle φ, which are experimentally
more practicable than $\underset{\sim}{k}_{\parallel}$). The observable quantities are usually presented as functions of one of the independent variables with the other
two fixed, e.g. as energy spectra at fixed angles θ and φ, as is illustrated in Fig. 8 for the specular beam from the ferromagnetic Gd(0001)
surface. (For a non-magnetic surface, the asymmetries A_{ex} and A_u

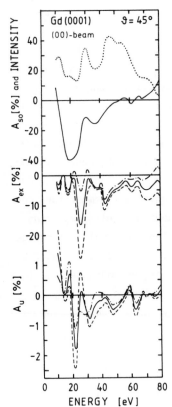

Fig. 8:

(SP)LEED specular beam from ferromagnetic
Gd(0001) for incidence at θ=45° and with
scattering plane parallel to mirror plane
and perpendicular to $\underset{\sim}{P}$ and $\underset{\sim}{B}$: Theoretical
results (from Tamura et al. 1984) for
spin-averaged I(....) and A_{so}(——) (cf.
eq.(16a)) in upper panel; and,in lower
panel, A_{ex} and A_u (cf. eqs.(16b,c)) for
top layer magnetization $M_1=M_b$ (bulk value)
(——), $M_1=1.5\ M_b$ (– – –) and $M_1=0.5\ M_b$
(–·–·)

(cf. eqs. (16b,c)) in the lower two panels would of course vanish). The spectra exhibit rich structure as a consequence of multiple scattering processes. (The increase in structural wealth in proceeding from a single crystal atom to an ordered mono-atomic layer and therefrom to a semi-infinite crystal has been explicitly demonstrated by numerical calculations (cf. Feder 1977 and Chapter 6 of this book)).

How can one extract the physical information coded in measured spectra of such complexity? Unlike in X-ray or neutron scattering, which is essentially "kinematic" (i.e. with negligible multiple scattering), the inverse scattering problem has not yet been solved. Instead, a systematic trial and error approach is well established, which proceeds as follows. Numerical calculations are performed using the full multiple scattering theory (outlined in Section 4.3.5) for physical model assumptions, which are varied within physically reasonable limits. One then looks for beams and regions in (E,θ,ϕ) space, for which the calculated quantities are selectively sensitive to changes in the model assumptions, i.e. respond strongly to one model characteristic (e.g. a geometrical structure parameter) and not (or weakly) to the others (e.g. parameters entering the effective potential construction). Comparison with the corresponding features in experimental data then determines the actual value of this particular model characteristic. As is well known from conventional LEED intensity analysis (cf. e.g. Jona et al. 1982, Marcus and Jona 1984), such determination must be based on a sufficiently large number of spectra and features. Spin-orbit-induced spin polarization or scattering asymmetry (A_{so}) spectra not only enlarge this basis, but also are generally more sensitive than (spin-averaged) intensity spectra, especially with regard to the scattering potential (cf. Feder 1981, Feder in ed. Marcus and Jona 1984, Chapter 6 of this Book, and refs. therein).

For ferromagnetic surfaces, the determination of magnetic properties by (SP)LEED is facilitated by the fact that the magnetic information (like the layer- and temperature-dependent (average) magnetization $M_n(T)$) is almost exclusively coded in A_{ex} and A_u if one employs the special diffraction geometry with $\underline{B}$ and $\underline{P}$ normal to the scattering plane (cf. Fig. 6). This is illustrated in Fig. 8, where an increase or decrease of the top layer magnetization by 50 % is seen to produce drastic changes in A_{ex} and A_u, whilest A_{so} and I are not affected. One can therefore first determine the non-magnetic properties by comparing calculated A_{so} and I spectra with their experimental counterparts, and subsequently deduce the magnetic properties from A_{ex} and A_u. The ordering of these two stages is important, since non-magnetic model changes (like geometrical surface reconstruction) in general affect A_{ex} and A_u, whereas magnetic model changes do not affect A_{so} and I. It should also be noted that this method allows the determination of $M_n(T)$ for any experimentally chosen temperature T even if $M_n(T=0)$ is not known. It is of course most interesting to then extrapolate to $M_n(T=0)$.

Although A_u, the asymmetry produced for unpolarized incident electrons by the joint action of spin-orbit coupling and exchange (cf. eq.(16c) and subsequent explanation), is generally about an order of magnitude smaller than the exchange-induced A_{ex}, it has actually been observed experimentally (Alvarado and Weller 1984a). Its theoretically predicted sensitivity to the top layer magnetization (cf. Fig. 8 and Tamura et al. 1984) therefore suggests that unpolarized electrons, which are experimentally less demanding than polarized ones, may provide a viable tool for stuying surface ferromagnetism. For polarized incident electrons, one has A_u as a magnetization-sensitive observable in addition to A_{ex} and can expect enhanced accuracy and reliability.

Since the calculations of A_{ex} (and/or A_u), which are required for the determination of the layer- and temperature-dependent magnetization $M_n(T)$ from experimental A_{ex} data, demand some theoretical competence and computing power (although becoming now feasible on Perconal Computers (Pendry 1985)), it is of interest to know whether A_{ex} is proportional to some average surface magnetization $\bar{M}_s(T)$. If yes, the temperature dependence of this $\bar{M}_s$ is directly given by the A_{ex} data. Unfortunately, it was found both theoretically (Feder and Pleyer 1982) and experimentally (Kirschner 1984) that such proportionality does not exist in general. For counter-examples see e.g. in Fig. 8 the small A_{ex} peak near 18 eV, the A_{ex} values near 80 eV (or A_u around 23 eV). Such behaviour can be traced back to interlayer multiple scattering (Feder and Pleyer 1982). If, however, the layer magnetizations within the information depth of LEED (typically 4 to 8 atomic layers) scale with temperature like the top layer magnetization M_1, A_{ex} becomes in very good approximation proportional to M_1 (the leading term in an expansion in odd powers of M_1). This is, due to the divergence of the magnetic correlation length (cf. e.g. Chapter 3 of this book) at the Curie temperature T_c, the case in the immediate vicinity of T_c (provided that T_c is the same in the bulk and at the surface, which is for example the case for Ni, but not for Gd (see Chapter 7)). The critical behaviour of M_1 can therefore be directly read off experimentel A_{ex} data (see Chapter 7 of this book and references therein).

Finally, attention is drawn to some practical aspects of measuring and of calculating A_{so} and A_{ex}. Experimentally, the definitions eqs.(16a,b) in terms of the four basic intensities I_σ^μ require absolute intensity measurements for opposite magnetic field orientations ($\mu=+/-$). Slight deflections of the primary beam by stray magnetic fields may therefore lead to errors. This is avoided by measuring instead (cf. Alvarado et al. 1982) A_u (using unpolarized electrons) and two new basic asymmetries

$$A^\mu = (I_+^\mu - I_-^\mu) / (I_+^\mu + I_-^\mu) \quad \text{with } \mu = +/- \ , \tag{51}$$

and therefrom calculating new asymmetries defined as

$$\tilde{A}_{so} = (A^+ + A^-)/2 \ , \quad \tilde{A}_{ex} = (A^+ - A^-)/2 \ , \tag{52}$$

$$\tilde{\tilde{A}}_{so} = \tilde{A}_{so} + A_u \tilde{A}_{so} \ , \quad \tilde{\tilde{A}}_{ex} = \tilde{A}_{ex} + A_u \tilde{A}_{so} \ . \tag{53}$$

Comparison with eqs.(16a,b,c) shows that $\tilde{A}_{so} = \tilde{\tilde{A}}_{so} = A_{so}$ in the absence of ferromagnetism, and $\tilde{A}_{ex} = \tilde{\tilde{A}}_{ex} = A_{ex}$ in the absence of spin-orbit coupling. If both spin-dependent interactions are present, $\tilde{A}_{so}$, $\tilde{\tilde{A}}_{so}$ and $\tilde{A}_{ex}$, $\tilde{\tilde{A}}_{ex}$ are approximations to A_{so} and A_{ex}, respectively, which are very good for Fe and Ni, as was plausibly suggested by Alvarado et al. (1982) and numerically verified by Tamura et al. (1984). For Gd, for which both spin-orbit coupling (Z=64) and magnetic exchange ($7\mu_B$) are much stronger, $\tilde{\tilde{A}}_{so}$ and $\tilde{\tilde{A}}_{ex}$ are still very good approximations (even if the scattering plane is not a mirror plane) (cf. Tamura et al. 1984). The same holds for the approximate theoretical quantities $A_{so}^{(o)}$ and $A_{ex}^{(o)}$ calculated by simpler theories: relativistically without magnetism, and non-relativistically with magnetism (i.e. two decoupled Schrödinger equations for V^+ and V^-), respectively. Since the above-discussed trial and error procedure requires calculation of A_{ex} for a substantial number of layer magnetizations M_n (at a given temperature), this finding permits a substantial reduction in computing requirements (time and storage).

4.4 Inelastic Scattering and Secondary Electron Emission

There is a wide variety of mechanisms and processes, by which electrons of energy E incident on a surface may lose energy, and at the end of which electrons of energy E' emerge from the surface. The vastness of this field is reflected by the fact that extensive monographs have been devoted to special sub-areas: e.g. one-electron and plasmon excitation by high-energy (up to 10 keV) electrons (Raether 1980 and Sturm 1982), or excitation of surface lattice vibrations by low-energy electrons (Ibach and Mills 1982). The present Section focuses - after some brief general remarks - on outlining key concepts and theoretical state-of-the-art for some phenomena, which are of particular interest in the context of polarized electrons: electron-hole pair and magnon excitation in ferromagnets, the resulting mean free path, and true secondary electron emission.

4.4.1 Loss Mechanisms and Spin-Dependence

The intensity of electrons of energy E'<E and momentum $\underline{k}'$, which emerge from a surface in response to an incident electron beam of energie E and momentum $\underline{k}$ (cf. Fig. 1), is schematically shown in Fig. 9 as a function of E'. Going from the elastic intensity at E = E', the

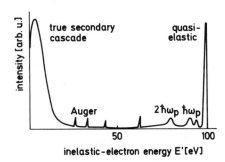

Fig. 9:

Schematic spectrum of inelastic electrons generated by a primary beam of kinetic energy E = 100 eV incident on a (perfectly or less ordered) surface. ($\hbar\omega_p$ is a plasmon energy).

"quasi-elastic" peak, which comprises electrons scattered by thermal lattice vibrations (phonons), is followed by a low-intensity "background" arising from (single and multiple) electron-hole pair excitation, superimposed on which are discrete peaks associated with (bulk and surface) plasmon excitation and Auger processes (involving core electrons). Finally, below about 10 eV, there is the true secondary electron emission "cascade" maximum. The spectral features fall into two categories according to the behaviour of their position E' when the primary energy E is varied: (a) if E' is correlated with E by a fixed difference w = E - E' (or multiples thereof), they are due to inelastic scattering by particular elementary excitations (loss processes) of energy w; (b) if the position E' is (nearly) independent of E, the feature is due to "true" secondary electrons emitted by the surface system. In the case of single-crystal surfaces, "extended" elementary excitations of energy w can be further specified by their surface-parallel momentum $\underline{q}_{\parallel}$, and the inelastically scattered electron of energy E' has surface-parallel momentum

$$\underline{k}'_{\parallel} = \underline{k}_{\parallel} - \underline{q}_{\parallel} + \underline{g} \, , \tag{54}$$

where g is a two-dimensional reciprocal lattice vector. One thus has "Inelastic Low-Energy Electron Diffraction" (ILEED) (for details see Duke 1974 and refs. therein). Denoting, for an incident beam with $\underset{\sim}{k}_{\parallel}$ and current $j_o(E)$, by $j_{e\ell}(E)$ the angle-integrated (or, for a periodic surface, the g-summed) elastically back-scattered current, by $j_{in}(E')$ the angle-integrated inelastic current (including true secondaries), and by $j_c(E)$ the total current "absorbed" by the solid (target current), current conservation requires

$$j_{e\ell}(E) + \int_0^E j_{in}(E')dE' + j_c(E) = j_o(E). \tag{55}$$

Spin polarization of the inelastic electrons (for an unpolarized primary beam) and a dependence of $j_{in}(E', \underset{\sim}{k}')$ on the primary-beam polarization (inelastic scattering asymmetry) may originate (a) from an intrinsic spin dependence of the relevant loss process(es) and (b) from spin-dependent elastic scattering events (due to spin-orbit or magnetic exchange interaction) (cf. Section 4.3). The latter effect can be understood from a multiple scattering expansion involving various sequences of elastic and inelastic events(cf. e.g. Duke 1974) (with the special two-step cases of "elastic diffraction before loss" and "loss before elastic diffraction"). Also, it is plausible in the approximation of a "golden-rule" transition between two (SP)LEED states. Experimentally, such an effect, arising from spin-orbit coupling, has been observed by Ravano and Erbudak (1982) (from Au(110)) and by Alvarado (1982) (from W(001)). If interpreted via a quantitative theory, such experiments promise a deeper insight into the interplay between elastic and inelastic events.

We now turn to the intrinsic spin dependence of individual loss mechanisms. Electron-hole pair excitation is, by virtue of the Pauli principle, spin-dependent already in non-magnetic materials. (It will be discussed in some detail in Section 4.4.2). Phonon excitation (cf. Ibach and Mills 1982) is hardly spin-dependent even for ferromagnets. The same is plausible for plasmons, since they are collective excitations involving both majority and minority electrons. (This conjecture by Feder (1979) was corroborated theoretically by Helman and Baltensperger (1980) and experimentally by Siegmann et al. (1981)). Excitation of (bulk and surface) magnons is of course strongly spin-dependent (see Section 4.4.3). Spin-exchange scattering in a paramagnetic surface layer may reduce the spin polarization of incident or emitted electrons (cf. Feuchtwang et al. 1978, Siegmann et al. 1984 and refs. therein). For ferromagnets, Auger emission involving a core hole and two 3d electrons exhibits substantial spin polarization (see Chapter 9 of this book).

4.4.2 Electron-Hole Pair Excitation

This aspect of the electron-electron interaction problem in solids has, in the framework of many-body perturbation theory (cf. e.g. textbooks by Fetter and Walecka (1971), Mahan (1981) and Inkson (1984) and refs. therein), a long history of research interest and successively improved approximations, mainly in the context of the interacting homogeneous electron gas. In view of the complexity and the width of this field, it must suffice here to briefly address some basic features and recent developments, which are of particular relevance for polarized-electron scattering at surfaces.

Electron-hole pair excitation by a "hot" electron (of energy $E_1 > E_F$) is, to first order in an effective Coulomb interaction V, symbolically represented by the diagrams in Fig. 10. Subject to conservation of energy and other quantum numbers, the incident (quasi-) electron generates a (quasi-) hole (with $E_2 < E_F$) and two (quasi-) electrons (with $E_F < E_3, E_4 < E_1$). The corresponding scattering amplitude is (with constants and a factor 2, for two equivalent diagrams with reversed vertices, implicitly included)

$$\tilde{A} = [<34|V|12> - <43|V|12>] \, \delta(1+2-3-4) \, \delta(E_1+E_2-E_3-E_4), \qquad (56)$$

which is manifestly antisymmetric under the exchange of the two out-

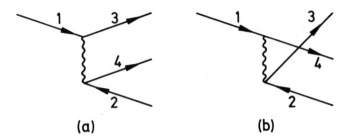

(a)   **(b)**

Fig. 10:
Electron-hole excitation by (a) direct and (b) exchange
scattering. The labels comprise sets of quantum numbers
(including spin) characterizing the individual (quasi-)
particles. The wiggly line represents an effective Coulomb
interaction.

going electrons (3 and 4). This expression is still general with regard to the quasi-particle states. If they are, for example, solutions
of the Dirac equation for a semi-infinite crystal, it corresponds to a
distorted-wave first Born approximation including exchange. If they
are (in the absence of spin-orbit coupling) eigenstates of σ_z (with
spin labels $\sigma_1 = +/-$, $\sigma_2 = +/-$, etc) and if V contains no spin-mixing
parts, spin is conserved at each vertex. Denoting the two integrals in
eq.(56) by f and g, respectively, and defining $A = [...]$, we then have

$$A = f(1234) \, \delta_{\sigma_1 \sigma_3} \, \delta_{\sigma_2 \sigma_4} - g(1234) \, \delta_{\sigma_1 \sigma_4} \, \delta_{\sigma_2 \sigma_3} \, . \tag{57}$$

For a given spin alignment σ_1, the outgoing electron 3 may thus have:
(a) $\sigma_3 = \sigma_1$, i.e. there is spin-conserving or "non-flip" scattering;
this can be realized either by $\sigma_2 = \sigma_1$, with a "non-flip" amplitude
$A_{nf1} = f-g$, or by $\sigma_2 = -\sigma_1$, with $A_{nf2} = f$; (b) $\sigma_3 = -\sigma_1$, i.e. "spin-flip"
scattering, with a "flip" amplitude $A_f = -g$, provided that $\sigma_2 = \sigma_3 = -\sigma_1$.
(The terminology "spin-flip scattering", which has become generally
accepted, is somewhat unfortunate, since spin is actually conserved at
the vertices, and the apparent flip is only in the eye of the beholder).
The non-flip amplitude $A_{nf1} = f-g$ corresponds to triplet scattering

($\sigma_2 = \sigma_1$) and therefore is - due to the Pauli principle - at very low E_1 substantially reduced compared to the singlet amplitude $f + g$ (cf. Wolff 1954, Bringer et al. 1979, Penn 1980).

The scattering rates, which correspond to these amplitudes, are, however, not readily observable, since they require detection of the two outgoing electrons (3 and 4) (or of one electron and the hole). While this has been achieved for high-energy electrons (above about 10 keV) with large energy losses ("Möller scattering", see e.g. Kessler 1976 ; "(e,2e) spectroscopy", see e.g. Ritter 1985), it is less feasible for low-energy electrons and small energy loss (in particular, if the second electron has not enough energy normal to the surface to emerge into the vacuum). One therefore sums over the possible states 2 and 4, obtaining from eqs.(56,57) and the above spin considerations the following inelastic scattering rates $R^{\sigma_1 \sigma_3}$ from state 1 into state 3: for $\sigma_3 = \sigma_1 = :\sigma$ the "non-flip" rate (for up-spin $\sigma=+$ and down-spin $\sigma=-$)

$$R^{\sigma\sigma}(1,3) = F(3,\sigma)\int d2 \int d4 [(1-F(2,\sigma))F(4,\sigma)|f-g|^2 +$$

$$+ (1-F(2,\bar\sigma)) F(4,\bar\sigma) |f|^2] \, \delta(1+2-3-4) \quad , \tag{58}$$

where $\bar\sigma=-\sigma$, $F(i,\sigma)$ is the Fermi occupation function (ensuring that states 3 and 4 are empty and 2 is occupied prior to the scattering event), and the arguments (1234) in f and g have been dropped. For $\sigma_3 = -\sigma_1 = \bar\sigma$, we have the "flip" rate

$$R^{\sigma\bar\sigma}(1,3) = F(3,\bar\sigma) \int d2 \int d4 \, (1-F(2,\bar\sigma))F(4,\sigma)|g|^2 \delta(1+2-3-4) \quad . \tag{59}$$

If $\sigma_1 = \sigma$ is given (polarized incident beam) and σ_3 is not detected, an appropriate derived quantity is the asymmetry

$$A_{ex} = [(R^{++} + R^{+-}) - (R^{-+} + R^{--})]/(R^{++} + R^{+-} + R^{-+} + R^{--}) \quad . \tag{60}$$

For unpolarized primary electrons, the spin polarization of the scattered electrons is

$$P_{ex} = [(R^{++} + R^{-+}) - (R^{+-} + R^{--})]/(R^{++} + R^{-+} + R^{+-} + R^{--}) \ . \tag{61}$$

If both ingoing and outgoing spins are determined, the four scattering rates (eqs.(58,59)) are known individually (cf. Chapter 8 of this book). If there is no spin alignment of occupied states (i.e. no magnetic order in the solid; or no experimentally produced orientation in atoms, cf. e.g. Raith 1984), $R^{\sigma_1 \sigma_3} = R^{\bar{\sigma}_1 \bar{\sigma}_3}$ and therefore $A_{ex} = 0$ and $P_{ex} = 0$.

Rather than by focusing on an inelastically scattered electron (3 in the above), inelastic events may be studied by "detecting" the hole (2 in Fig. 10) (e.g. from its radiative or Auger decay). This is, for example, done in "appearance potential spectroscopy"(APS), in which a core hole and two conduction band electrons are produced (i.e. an inverse Auger process) (see e.g. Kirschner 1984 and refs. therein). The above scattering rates are then modified by integrating over the two electrons (3 and 4) rather than one electron and the hole (4 and 2).

Evaluation of the scattering rates (eqs.(58,59)) requires calculation of the direct and exchange amplitudes f and g, i.e. the matrix elements in eq.(56). Details depend on the case to be studied (material, primary electron energy and loss energy, localized or itinerant nature of hole) and on the accuracy sought. The simplest case is the idealized material "jellium", an infinitely extended homogeneous electron gas with a uniform positive background (instead of discrete ion cores on lattice sites). The single-particle states are then plane waves, characterized by momenta p_i (with $i = 1,-,4$), and the amplitudes become

$$f = \tilde{V}(|p_1 - p_3|), \quad g = \tilde{V}(|p_1 - p_4|) \ , \tag{62a}$$

$$\text{with} \quad \tilde{V}(|p|) = (4\pi e^2)/[|p|^2 \ \epsilon(|p|,\omega(p))] \ , \tag{62b}$$

i.e. a Coulomb potential screened by a dielectric function ε. Electron-
hole pair scattering rates in this approximation were, for example,
calculated (with different approximations for ε) for non-magnetic
jellium by Ritchie and Ashley (1965) and by Penn (1980), and, for a
ferromagnetic jellium model with a spin-split "window" (mimicking
d-bands) by Yin and Tosatti (1981), Glazer and Tosatti (1984) and
Glazer et al. (1985). The latter authors, focusing in particular an
low primary energies (E_1 of the order of 10 eV) and low loss energies
($E_1 - E_3 = :w \lesssim 5$ eV), obtained interesting model results regarding spin-
flip rates $R^{\sigma\bar{\sigma}}$ (cf. eq.(59)) associated with opposite-spin pair excitations in the d band (in Fig. 10: $\sigma_2 = \bar{\sigma}$, $\sigma_4 = \sigma$) ("Stoner excitations"].
R^{+-} and R^{-+} as functions of w and $q = p_1 - p_3$ probe (approximately in
the sense of a joint density of states, as indicated by eq.(59) when
using p-independent matrix elements) the "minority-hole/majority-elec-
tron" and "majority-hole/minority-electron" pairs, respectively.
Fig. 11 schematically depicts the Stoner pair-excitation continuum

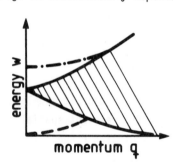

Fig. 11:

Elementary excitations in transition
metal ferromagnets: opposite-spin
pair (Stoner) excitations (shaded
area), acoustic (- - -) and optical
(-·-·) magnons.

in (q,w) space. (For a general introduction see e.g. White (1970), for
quantitative results on Fe and Ni: Cooke et al. (1980) and Cooke (1985)).
For further discussion and pioneering experimental electron scattering
results see Chapter 8 of this Book.

Before leaving electron-hole pair excitation, two recent theoretical developments are briefly mentioned. One is a qualitative discussion of the spin dependence of 3p→3d excitation in Ni and Fe and its
relation to elastic scattering (in terms of a "resonance"), Auger emis-

sion and a "satellite" observed in photoemission (Nesbet 1985). The other is an appealing response-function approach to (small w, small q) bulk and surface pair excitation processes, so far developed for non-magnetic materials (Persson and Zaremba 1985, and refs. therein). The importance of pair-excitation for the mean free path of "hot" electrons and for the secondary electron cascade will be discussed in Sections 4.4.4 and 4.4.5.

4.4.3 Spin Fluctuations

A second type of inelastic electron scattering process, which is naturally spin-dependent, involves collective spin excitations (spin fluctuations) in magnetically ordered materials. At temperatures well below T_c, these excitations are nearly independent spin waves (magnons). Schematic magnon dispersion curves ($w \propto q^2$) have been shown in Fig. 11. For recent results and references on magnons in bulk transition metals see e.g. Cooke et al. 1985. In the surface region, spin fluctuations generally differ - in line with other magnetic properties (see Chapter 3 of this book) - from those in the bulk (see e.g. Castiel 1976, Demangeat et al. 1977, Demangeat and Mills 1978, Lêvy 1981, Mathon 1981, Selzer and Majlis 1982).

Scattering of an electron (in the absence of spin-orbit coupling) from a state $|k\sigma>$ of energy E, where k subsumes the relevant spatial quantum numbers and $\sigma = +/-$ gives the spin alignment with respect to the majority spin direction of the ferromagnet, to a state $|k'\sigma'>$ of energy E', along with the transition of the solid from an initial state $|i>$ to a final state $|f>$, is described in the first (distorted-wave) Born approximation by the scattering rates

$$R^{\sigma\sigma'} = (\frac{E}{E'})^{1/2}(\frac{m}{2\pi})^2 \sum_{i,f} |<k'\sigma'f|V|k\sigma i>|^2 \ W(i)\delta(\epsilon-E_i+E_f) \quad , \qquad (63)$$

where $\epsilon = E'-E$ is the energy transfer and W(i) the probability of finding the solid in the initial state $|i>$. Since dipole-dipole coupling is, for electron energies $E << mc^2$, negligible compared to the exchange.

interaction, V contains only the latter. Associating with each lattice site $\underset{\sim}{R}$ a spin operator $\underset{\sim}{S}_R$, it may be approximated (cf. Mills 1967 and refs. therein) by the local form

$$V(E,\underset{\sim}{r}) = \underset{\sim}{\sigma} \cdot \underset{R}{\Sigma} \; \underset{\sim}{S}_R \; B(E,\underset{\sim}{r}-\underset{\sim}{R}) \quad , \tag{64}$$

where the Pauli matrix $\underset{\sim}{\sigma}$ represents the spin of the "hot" electron and $B(E,\underset{\sim}{r}-\underset{\sim}{R})$ is the magnitude of the effective magnetic field (cf. eq.(3)) in the cell around $\underset{\sim}{R}$. The scalar products in eq.(64) can be written as

$$\underset{\sim}{\sigma} \cdot \underset{\sim}{S}_R = \sigma^z S_R^z + \frac{1}{2} (\sigma^+ S_R^- + \sigma^- S_R^+) \quad , \tag{65}$$

where $\sigma^\pm = \sigma^x \pm i\sigma^y$ and $S_R^\pm = S_R^x \pm iS_R^y$ are the usual spin raising and lowering operators. Substitution into eq.(63) firstly produces, due to $\langle\sigma'|\sigma^z|\sigma\rangle = \delta_{\sigma'\sigma}$, two "non-flip" rates R^{++} and R^{--}. Conservation of spin implies $|f\rangle = |i\rangle$ and therefore $E_f = E_i$, i.e. this scattering contribution is elastic. For $T \ll T_c$, it is already incorporated in the elastic multiple scattering formalism described in Section 4.3, and its interference with the strong Coulomb scattering by the potential $V(E,\underset{\sim}{r})$ (cf. eq.(3)) actually renders the Born approximation quantitatively inadequate. The other two terms from eq.(64) lead, in eq.(63), to two "spin-flip" rates R^{+-} and R^{-+}, which lower and raise E_i by destroying or creating a spin fluctuation of energy $|\varepsilon|$, respectively. They are thus clearly inelastic. If the two electron states pertain to a semi-infinite periodic system, i.e. have $\underset{\sim}{k}_\parallel$ and $\underset{\sim}{k}'_\parallel = \underset{\sim}{k}_\parallel + \underset{\sim}{q}_\parallel$ as good quantum numbers, the flip rates $R^{\sigma\bar{\sigma}}$ are proportional to the dynamic structure factor

$$\widetilde{S}^{\sigma\bar{\sigma}}(\underset{\sim}{q}_\parallel,\varepsilon) = \int\limits_{-\infty}^{+\infty} dt \; e^{i\varepsilon t} \langle S^\sigma(\underset{\sim}{q}_\parallel,0) \; S^{\bar{\sigma}}(\underset{\sim}{q}_\parallel,t)\rangle_T \quad , \tag{66}$$

which characterizes the spin fluctuation excitation spectrum. For details see Mills (1967) and Selzer and Majlis (1982); for the low-temperature limit in the plane-wave approximation see also DeWames and Vredevoe (1967). Since for $\varepsilon \neq 0$ also $q_\parallel \neq 0$, the scattering angles deviate from the elastic ones ($\underset{\sim}{k}'_\parallel = \underset{\sim}{k}_\parallel + \underset{\sim}{g}$), i.e. the inelastically scattered electrons do not appear in the LEED beam but in the "diffuse background" in-between.

Spin polarization effects are obvious from the above. In the spin-wave regime (well below T_c), the association of the flip rate R^{-+} (R^{+-}) with magnon creation (annihilation) implies that, for an unpolarized incident beam, scattered electrons with $E'<E$ ($E'>E$) are completely polarized with spin up (down). Further, $R^{-+}>R^{+-}$, with the latter vanishing for $T{\to}0$ (neglecting zero-point magnons). The magnon-induced diffuse background in LEED is therefore highly spin-polarized. In principle, this mechanism could be used for a source of polarized electrons, but the low scattering rates presumably render such a source inferior to already existing ones (see Chapter 5). A more promising application concerns the determination of magnon dispersion relations, especially for surface magnons. Employing an energy resolution of the order of 10 meV (as currently used in electron energy loss spectroscopy for studying surface lattice vibrations; see e.g. Ibach and Mills 1982), one might think of measuring, for $E'<E$, $R^{-+}(q_\parallel, E-E')$ simply using unpolarized electrons without spin analysis, since $R^{+-}(q_\parallel, E-E')=0$. It is, however, important to discriminate against phonon scattering contributions I_p, which occur for the same energies (of the order of 10 - 100 meV). Since I_p is - except at rather special energies and diffraction conditions - little dependent on the spin direction of the primary beam, using a polarized beam (without subsequent spin analysis) one obtaines $I^+ = I_p$ and $I^- = R^{-+} + I_p$, and hence $R^{-+}(q_\parallel, E-E')$. Further discrimination against phonons is possible on the grounds of the opposite behaviour of magnon and phonon scattering on decreasing E and T: the rate for magnon creation increases (cf. the energy dependence of the exchange interaction, i.e. of the effective field B(E,r) in eq.(3) and eq.(64)), whilest that for phonons decreases (cf. Debye - Waller factor).

4.4.4 Imaginary Potential and Mean Free Path

The effective one-electron potential V^σ (with $\sigma = +/-$ for spin up/down) (cf. eq.(7)) has as an important ingredient an imaginary part V_i^σ, which is associated with the quasi-particle life-time τ^σ (cf. discussion, eq.(13) and Fig. 4 in Section 4.2.2). In terms of scattering theory, it describes the removal of flux from the elastic channel into (open) inelastic channels and therefore determines the inelastic mean free path λ^σ (cf. eq.(31)). We now consider the relative contributions of different types of inelastic channels, i.e. loss mechanisms, to λ^σ with an emphasis on the resulting spin dependence in the case of ferromagnets.

The rate of removal $(\tau^\sigma)^{-1}$ of flux from a state $|Ek\sigma\rangle$, where k subsumes relevant quantum numbers (in addition to E and σ) is the sum of the rates $(\tau_\ell^\sigma)^{-1}$ due to loss mechanisms of type ℓ. Denoting by $R_\ell^{\sigma\sigma'}(Ek,E'k')$ the scattering rate from $|Ek\sigma\rangle$ to $|E'k'\sigma'\rangle$ for loss mechanism ℓ (see e.g. eq.(59)), we then have

$$(\tau^\sigma(E,k))^{-1} = \sum_\ell \sum_{\sigma'} \int_{E'<E} dE' dk' \, R_\ell^{\sigma\sigma'}(Ek,E'k') \quad . \tag{67}$$

Its spin dependence is conveniently characterized by an asymmetry (analogous to eq.(60))

$$A_\lambda(E,k) = [\tau^+(E,k)^{-1} - \tau^-(E,k)^{-1}]/[\tau^+(E,k)^{-1} + \tau^-(E,k)^{-1}] \, , \tag{68}$$

which is positive (negative) if majority (minority) electrons are removed more strongly, i.e. have the shorter mean free path (recall eq.(13): $\lambda^\sigma(E) = \tau^\sigma(E)\sqrt{2}(E+V_{or}))$. From the previous Sections it is clear that a spin dependence of λ^σ may originate from magnon and electron-hole pair excitation and get reduced by the spin-independent excitation of phonons and plasmons. For electron-magnon scattering, R_m^{-+} dominates and consequently $\lambda_m^-(E) << \lambda_m^+(E)$, giving a negative contribution to A, which decreases with increasing E. Model calculations gave λ_m^- between 25 Å and 10^4 Å for a Heisenberg ferromagnet (De Wames and Vredevoe 1969), and $V_{im}^- = 0.15$ eV and $V_{im}^+ < 0.01$ eV for Ni (Kleinman 1978). Comparison of these values to typical total λ and V_i values (cf. Fig. 4) suggests that the influence of magnon creation on the mean free path is insignificant at energies above the vacuum threshold and thus for elastic scattering and energy loss spectroscopy. It may, however, be important in the vicinity of the Fermi energy, i.e. for the lower state in photoemission and its inverse (cf. Sections 4.5 and 4.6).

Electron-hole pair excitation is the dominant loss mechanism at energies below the threshold of plasmon excitation (typically 20 eV for bulk plasmons). The resulting spin dependence of the mean free path λ_e^σ has been semi-quantitatively understood by theoretical models involving various simplifying assumptions. In an early heuristic model (Bringer et al. 1979, Feder 1979; see also Feder 1981), the Pauli principle was assumed to permit only singlet scattering, i.e. to make the exchange scattering amplitude g (cf. eq.(57)) equal to the direct amplitude f (and hence the triplet scattering amplitude f-g=0): in a ferromagnet, a "hot" spin-up (spin-down) electron is thus scattered only by minority (majority) electrons; instead of then evaluating the integrals in eqs. (58,59,67), the mean free path ratio was assumed as $\lambda^+/\lambda^- = n^+/n^-$, where $n^+(n^-)$ is the number of spin-up (spin-down) valence electrons (per atom, i.e. $\lambda^- < \lambda^+$. This implies (cf. eq.(68)) a negative asymmetry A_λ of the magnitude of the bulk spin polarization at all energies. The decrease of g with increasing E was taken into account in a higher-energy atomic approximation (Matthew 1982), leading above 100 eV to a negative A_λ of about two orders of magnitude below the bulk polarization. Evaluation of the respective integrals in a plane-wave approximation (Yin and Tosatti 1981) gave, at all energies, negative asymmetries below the bulk polarization. (In a plane-wave calculation by Rendell and Penn (1980), spin flip processes (cf. eq.(59)) were completely neglected, which is probably the cause for the wrong sign of the asymmetry obtained for Fe and Co below about 60 eV). A negative A_λ was also suggested for 3p→3d excitation (Nesbet 1984).

The A_λ resulting from all concomitant loss processes (cf. eq.(68))

(with spin-independent ones, like plasmon excitation, reducing the magnitude of A_λ) can also be determined empirically by comparing experimental electron spectroscopy data with corresponding theoretical results obtained for various assumed loss rates. This has, for example, been done by spin-polarized LEED, yielding A_λ from -0.04 at 30 eV to -0.02 at 130 eV for Fe (Tamura and Feder 1982) (cf. bulk polarization 0.26) and $|A_\lambda| \leq 0.003$ for Ni between 70 and 90 eV. A direct extraction of A_λ from experimental data (e.g. inelastic scattering cross sections) is not quantitatively feasible due to the influence of elastic events on the observed intensities at E'<E.

4.4.5 True Secondary Electron Emission

The "cascade" of successive pair excitation events, which is triggered by an incident primary electron, leads to a large peak at the low-energy end of inelastic scattering spectra (see Fig. 9). (For references, cf. e.g. Chung and Everhart 1974). In the following, a few key aspects are outlined with an emphasis on spin dependence.

In principle, the energy- and angle-resolved SEE current may be obtained from a steady-state collision-rate equation. Denoting by $N^\sigma(Ek)$ the occupation probability of the quantum-mechanical state $|Ek\sigma>$ (neglecting spin-orbit coupling), where k comprises again a set of relevant quantum numbers, the rate of scattering out of a particular $|Ek\sigma>$, with $E<E_o$ (the primary energy), equals the rate of scattering into this state:

$$N^\sigma(Ek) \int_{E_F}^{E} dE' \int dk' [\sum_{\sigma'} R^{\sigma\sigma'}(Ek, E'k')(1-N^{\sigma'}(E'k'))] =$$

$$= (1-N^\sigma(Ek)) \int_{E}^{E_o} dE' \int dk' \sum_{\sigma'} [N^{\sigma'}(E'k') R^{\sigma'\sigma}(E'k', Ek)] \qquad (68)$$

where $\sigma = +/-$ and $R^{\sigma\sigma}$ and $R^{\sigma\bar\sigma}$ are "non-flip" and "flip" rates for scattering from $|Ek\sigma>$ to $|E'k'\sigma>$ and $|E'k'\bar\sigma>$ (cf. eqs.(58,59)). This set of coupled integral equations has to be solved with the boundary condition

of a given $N^\sigma o(E_o k_o)$ corresponding to the incident beam. Using real-
istic states $|Ek\sigma>$ (e.g. time-reversed LEED states $|Ek_{||}\sigma>$ in the case
of two-dimensionally periodic systems, cf. Section 4.3.5) appears,
however, prohibitively complicated at present (even for non-magnetic
materials), and one has to resort to approximations.

As was shown by Feder and Pendry (1978) for non-magnetic systems,
the observed angle-resolved SEE current $I(Ek_{||})$ can be represented as a
product of a slowly varying "smooth" cascade function $C_o(E)$ and a
"fine structure function" $F(Ek_{||})$, which for periodic systems is given
as $F(Ek_{||}) = 1 - \sum_g R_g(E,k_{||})$, where R_g is the elastic current reflection
coefficient for the gth LEED beam. This relation indicates some correlation of F with the bulk band structure (cf. Kleinherbers et al. 1984
and Feder et al. 1985). For amorphous systems, F can be written simi-
larly, with g replaced by a continuous $k_{||}'$, but is much smoother (to
the extent of being indiscernible when comparing with experimental
data). The cascade function $C_o(E)$ can approximately be taken from a
homogeneous electron gas calculation (e.g. Chung and Everhart 1974).
This approach led to very good agreement with experiment for SEE from
W(001) (Schäfer et al. 1981), Cu(001) and Cu(001)c(2x2) (Kleinherbers
et al. 1984, Feder et al. 1985).

For ferromagnets, the extension of the above is straightforward
with regard to the fine structure function $F^\sigma = 1 - \sum_g R_g^\sigma$, where the R_g^σ
(with $\sigma = +/-$) are obtained separately for the spin-up/down potentials
V^+ and V^- (cf. eq.(7)). The cascade function also becomes spin-dependent, $C_o^\sigma(E)$, and can e.g. be obtained by solving eq.(68) for a ferromagnetic modified electron gas with the same magnetization as the real
solid (cf. Glazer et al. 1985). Since results are presented in Chapter
9, it may suffice here to mention two salient features. Firstly, the
SEE spin polarization $P(E) = [C_o^+(E) - C_o^-(E)]/[C_o^+(E) + C_o^-(E)]$ equals the
bulk polarization (e.g. about 5 % for Ni) at energies above about 10 eV
and rises by about a factor two for E→0. This behaviour is explained
by an increasing importance of the (exchange-induced) spin-flip rate
$R^{-+}(E,E')$ (cf. eq.(59)), which removes minority electrons at E, and

replaces them by majority electrons at $E' < E$ (for $E' > E_{vac}$). A second finding concerns the intensity asymmetry of $(C_0^+ + C_0^-)$ upon reversal of the primary-beam polarization: for typical primary energies of the order of 100 eV, it is practically zero, because the exchange amplitude g and hence the flip rates $R^{\sigma\bar{\sigma}}$ (eq.(59)) practically vanish and the non-flip rates R^{++} and R^{--} (eq.(58)) are the same.

In concluding this Section on inelastic scattering, a complementary and experimentally simpler method should be mentioned. By virtue of current conservation (cf. eq.(55)), the current j_c absorbed by the crystal at primary energy E contains the combined information of elastic scattering and energy-integrated inelastic scattering (including true secondary electrons). The spin dependence of its zero crossing (at energies above 100 eV) is the basis of a recent spin polarization detector (cf. Chapter 5.2.3). In the kinetic energy range from 0 to about 70 eV, separation - by means of the second energy derivative - of the slowly varying inelastic part from the more rapidly varying elastic part has revealed correlations with the quasi-particle bulk band structure and permitted a determination of the energy dependence of the real and imaginary inner potential parts for Fe (Tamura et al. 1985).

4.5 Photoemission

Within the past ten years, angle- and energy-resolved photoemission has developed into a very widely used powerful technique for the study of the electronic structure of solids and their "clean" and adsorbate-covered surfaces. Details and references to a vast body of earlier original literature may be found in the "classical" volumes by Feuerbacher et al. (1978) and Cardona and Ley (1978), and in review articles by Williams et al. (1980), Plummer and Eberhardt (1982), Wandelt (1982), Fadley (1984), and Courths and Hüfner (1984). While recent experimental and theoretical results obtained by spin-resolved photoemission will be reviewed in Chapters 10, 11, 12 and 14 of this book, the present Chapter provides, after outlining some basic concepts, a survey of the theoretical foundation, with an emphasis on physical ideas rather than details of formalism.

4.5.1 Basic Concepts

Photoemission may be viewed as a scattering phenomenon by a surface system (cf. Fig. 1) with (monochromatic) electromagnetic radiation (i.e. photons of energy $h\nu$, momentum $\hbar q$ and a polarization described by the electric field vector $\underline{E}$) in the input channel and electrons (of energy $E'<h\nu$, momentum $\hbar k'$ and spin polarization vector $\underline{P}$) in the output channel. The basic mechanism "inside" the system is depicted in Fig. 12. The photon imparts its energy $h\nu$ to an electron of

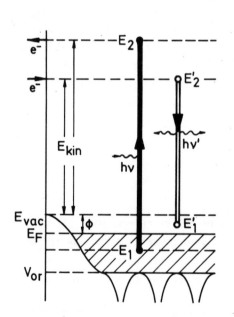

Fig. 12:

Schematic representation of the (valence band) photoemission process (involving the filled vertical line) and its inverse (involving the empty vertical line) in an energy level diagram in the surface region. The real part of the effective potential is as in Fig. 2, and the hatched area indicates the occupied levels (below the Fermi energy E_F).

energy $E_1<E_F$, raising it into a previously empty state of energy $E_2>E_{vac}$. The current associated with the extension of the latter state into the vacuum region is then detected as the "photo-current". In a classical picture one might visualize the entire process as a sequence of five events: (1) penetration of the radiation through the surface, (2) propagation of the radiation inside the solid, (3) excitation of an electron from E_1 to E_2, (4) propagation of this electron towards the surface, (5) penetration through the surface into the vacuum. The sequence of separate events (3) (specified as excitation of a "bulk" electron), (4) and (5) is traditionally referred to as "three-step model" of photoemission. Viewing (3) to (5) as a single event of excitation from the bound half-space state $|1\rangle$ with E_1 to the scattering state $|2\rangle = |E_2, \underset{\sim}{k}_2''\rangle$, where k_2'' is the momentum of the photoelectron parallel to the surface, is referred to as "one-step model".

The above suggests as a first approximation for the photon current I the "golden rule" form

$$I(E_2, \underset{\sim}{k}_2'') \propto \sum_1 |\langle 2|\underset{\sim}{A} \cdot \underset{\sim}{p}|1\rangle|^2 \delta(E_2 - h\nu - E_1) , \qquad (69)$$

where $\underset{\sim}{A}$ is the classical magnetic vector potential and $\underset{\sim}{p}$ the momentum operator. The δ-function expresses conservation of energy, which implies for the kinetic energy E_{kin} of the photo-electron (cf. Fig. 12): $E_{kin} = E_1 + h\nu - E_{vac}$. The approximate form eq.(69) comprises a variety of further approximations depending on the description of the initial states $|1\rangle$ and the final states $|2\rangle$. The most accurate (nonrelativistic) approximation is obtained if both are calculated as solutions of Schrödinger's equation for the semi-infinite surface system. $|2\rangle$ is then just a time-reversed LEED state (cf. Section 4.3). For the subtlety of using the time-reversed state, i.e. ingoing scattered-wave asymptotics rather than outgoing ones (as in LEED), see e.g. Breit and Bethe (1954).

The dependence of the photo-electron current on the energy, direction and polarization of the incident electro-magnetic radiation and on the energy and direction of the emitted electrons provides most extensive information on the electronic structure both of the "bulk" solid and of the surface region. In particular, it reveals the band

structure of the infinite solid, the two-dimensional band structure of
surface states on clean surfaces, the energy levels of atoms or molecules adsorbed on surfaces, the symmetry-type of electronic wave functions and - via diffraction in the surface region - the geometrical
arrangement of adsorbed atoms and molecules.

Further and unique insight into the electronic structure of solids and into the photoemission process can be obtained by analyzing
the spin orientation of the emitted electrons. That the latter should
be (partially) spin-polarized, appears most plausible if the initial
states |1> have a preferential spin orientation, i.e. if the ground
state of the system exhibits long-range magnetic order. This comprises,
in addition to bulk ferromagnetism (like in Fe, Co, Ni or Gd), ferromagnetic ordering at surfaces of non-ferromagnetic materials (like V,
Cr, Si) (cf. Chapter 1 of this book), ferromagnetic epitaxial films a
few monolayers thick, and possibly (more generally) ground states of
Fermi liquids with spontaneously broken parity (cf. Akhiezer and
Chudnovskii 1978). A second category of spin-polarized photo-electrons
is associated with spin-orbit interaction in the initial or the final
state (without requiring a direct coupling of the electron spin to the
radiation field). The effect is most readily understood for a single
atom (for which it was originally predicted by Fano (1969)) or "crystal
atom": the wave functions involve the two-component spin angular functions χ_κ^μ (cf. eqs.(29,30)); for circularly polarized radiation, the dipole selection rule $\Delta\mu = \pm 1$ (in addition to $\Delta\ell = \pm 1$, and with the re-
striction $|\mu| \leq j$, where $j = \ell \pm \frac{1}{2}$ characterizes the total angular momentum)
then restricts excitation processes such that the final state electrons
are (partically) spin-polarized. (A simple illustrative example is given in Chapter 10.1).

4.5.2 General Theoretical Framework

A complete microscopic theoretical description of the photoemission process from solids has to take account of the lattice periodicity
and its breaking by the surface as well as of the electron-electron

interaction. This many-body problem, formidable already when disregar-
ding the electron spin, is further complicated if spin-orbit coupling
and magnetic ordering of the solid are to be considered. A general solution has been given by Schaich and Ashcroft (1971) in a quadratic
response approximation in terms of a three-current correlation function. An equivalent, but physically more transparent formulation has
been derived by Caroli et al. (1973) with the aid of a non-equilibrium
many-body theory due to Keldysh (1965). Since these approaches and related ones have been reviewed earlier (cf. Williams et al. 1980,
Ashcroft in Feuerbacher et al. 1978 and Caroli et al. 1978), it may
suffice here to recall, in the illustrative language of Feynman diagrams, some central results. As was pointed out by Caroli et al.(1978),
this language provides, without any computation, a rather large amount
of physical insight by interpreting the various classes of diagrams.

For a system consisting of non-interacting electrons in a simi-
infinite lattice-periodic static electromagnetic field, the photocurrent, i.e. the current I of electrons of energy E and surface-parallel
momentum $k_{\parallel}$ outside the system generated by an electron-photon interaction H' is respresented by the diagram shown in Fig. 13, where G_{+} and
G_{-} denote retarded and advanced single-particle Green functions, and

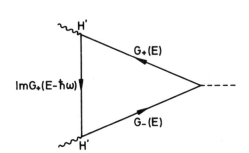

Fig. 13:
Diagram respesenting elastic
photoemission (cf. eq.(70)).
G_{+} and G_{-} are one-electron
Green functions, ∿∿ symbo-
lizes the electron-photon interaction and - - - the photocurrent.

$E-\hbar\omega$ is the energy of the "hole". If electron spin is ignored, this
diagram describes the current

$$I(E\underset{\sim}{k}_\parallel) = (-1/\pi) \ <E\underset{\sim}{k}_\parallel |G_+(E)H' \ \text{Im} \ G_+(E-\hbar\omega)H'^+G_-(E)|E\underset{\sim}{k}_\parallel> \ , \qquad (70)$$

where $<r|E\underset{\sim}{k}_\parallel> = \exp \ (i\underset{\sim}{k}_\parallel \cdot \underset{\sim}{r})\delta(z-Z)$ with Z specifying the position of a surface-parallel observation plane outside the solid (cf. Pendry 1976 and Hopkinson et al. 1980), and the Green functions are those for a Schrödinger Hamiltonian. The non-relativistic interaction H' with the radiation field is (neglecting the diamagnetic part)

$$H' = -(e/2mc) \ (\ \underset{\sim}{P} \cdot \underset{\sim}{A} + \underset{\sim}{A} \cdot \underset{\sim}{P} \ ), \text{ with } \underset{\sim}{P} = -i\hbar\nabla \ , \qquad (71)$$

where the radiation (or Coulomb) gauge has been used, i.e. the electric field vector is $\underset{\sim}{E} = -(1/c) \ \partial_t \underset{\sim}{A}$. For a plane wave field, the electric field vector amplitude $\underset{\sim}{E}_0$ is thus $(i\omega/c)\underset{\sim}{A}_0$, i.e. the photon polarization, which is primarily characterized by $\underset{\sim}{E}_0$, enters - via $\underset{\sim}{A}_0$ - directly into the formalism. The radiation gauge is therefore particularly convenient for photoemission purposes. Returning to the diagram in Fig. 13, we note that it corresponds to "elastic" photoemission, i.e. the photoelectron leaves the solid at the same energy to which it was excited by the radiation field. For the non-interacting electron system, Fig. 13 or eq.(70) give in fact the complete photocurrent. Further, eq.(70) then reduces to the familiar "golden rule" form (eq.(69)) (cf. e.g. Feibelman and Eastman 1974, Pendry 1976, Liebsch 1979).

Interaction processes like electron-electron, electron-phonon or electron-magnon scattering are taken into account by means of four classes of diagrams obtained by renormalizing the "bare" diagram (Fig. 13) (cf. Caroli et al. 1973 and 1978): (a) Separate renormalization of the three Green functions, via (complex) self-energies; this class, still corresponding to an expression analogous to eq.(1), contributes only to the elastic photocurrent. (b) Renormalization of the I-vertex, as illustrated by the two diagrams involving electron-hole pair excitation (Fig. 14), can semi-classically be interpreted as

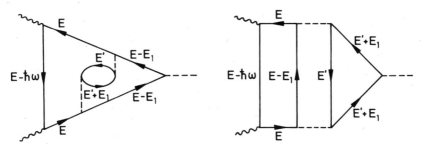

Fig. 14:
Diagrams representing inelastic contributions to the photocurrent.
(Notation as in Fig. 13, with additional internal broken lines
symbolizing the effective electron-electron interaction).

"loss after excitation", i.e. "inelastic" parts of the photocurrent.
(c) Renormalization terms of the H' vertices and (d) three-particle
vertex corrections can neither be regarded as elastic nor as inelastic,
defying this semi-classical dichotomy (cf. Caroli et al. 1978). As a
consequence of the integrations involved, contributions of classes
(b), (c) and (d) to the photocurrent can generally be expected to be
slowly varying as functions of the detected energy compared to the
elastic part described by class (a). Quantitative photoemission calculations must therefore primarily focus on the latter, i.e. an independent quasi-particle picture, and evaluate them as rigorously as possible.

To this end, several nonrelativistic formalisms (ignoring spin-
orbit coupling) have been developed, which treat exactly the multiple
scattering of the photoelectron (time-reversed $\underline{L}$ow-$\underline{E}$nergy $\underline{E}$lectron
$\underline{D}$iffraction (LEED) state) and of the "hole" by the semi-infinite
lattice (cf. above-cited reviews). Amongst these, a formalism due to
Pendry (1976) is distinguished by a scattering theoretical treatment
of the "hole", which permits the calculation of the renormalized
$G_+(E-\hbar\omega)$, i.e. an a priori incorporation of the hole lifetime. A relativistic extension of this method to include both spin-orbit coupling
and ferromagnetic order of the ground state has recently been formula-

ted (Ackermann and Feder 1985 a) and successfully applied to Pt(111) (Ackermann and Feder 1985b) and ferromagnetic Ni(001) (Ackermann 1985, Ackermann and Feder 1986). A similar formalism restricted to non-magnetic systems was about simultaneously and independently developed and applied to Pt(111) (Ginatempo et al. 1985). (The results for Pt(111) are, together with experimental data, reviewed in Chapter 11 of this book).

In the following, I first outline the relativistic formulation of elastic photoemission from semi-infinite systems, which incorporates both spin-orbit coupling and a ferromagnetic ground state. Simpler spin-dependent approaches, which have earlier been developed and applied, are therefrom obtained by introducing special assumptions and will subsequently be discussed in more detail.

4.5.3 Relativistic One-Step Theory for Ferromagnets

This theory (developed by Ackermann and Feder (1985a)) can be schematically characterized by the "elastic" diagram in Fig. 13, with an appropriate spin-dependent interpretation of its ingredients. Its basis is thus a spin-dependent generalization of the expression given in eq.(70). Firstly, the state $|E\underline{k}_\parallel>$ at the detection plane is replaced by two plane-wave states $|E\underline{k}_\parallel s> = |E\underline{k}_\parallel>|s>$ with spin projections $s = \pm\frac{1}{2}$ (with respect to some preferential direction $\hat{\underline{n}}$), i.e. (cf. e.g. Rose 1961)

$$|s> = \begin{pmatrix} \chi^s \\ c(\underline{\sigma}\underline{k})/(E+mc^2)\chi^s \end{pmatrix}, \text{ with } \underline{k} = (\underline{k}_\parallel, k_z), \tag{72}$$

where $\underline{\sigma}$ comprises Pauli spin matrices (see e.g. eq.(2)) and χ^s are the two Pauli basis spinors. Since the photocurrent consists in general of an incoherent superposition of electrons in different spin states, it has to be described by a spin density matrix $\rho(E,\underline{k}_\parallel)$, from which the spin-averaged intensity $\bar{I}$ and the spin polarization vector $\underline{P}$ are obtained in the usual way (as tr$[\rho]$ and tr$[\underline{\sigma}\rho]$/tr$[\rho]$). If spin analysis is performed with respect to some chosen direction $\hat{\underline{n}}$, the spin-up and spin-down currents j_s are consequently

$$j_s = (1 + s \, \underset{\sim}{P} \cdot \hat{n}) \bar{I}/2 \, , \text{ with } s = +/- \tag{73}$$

In terms of the basis states $|s>$, ρ is given as

$$\rho(E, \underset{\sim}{k}_\parallel) = \underset{ss'}{\Sigma} \, |s> \, I_{ss'}(E, \underset{\sim}{k}_\parallel) <s'| \, ; \quad s,s' = +/- \, ; \tag{74}$$

with matrix elements $I_{ss'}$, defined in formal analogy to eq.(70):

$$I_{ss'}(E, \underset{\sim}{k}_\parallel) = -\frac{1}{\pi} <E\underset{\sim}{k}_\parallel s| \, G_+(E)H' \text{Im } G_+(E-\hbar\omega) \, H'^+ G_-(E) |E\underset{\sim}{k}_\parallel s'> \, . \tag{75}$$

The relativistic retarded and advanced single-particle Green functions $G_+(E)$ and $G_-(E)$ (for the photoelectron) and the retarded $G_+(E-\hbar\omega)$ (for the hole below E_F) are (4x4) matrices. Assuming, in view of computational tractability, the local effective electron-ferromagnet Hamiltonian H introduced in Section 4.2.1 (eq.(3)), they are formally defined as

$$G_\pm(E') = \lim_{\varepsilon \to 0^-} (E' - H(E) \pm i\varepsilon)^{-1} \, , \tag{76}$$

where E' stands for E or E-$\hbar\omega$.

The interaction H' (in eq.(75)) with the radiation field is treated classically, as in non-relativistic theories (cf. eq.(71)). For a monochromatic plane-wave field (with frequency ω and wave vector $\underset{\sim}{q}$), its relativistic form is (cf. standard textbooks on quantum mechanics)

$$H' = -e\underset{\sim}{\alpha} \cdot \underset{\sim}{A}_0 \exp (i(\underset{\sim}{q}\underset{\sim}{r}-\omega t)) \, , \tag{77}$$

where $\underset{\sim}{\alpha}$ is the three-vector consisting of the 4x4) matrices α_k defined in eq.(4), and the radiation gauge (cf. above) has been used. Inside a solid and in its surface region, the vector potential amplitude $\underset{\sim}{A}_0$ is in general modified by the response of the (inhomogeneous) electron gas, which leads to lattice-periodic local field corrections and to near-surface induced fields of short wavelength. Since these fields and their possible effects on photoemission ("Surface photo-electric effect", break-down of usual momentum and dipole selection rules) have been extensively reviewed (Feibelman 1982), it may suffice here to note that fortunately, for a wide range of photoemission situations (choice of material, geometry, photon energy and polarization) a classical macroscopic description of $\underset{\sim}{A}_0$ as determined by the macroscopic complex dielectric constant $\varepsilon(\omega)$ and Fresnel's formulae (cf. e.g. Born and Wolf 1980) is adequate (cf. Goldmann et al. 1983), i.e. $\underset{\sim}{A}_0$ can be taken as a constant vector normal to the internal direction of propagation $\hat{g}$. Its two Cartesian components A_{o1} and A_{o2} characterize the (pure) state of polarization of the plane wave field (cf. e.g. Born and Wolf 1980, Jackson 1962, 1975): real values correspond to linear polarization; a phase factor $\exp(i\theta)$ indicates elliptical polarization with the special case of circular polarization described by $A_{o2} = \pm i A_{o1}$, where $+(-)$ corresponds to positive (negative) helicity or, in the classical optics terminology, "left-handed" ("right-handed"). Since the polarization of the radiation field plays a very decisive role in spin-polarized photoemission - as will be discussed later -, it is, however, important to note that even classical transmission through a surface generally alters this polarization. In the case of metals - described by a complex index of refraction -, linearly polarized incident radiation generally leads to elliptically polarized radiation inside the solid, unless θ_p, the photon angle of incidence, is zero or - for $\theta_p > 0$ - $\underset{\sim}{A}_{inc}$ is perpendicular (s-polarized) or parallel (p-polarized) to the plane of incidence; circular polarization is - at arbitrary frequency ω - conserved only for $\theta_p = 0$. Normal incidence has the further advantage, that the above-mentioned near-surface induced field will be minimal (cf. Feibelman 1982).

So far, the radiation field has been assumed to be in some completely polarized state characterized by $\underset{\sim}{A}_0$. The extension to unpolarized or partially polarized radiation is, however, very straightforward, since such radiation can be represented as an incoherent superposition of two completely polarized "basis" fields. The resulting photocurrent spin density matrix (eqs.(74,75)) is obtained by summing, with the appropriate weights, the density matrices produced by two "basis" fields (e.g. linearly s- and p-polarized).

The actual evaluation of the photocurrent spin density matrix elements (eq.(75)) is facilitated if the effective one-electron potential $V(E,\underline{r})$ and the effective magnetic field $\underline{B}(E,\underline{r})$, which determine H (eq.(3)) and thence $G_\pm$(eq.(76)), are shape-restricted to the muffin-tin form (cf. Fig. 3 in Section 4.2.1), and, moreover, periodic parallel to the surface. The relativistic multiple scattering (layer-KKR) approach, which has been described in Section 4.3, can then be employed, and the calculation can proceed in analogy with the non-relativistic formalism mentioned in conjunction with eq.(70) (Pendry 1976 and Hopkinson et al. 1980). Since a presentation of the formalism, which culminates in the evaluation of complicated matrix element expressions, seems too technical in the present context, the interested reader is referred to the original articles (Ackermann and Feder 1985, Ackermann 1985). Instead, we now proceed to deriving various approximations and specializations, which provide more physical insight.

4.5.4 "Golden Rule" Form and Three-Step Model

While the general quasi-particle expression (eqs.(74,75)) for the photocurrent density matrix is advisable as a basis for obtaining realistic numerical results, it is still instructive to consider approximate forms, which historically preceded the general expression. In particular the time-honoured "three-step model" (cf. Williams et al. (1980) and references therein), in which the photoemission process is decomposed into (1) excitation from an occupied to an unoccupied bulk (Bloch) state, (2) transport to the surface and (3) transmission through the surface, has been found useful for spin-polarized photoemission: in addition to ease of computation, this model provides deeper insight into the various physical mechanisms, which can produce spin polarization of the photocurrent.

The general expression eq.(75) can be cast into a "one-step model" "golden rule" form if one ignores the damping of the hole, i.e. the effect of an imaginary self-energy part on $G_+(E-\hbar\omega)$. For two-dimensionally periodic systems, initial states $|is>$ of energy E_i^s, where i stands for surface-parallel momenta $\underset{\sim}{k}_\parallel^i$ and s labels two linearly independent states (which in the absence of spin-orbit coupling are eigenstates of $\underset{\sim}{\sigma}\cdot\underset{\sim}{n}$ with opposite spin alignment), can then be constructed as linear combinations of Bloch waves matched at the surface, and one has the spectral representation

$$\text{Im } G_+(E-\hbar\omega) = -\pi\underset{is}{\Sigma} |is><is| \; \delta(E-\hbar\omega-E_i^s) \; . \qquad (78)$$

Next we identify in eq.(75) the time-reversed LEED states

$$|G,E\underset{\sim}{k}_\parallel s> := G_-(E)|E\underset{\sim}{k}_\parallel s> \quad , \qquad (79)$$

which are generated from the (outgoing) states at the detection plane $|E\underset{\sim}{k}_\parallel s>$ by the advanced Green function $G_-(E)$. In the presence of spin-orbit coupling, the $|G,E\underset{\sim}{k}_\parallel s>$ are like the $|E_i\underset{\sim}{k}_\parallel s>$, not eigenstates of $\underset{\sim}{\sigma}\cdot\hat{\underset{\sim}{n}}$, and s is merely a label to distinguish two independent states of opposite spin expectation values. Substitution of eqs.(78,79) into eq. (75) yields the spin density matrix elements

$$I_{ss'}(E,\underset{\sim}{k}_\parallel) = \underset{is''}{\Sigma} <G,E\underset{\sim}{k}_\parallel s|H'|is''><is''|H'^+|G,E\underset{\sim}{k}_\parallel s'> \times \delta(E-\hbar\omega-E_i^s) \; . \qquad (80)$$

In general, there are thus four different golden-rule matrix elements, between initial and final states with s, s', s" = +/-. The initial states comprise, in addition to (semi-)infinitely extended bulk (Bloch) states, surface states of arbitrary degree of localization in the surface region. The projections of the final states onto the crystal-half-space are either linear combinations of Bloch waves, propagating (with damping by inelastic processes) toward the surface, or of "gap states" decaying into the crystal. The resulting combination possibilities lead to the well-known distinction between "bulk-", "surface-state-" and "gap-photoemission" (cf. Williams et al. (1980) and references therein). This classification is also useful for discussing spin polarization effects.

In the case of bulk photoemission, i.e. Bloch wave initial and final states, eq.(80) can be cast into a "three-step-model" form by proceeding in spin-dependent analogy to the method well known in the spinless case (cf. e.g. Feibelman and Eastman 1974). The essential idea is the expansion of the initial and the final states inside the crystal in terms of Bloch waves $|E\underset{\sim}{k}s>$ and $|E_i^s\underset{\sim}{k}_is>$, respectively, where $\underset{\sim}{k} = (\underset{\sim}{k}_\parallel^B, k_z)$, with $\underset{\sim}{k}_\parallel^B = \underset{\sim}{k}_\parallel + \underset{\sim}{g}_\parallel$; the spin index s refers to 'spin up/down' in the absence of spin-orbit coupling and in general labels two states, which have the same $\underset{\sim}{k}_\parallel$ and E or E_i, but opposite spin polarization P expectation values. In the limit of very weak final state inelastic scattering, for a given E and $\underset{\sim}{k}_\parallel$, only one or a few Bloch wave pairs $|E\underset{\sim}{k}s>$, with $s = \pm\frac{1}{2}$, are important for the composition of the coherent final state. Matching at the surface the final state Bloch waves to the plane waves in vacuum then yields (2x2) matrices $T(E,\underset{\sim}{k})$ such that

$$|E\underset{\sim}{k}_\parallel s> = T(E,\underset{\sim}{k})|E\underset{\sim}{k}s> \quad . \tag{81}$$

If the Bloch waves are four-component spinors, $T(E,\underset{\sim}{k})$ is implicitly assumed to contain a projection onto the two large components, since scattering into plane wave electron states is completely specified by these components. The coherent final state reached from an initial state with Bloch wave vector $\underset{\sim}{k}_i$, spin label s_i and energy E_i^s is

$$|E\underset{\sim}{k}; \underset{\sim}{k}_is_i> = \underset{s}{\Sigma} <E\underset{\sim}{k}s|H'|\underset{\sim}{k}_is_i>|E\underset{\sim}{k}s> \quad , \tag{82}$$

with the corresponding statistical operator

$$\rho(E\underset{\sim}{k}; \underset{\sim}{k}_is_i) = |E\underset{\sim}{k}; \underset{\sim}{k}_is_i><E\underset{\sim}{k}; \underset{\sim}{k}_is_i| \quad . \tag{83}$$

From eq.(80), one thus eventually obtains the spin-dependent three-step model expression

$$\rho(E,\underset{\sim}{k}_{\parallel}) = \sum_{\underset{\sim}{ikg}} T(E,\underset{\sim}{k})D(E,\underset{\sim}{k})\rho(E\underset{\sim}{k};\underset{\sim}{k}_i s_i)T^+(E,\underset{\sim}{k}) \ \delta(E-\hbar\omega-E_i)\delta(\underset{\sim}{k}_{\parallel}+g_{\parallel}-k_{\parallel}^B), \quad (84)$$

where $D(E,\underset{\sim}{k})$ accounts for the finite escape depth, i.e. the probability that the photo-electron reaches the surface without inelastic scattering. D is a scalar for non-magnetic solids and, in general, a (2x2) matrix for ferromagnets. (cf. Section 4.5.7).

Eq.(84) shows that spin polarization of the photocurrent may arise from four distinct sources: initial bulk density of states, direct bulk transition matrix elements (cf. eqs.(82,83)), transport to the surface and transmission through the surface.

Before proceeding to a further discussion and evaluation of the above one-step (eq.(80)) and three-step (eq.(84)) model expressions, it should be noted that the hole damping (described by an imaginary part V_{i1}^S in the initial-state effective potential), which was assumed zero in deriving these expressions, can approximately be introduced by replacing the δ-function by the hole spectral function (cf. e.g. Liebsch 1979), or, more approximately, by convoluting the above zero-damping results with a Lorentzian of half-width V_{i1}^S (assuming V_{i1}^S as spatially homogeneous). This a posteriori approximation, in which multiple scattering in the initial state is calculated without damping, is in general good as long as V_{i1} is fairly small (i.e. especially near E_F). A comparison of results, for Ni(110), may be found in Chapter 14.2 of this book.

4.5.5 Nonrelativistic Limit and Electric Dipole Approximation

As equations (80) and (82) show, the central ingredient to the above relativistic one-step and three-step models consists of matrix elements of the type $<E|H'|E_i>$, where $|E>$ and $|E_i>$ are (time-independent) for-spinor eigenstates to the Dirac Hamiltonian H (cf. eq.(3)) (with V and $\underset{\sim}{B}$ representing the semi-infinite or the infinite solid) and H' has been re-defined as the spatial part of eq.(77). The possible origins of a spin dependence of such matrix elements become more transparent in the following nonrelativistic approximation. Denoting by $|\phi\rangle$ and $|\phi_i\rangle$ the upper two components of $|E\rangle$ and $|E_i\rangle$, respectively, the lower components $|\chi\rangle$ and $|\chi_i\rangle$ are given as (cf. e.g. Messiah 1965, Rose 1961)

$$|\chi_{(i)}\rangle = (E_{(i)} + mc^2 - eV(\underset{\sim}{r}))^{-1} c\underset{\sim}{\sigma}\underset{\sim}{p}|\phi_{(i)}\rangle \quad . \tag{85}$$

Approximating the first factor nonrelativistically to order $1/c^2$, substituting into the matrix element the upper and lower two-component spinors, and expressing the Dirac matrix $\underset{\sim}{\alpha}$ by Pauli matrices according to eq.(4), one obtains

$$\langle E|H'|E_i\rangle = - \frac{e}{2mc} \langle\phi|[\underset{\sim}{A}\cdot\underset{\sim}{p}+\underset{\sim}{p}\cdot\underset{\sim}{A}+h\underset{\sim}{\sigma}\cdot\underset{\sim}{B}_r]|\phi_i\rangle \quad , \tag{86a}$$

$$\text{with } \underset{\sim}{\tilde{p}} = \underset{\sim}{p} - \frac{1}{4m^2c^2} \underset{\sim}{p}^2\underset{\sim}{p} + \frac{\hbar}{4mc^2}(\sigma\times\nabla V(r)) \quad , \tag{86b}$$

where $\underset{\sim}{A} = \underset{\sim}{A}_o \exp(i q\underset{\sim}{r})$ is the magnetic vector potential of the radiation field, and $\underset{\sim}{B}_r = \nabla\times\underset{\sim}{A}$. Eq.(86) corresponds to an approximation to the order of $1/c^2$ if one approximates the upper two-component solutions $|\phi\rangle$ and $|\phi_i\rangle$ of the Dirac equation by eigenstates of the two-component Hamiltonian eq.(5). If local field corrections are ignored, the first two terms in eq.(86a) reduce to $-(e/mc)\underset{\sim}{A}\cdot\underset{\sim}{\tilde{p}}$. The third term in eq.(86a) represents a coupling of the spin magnetic moment $(e\hbar/2mc)\underset{\sim}{\sigma}$ to the radiation field. It is conceptually closely related to and of the same order of magnitude as the orbital magnetic dipole transition matrix element, which is obtained- together with the electric quadrupole element - in the $\underset{\sim}{A}\cdot\underset{\sim}{p}$ part as a contribution associated with the $(i q\cdot r)$ term of the exponential expansion. (Details may be found in a monograph by Bethe and Salpeter (1957)).

For $q\cdot r \ll 1$, which applies to visible and to ultraviolet radiation, the $\underset{\sim}{A}\cdot\underset{\sim}{p}$ part of the matrix element reduces to the electric dipole form $-(e/mc)\underset{\sim}{A}_o\cdot\langle\phi|\underset{\sim}{p}|\phi_i\rangle$. Of course, the electric dipole approximation can also be made in the fully relativistic expression:

$$<E|H'|E_i> \approx - e\underset{\sim}{A}_0 \cdot <E|\underset{\sim}{\alpha}|E_i> \quad . \tag{87a}$$

If $|E>$ and $|E_i>$ are eigenstates of the same effective Hamiltonian H_0, use of the Heisenberg equation of motion, $c\underset{\sim}{\alpha} = -i[\underset{\sim}{r},H_0]$, for the relativistic velocity operator $c\underset{\sim}{\alpha}$, transforms the above matrix element into

$$-i \ (E-E_i) \ \frac{e}{c} \ \underset{\sim}{A}_0 \ \cdot \ <E|\underset{\sim}{r}|E_i> \quad , \tag{87b}$$

the so-called "length form". In this form, nonrelativistic approximations are most straightforward: replacement of the four-spinors $<E|$ and $|E_i>$ by two-component spinor or scalar functions $<\tilde{E}|$ and $|\tilde{E}_i>$. In the Schrödinger approximation (for non-magnetic materials), a popular (cf. e.g. Williams et al. 1980) alternative form of the electric dipole matrix element is obtained with the aid of the commutator $[p,H_s] = \frac{\hbar}{i} V(\underset{\sim}{r})$, where H_s is the Schrödinger Hamiltonian:

$$<E|H'|E_i> \approx - \frac{e}{mc} \ (E-E_i)^{-1} \ \underset{\sim}{A}_0 \ \cdot \ <E|\nabla V(\underset{\sim}{r})|E_i> \quad . \tag{88}$$

This "acceleration form" is particularly convenient if the potential has the "muffin tin" shape, i.e. is spherically symmetric in spheres centred at the atomic nuclei and constant in the interstitial region: integrations have then to be carried out only over the atomic spheres. In the nonrelativistic approximation to order $1/c^2$ (cf. eq.(5)), the commutator $[p,H]$ is seen to be augmented by terms involving spatial derivatives of $\underset{\sim}{B}$, ΔV and ∇V. The analogue of eq.(88) becomes consequently more cumbersome; in particular, it explicitly involves spin-orbit coupling.

The theoretical framework, which has been presented so far, incorporates simultaneously spin-orbit coupling and ferromagnetic exchange coupling. Treating these two spin-dependent interaction mechanisms separately does not only facilitate numerical work, but also provides some direct insight into their role in producing spin-polarized photo-electrons.

4.5.6 Non-Magnetic Solids with Spin-Orbit Coupling

We now specialize to the theory of spin-polarized photoemission from non-magnetic solids, i.e. basically $\underline{B} = 0$ in the Dirac Hamiltonian (eq.3). In particular, general features and results are presented, which can be obtained by means of symmetry arguments (cf. also Section 4.3.2) without numerical calculations.

4.5.6.1 One-Step-Model Features and Selection Rules

As indicated above, a nonrelativistic approximation makes the physical origin of photoelectron spin polarization more transparent. Equations (86a,b) and (5) (with $\underline{B} = 0$) show that the electron spin, represented by the Pauli spin operator $\hbar\sigma/2$, features in two ways: firstly by a direct coupling of the electron spin magnetic moment to the radiation field ($\underline{\sigma} \cdot \underline{B}_r$ term in eq.(86a)), and secondly by spin-orbit coupling, which occurs both in the transition operator (cf. last term of eq.(86b)) and in the Hamiltonian (eq.(5)) determining the initial and the final states. Before investigating the effects of these two couplings, it is instructive to see what happens in their absence. The remaining Hamiltonian H_s is then spin-independent and its eigen-functions for a general $\underline{k}_{\parallel}$ are two-fold degenerate and can be written as direct products $|E\underline{k}_{\parallel}\rangle|s\rangle$ of scalar spatial functions $|E\underline{k}_{\parallel}\rangle$ and two-component spinors $|s\rangle$. Since the remaining interaction operator does not affect the $|s\rangle$, and since the Green functions G associated with H_s are scalar, the general "one step model" expression (eq.(74)) becomes

$$\rho(E,\underline{k}_{\parallel}) = \begin{pmatrix} I & 0 \\ 0 & I \end{pmatrix} \text{ , with } I := I_{++} = I_{--} \tag{89}$$

Using eq.(1), one obtains the spin polarization vector $\underline{P}(E,\underline{k}_{\parallel}) = 0$. This result extends to special $\underline{k}_{\parallel}$ of higher degeneracy, since the distribution of the general $k_{\parallel}$ is dense and P is a continuous function of $k_{\parallel}$. Of course, it holds a fortiori for the golden rule form eq. (80), and also for the bulk excitation part of the three-step model (cf. eq.(83)). Irrespective of the polarization of the radiation, there is thus no spin polarization in photoemission (from nonmagnetic solids) in the absence of the above two spin coupling mechanisms.

Let us next investigate effects of the $\underline{\sigma} \cdot \underline{B}_r$ coupling (eq.(86a)) in the absence of spin-orbit coupling. The eigenstates still have the form $|E k_{\parallel}\rangle |s\rangle$. Without loss of generality, the spin quantization axis (z-axis) can be chosen parallel to the direction $\hat{q}$ of propagation of the radiation. Then

$$\underline{\sigma} \cdot \underline{B}_r = \underline{\sigma} \cdot [\nabla \times \underline{A}] = \underline{\sigma} \cdot [iq \times \underline{A}] = iq \, (-\sigma_x A_y + \sigma_y A_x) \, , \tag{90}$$

where $\underline{A} = \underline{A}_0 \exp(iqr)$. By virtue of the Pauli matrices σ_x and σ_y, $\underline{\sigma} \cdot \underline{B}_r$ flips the spin of the eigenstates $|s\rangle$ of σ_z. For linearly polarized light (e.g. $A_y = 0$, A_x real; or vice versa), $|+\rangle$ and $|-\rangle$ are flipped with equal probability. If the spin-independent $(\underline{A}p + p\underline{A})$ part of H' does not contribute, i.e. has zero matrix elements, one sees from eq. (74) that the photo-current density matrix ρ assumes the diagonal form of eq.(89), which implies $\underline{P} = 0$. If the $(\underline{A}p + p\underline{A})$ part contributes, ρ formally acquires an interference term proportional to σ_y or to σ_x, which, if non-zero, would imply $P_y \neq 0$ or $P_x \neq 0$. For circularly polarized light $(A_y = +iA_x$ or $-iA_x)$ eq.(90) yields

$$\underline{\sigma} \cdot \underline{B}_r = 2qA_x \begin{pmatrix} 0 & 1 \\ 0 & 0 \end{pmatrix} \text{ or } -2qA_x \begin{pmatrix} 0 & 0 \\ 1 & 0 \end{pmatrix} \, , \tag{91}$$

i.e. radiation of positive (negative) helicity flips only $|-\rangle$ $(|+\rangle)$. The $\underline{\sigma} \cdot \underline{B}_r$ term by itself then leads to

$$\rho = \begin{pmatrix} I & 0 \\ 0 & 0 \end{pmatrix} \text{ or } \begin{pmatrix} 0 & 0 \\ 0 & I \end{pmatrix} \, , \tag{92}$$

which means $P_z = +1$ or -1, i.e. 100 % spin polarization. However, the

(A_p + pA) part of H' adds equal contributions to the diagonal elements of ρ and thereby reduces the spin polarization. The $\sigma \cdot B_r$ interaction is thus seen to entail, in principle, spin polarization of the photocurrent, the magnitude of which depends on the ratio of the strengths of the spin-flipping $\sigma \cdot B$ and of the spin-conserving $A \cdot p$ matrix elements. This ratio R has been estimated by Feuchtwang et al. (1978) for two model cases: (a) single atom with plane-wave $\exp(ik_f \cdot r)$ final state, where typically $R = q/k_f \ll 1$ (e.g. for photon energy 100 eV and photoelectron kinetic energy 90 eV: $R \approx 0.02$); note, however, that approximation of the final state by a plane wave (i.e. no scattering by the atomic potential) is inadequate for kinetic energy below a few hundred eV, and that this estimate of R ist therefore not valid for photoemission by visible or ultraviolet radiation; (b) infinite solid with the final Bloch state expanded in powers of q by "$k \cdot p$ perturbation theory" to second order; this leads to $R \approx \hbar \omega / (mc^2)$ (e.g. for $\hbar \omega = 100$ eV, R = 0.0002). As pointed out by Feuchtwang et al. (1978), this generally very small ratio R may substantially increase in favour of the spin-flip transitions for special geometrical configurations, for which the $A \cdot p$ transitions are very weak. One should bear in mind, however, that the associated photocurrent is then very weak. The generally small value of R is also plausible from a multipole expansion of the matrix elements. As mentioned above, the leading term of the $\sigma \cdot B$ transition is of the magnetic dipole type (M1 in the terminology of atomic spectrosopy), while the leading term of the $A \cdot p$ transition is the electric dipole (E1) matrix element. In the absence of spin-orbit coupling, it follows further from the orthogonality of the initial and final states (for photo-excitation of atoms cf. e.g. Condon and Shortley 1977) that M1 transitions are "forbidden" and $\sigma \cdot B_r$ can consequently at best contribute via higher multipoles. Although quantitative studies still seem desirable, one can thus expect spin polarization effects due to $\sigma \cdot B_r$ to be generally very small (less than about 0.1 %) in ultraviolet photoemission.

In contrast to the $\underset{\sim}{g} \cdot \underset{\sim}{B}_r$ coupling, the spin-orbit interaction influences already the dominant electric dipole transitions ($q = 0$, $\underset{\sim}{A} = \underset{\sim}{A}_0$), as is clear from equations (86,5). Some general information on spin polarization effects can therefore be obtained by considering the electric dipole selection rules, i.e. the dipole-allowed transitions, in the presence of spin-orbit coupling. If V is a central field, describing an atom or a single muffin tin sphere in a crystal, eigenstates of the Dirac or Pauli Hamiltonian, $|E\ell j\mu\rangle$, are characterized by the orbital and total momentum quantum numbers ℓ and j, with $j = \ell \pm \frac{1}{2}$, and the magnetic quantum number $\mu = -j, \ldots, j$. (Note that these are eigenstates of L^2 only in the Pauli, but not in the Dirac case). For matrix elements $\langle E\ell'j'\mu'|O|E_i\ell j\mu\rangle$ to be non-vanishing, necessary conditions are (cf. e.g. Bethe and Salpeter 1957, Condon and Shortley 1977)

$$\ell' = \ell \pm 1 \quad \text{and} \quad j' = j, j \pm 1 \quad \text{for } 0 = x,y,z \tag{93a}$$

$$\mu' = \mu \quad \text{for } 0 = z; \; \mu' = \mu \pm 1 \quad \text{for } 0 = x \pm iy \; ; \tag{93b}$$

$(x + iy)$ and $(x-iy)$, arising from $\underset{\sim}{A}_0 \cdot \underset{\sim}{r}$, correspond to circularly polarized electromagnetic radiation of positive and of negative helicity, respectively, propagating in the positive z-direction.

If V describes a semi-infinite single crystal, dipole photoemission can only occur if dipole operator matrix elements between initial and final states of the same surface-parallel momentum $\underset{\sim}{k}_{\|}$ are non-vanishing. Specialization from the general form eq.(75) to the one-step model golden rule form eq.(80) is not necessary to obtain the following results; it makes, however, the presentation more transparent. Dipole selection rules are obtained (cf. e.g. Hermanson 1977) by noting that (a) the one-electron final state is invariant under crystal symmetry operations that leave $\underset{\sim}{k}_{\|}$ unchanged, and (b) the symmetry of dipole-allowed initial states is then the same as that of the dipole operator. Since $\underset{\sim}{k}_{\|}$ (e.g. in a mirror plane) and the dipole

operator (e.g. linearly polarized light with $\underset{\sim}{A}$ parallel or perpendicular to the surface) can be controlled by the experimentor, selection
rules have proven to be most valuable for identifying initial states
and their symmetries (cf. e.g. Goldmann et al. 1983, Plummer and Eberhardt 1982 and references therein). If spin-orbit coupling is ignored,
the symmetry operations can be confined to coordinate space, i.e. to
"single-groups"; spin-orbit coupling connects symmetry operations in
coordinate and in spin space, i.e. requires the use of "double groups"
(cf. e.g. Cornwell 1969). In the special case of emission normal to
the surface, the final state (both inside and outside the crystal) is
totally symmetric under point group operations about the surface normal. The initial state symmetries, which are consequently allowed for
$\underset{\sim}{A} \parallel$ x,y,z - with z chosen normal to the surface -, are illustrated in
Table 1 for three low-index surfaces of cubic crystals (with a primi-

crystal face	point group	final state symmetry without s.o.	with s.o.	$\overset{A}{\underset{\sim}{}}$ parallel to	initial state symmetry without s.o.	with s.o.
(001)	C_{4v}	Δ_1	Δ_6	x: (100)	Δ_5	$\Delta_6\ \Delta_7$
				y: (010)	Δ_5	$\Delta_6\ \Delta_7$
				z: (001)	Δ_1	Δ_6
(111)	C_{3v}	Λ_1	Λ_6	x: ($\bar{1}$10)	Λ_3	$\Lambda_4\ \Lambda_5\ \Lambda_6$
				y: ($\bar{1}\bar{1}$2)	Λ_3	$\Lambda_4\ \Lambda_5\ \Lambda_6$
				z: (111)	Λ_1	Λ_6
(110)	C_{2v}	Σ_1	Σ_5	x: (001)	Σ_3	Σ_5
				y: (1$\bar{1}$0)	Σ_4	Σ_5
				z: (110)	Σ_1	Σ_5

Table 1:

Dipole-allowed initial state symmetries in photoemission normal to
low-index faces of primitive cubic crystals. In the absence ("without
s.o.") and in the presence ("with s.o.") of spin-orbit coupling, the
irreducible representations of the point groups are given in single
and in double group notation (cf. e.g. Cornwell 1969), respectively.
The results "without s.o." are taken from Hermanson (1977), those
"with s.o." from Wöhlecke and Borstel (1981a,b).

tive lattice). Due to the lowering of symmetry by the spin-orbit interaction, the selection rules with spin-orbit coupling are weaker,
i.e. less restrictive, than those without. This difference is seen to
be most drastic for (110) surfaces: In the absence of spin-orbit coupling, the initial state excited by A_x, A_y and A_z is characterized by
the single-group representations Σ_3, Σ_4 and Σ_1, respectively, whereas
with spin-orbit coupling table 1 gives only the trivial statement that
the initial state belong to Σ_5, the only representation of the double
group. As pointed out by Borstel et al. (1981), there is, however, an
interesting consequence of this "relaxation" of selection rules: states
(e.g. those of Σ_2 symmetry for (110) surfaces), from which no dipole
transitions may occur without spin-orbit coupling, can contribute to
the photocurrent as a consequence of being mixed with "allowed" spatial
wave functions by the spin-orbit interaction. Since spin-orbit coupling,
while large for large-Z materials, is present also at very small atomic
numbers Z, the interesting question arises as to the conditions, under
which it may be neglected and experimental photoemission results may
consequently be interpreted with the aid of the more restrictive "ordinary" selection rules. This issue has been vividly discussed by Borstel
et al. (1981), Goldmann et al. (1982) and Przybylski et al. (1983).
Since off-normal emission, which always contributes to experimental
data due to finite angular resolution, also involves reduced symmetry
and can produce similar effects, quantitative one-step calculations are
necessary to identify the actual (dominant) origin of observed features
(cf. Ackermann 1985 and Ackermann and Feder 1985). (A semantic remark:
selection rules in the absence (presence) of spin-orbit coupling are
usually referred to as "non-relativistic" ("relativistic")).

The spin polarization of the photo-electrons can be studied with
the aid of relativistic dipole selection rules, as will be discussed
below. Some general information may, however, already be obtained -
without recourse to matrix elements - from simpe spatial symmetry arguments as follows. Let us again consider the case of radiation incident and of electrons emitted perpendicular to a centrosymmetric crystal surface, the normal direction to which is parallel to an n-fold

(with n≥2) rotation axis of the crystal. (The surfaces in Table 1 are important special cases in this category). For arbitrary polarization of the radiation, the photocurrent spin density matrix ρ (eq.(74)) must be invariant under the representation R_n of the basic rotation, i.e. $\rho = R_n \rho R_n^+$. Choosing the z-axis as the (inward-directed) surface normal, this implies that ρ is diagonal and therefore (cf. eq.(1)) $P_x = P_y = 0$, i.e. there is no spin polarization component parallel to the surface. Wheter P_z vanishes or not, can be seen to depend on the polarization of the radiation. To this end, it is useful to represent the real crystal not by a semi-infinite model but rather by a "slab" of finite extension d in the z-direction such that d/2 is much larger than the photo-electron mean free path and that an origin placed in the central mono-atomic layer is an inversion centre of the slab. Light incident from the left (right) then induces photoemission only at the left (right) surface . For linearly polarized light (with real $\underline{A}$ and $-\underline{A}$) incident from both sides, the entire set-up is invariant under spatial inversion. This invariance must extend to the photo-electron polarization vector $\underline{P}_z$. Since inversion leaves $\underline{P}_z$ unchanged, the electron helicity at the left-hand surface is opposite to that at the right-hand surface, which constitutes a parity violation unless $\underline{P}_z = 0$. As unpolarized light can be regarded as an incoherent superposition of linearly polarized light in two orthogonal states, it follows that there is also no spin polarization obtainable by unpolarized light. For circularly (or elliptically) polarized light of the same helicity incident on both sides, the set-up is no longer invariant under inversion (since inversion changes the photon helicity) and the photoelectrons may acquire a non-vanishing $P_z^{left} = -P_z^{right}$, i.e. the same helicity at both sides. For light of opposite helicity incident on the two sides, the set-up is inversion-invariant and $P_z^{left} = P_z^{right}$, i.e. the electron helicities on the two sides are opposite; since the two light helicities are opposite, there is no parity violation provided that an interchange of the light helicities interchanges the electron helicities. This implies again zero electron spin polarization for unpolarized light, which can also be viewed as an incoherent superposition of

two circularly polarized states of opposite helicity. Focusing on only
one of the surfaces (of a centrosymmetric crystal, with an at least
two-fold rotation axis normal to the surface), we thus have - for normal incidence and emission - the results that (a) linearly polarized
and unpolarized light cannot produce spin-polarized photo-electrons,
(b) circularly (or elliptically) polarized light may lead to a non-
zero photo-electron spin polarization normal to the surface and (c)
reversal of the photon helicity reverses the electron spin polarization vector. These findings are the same as in photoemission from a
single atom (cf. e.g. Kessler 1976, 1985). With regard to (a), it is
important to bear in mind that the direction of electron emission has
been chosen (anti)parallel to the direction of light propagation and
normal to the surface. As will be illustrated later with the aid of
the three-step model, for off-normal emission linearly polarized or
unpolarized light generally leads to spin polarization. The same will
happen if there is neither an at least two-fold rotation axis nor a
set of at least two mirror planes normal to the surface. The existence
of such a mirror plane implies, by the way, most directly the above
statement (c), since the mirror operation reverses the photon helicity
and changes P_z into $-P_z$.

The above results, derived for precisely normal emission, remain
valid for the experimentally more realistic situation of emission in a
cone centred around the surface normal, since deviations occurring at
individual $\underset{\sim}{k}_{\parallel} \neq 0$ cancel in the $\underset{\sim}{k}_{\parallel}$ summation. An important feature of
the above arguments is that they are based only on spatial symmetries:
in contrast to proofs involving time reversal (see below), they are
valid also in the presence of electron or hole damping, i.e. of imaginary self-energy parts in the Hamiltonians.

So far, conditions have been identified, under which spin polarization of the photocurrent may occur. More detailed information on
its actual occurrence and on its sign may be obtained by examining
electric dipole transition matrix elements (for circularly polarized
light) and the symmetry properties of the wave functions involved. Since

this subject has recently been reviewed (Wöhlecke and Borstel 1984), a brief outline together with some typical results may suffice. Let us return to the case of normal emission from cubic crystal surfaces induced by normally incident circularly polarized light. The symmetries of the final state and of possible initial states (with spin-orbit coupling), which can be excited by the operator $x \pm iy$ (or, in terms of spherical harmonics, $Y_1^{\pm 1}$), have already been given in Table 1. The essential idea is now to express final and initial states as linear combinations of the basis functions of the respective extra irreducible representations of the double group. These basis functions can be classified with the aid of the underlying single group labels and expressed as superpositions of products of spherical harmonics and Pauli spinors, as is illustrated in Table 2 for (001) and (111) surfaces. For example, for a (001) surface we thus have as the final state

$$|E,\sigma,\Delta_6> = a_6^1|E,\sigma,\Delta_6^1> + a_6^5|E,\sigma,\Delta_6^5> \qquad (94a)$$

and as a possible initial state

$$|\tilde{E},\tilde{\sigma},\Delta_7> = a_7^5|\tilde{E},\tilde{\sigma},\Delta_7^5> + a_7^2|\tilde{E},\tilde{\sigma},\Delta_7^2> + a_7^{2'}|\tilde{E},\tilde{\sigma},\Delta_7^{2'}> \quad , \qquad (94b)$$

where Δ_i^k characterizes the basis functions and the a_i^k are the respective linear combination coefficients. The matrix element of the operator $Y_1^{\pm 1}$ between final and initial state then has the form

$$\sum_{f=1,5} \sum_{i=5,2,2'} (a_6^f)^* a_7^i <E,\sigma,\Delta_6^f |Y_1^{\pm 1}| \tilde{E},\tilde{\sigma},\Delta_7^i> \quad . \qquad (95)$$

Using the special basis function parts given in Table 2, the basic matrix elements in eq.(95) can be easily further evaluated: only those may be nonvanishing, for which initial and final state contain the same Pauli spinor and the difference between final and initial orbital magnetic quantum number m is +1 (for Y_1^{+1} light) or -1 (for Y_1^1 light) and $\Delta \ell = \pm 1$. For some pairs (f,i), the basic matrix elements

(111)-surface C_{3v}	
Λ_6^1	$Y_\ell^0|\sigma\rangle$
Λ_6^2	$(Y_3^3+Y_3^{-3})|\sigma\rangle$
Λ_6^3	$Y_\ell^{\sigma 1}|\bar\sigma\rangle$
Λ_4^3	$Y_\ell^{-1}|-\rangle+iY_\ell^1|+\rangle$ $-Y_\ell^{-2}|-\rangle+iY_\ell^2|+\rangle$
Λ_5^3	$iY_\ell^{-1}|-\rangle+Y_\ell^1|+\rangle$ $-iY_\ell^{-2}|-\rangle+Y_\ell^2|+\rangle$

(001)-surface C_{4v}	
Δ_6^1	$Y_\ell^0|\sigma\rangle$
Δ_6^5	$Y_\ell^{\sigma 1}|\bar\sigma\rangle$, $Y_3^{\sigma 3}|\bar\sigma\rangle$
Δ_7^5	$Y_\ell^{\bar\sigma 1}|\bar\sigma\rangle$ $Y_3^{\bar\sigma 3}|\bar\sigma\rangle$
Δ_7^2	$(Y_\ell^2+Y_\ell^{-2})|\sigma\rangle$
$\Delta_7^{2'}$	$(Y_\ell^2-Y_\ell^{-2})|\sigma\rangle$

Table 2:

Angular momentum parts (up to orbital angular momentum $\ell = 3$) of basis functions of irreducible double group representations (in BSW notation, with the subscript indicating the double group representation and the superscript the underlying single group representation) of the point groups C_{4v} and C_{3v} relevant for normal emission from cubic (001) and (111) surfaces, respectively. $|\sigma\rangle$, with $\sigma = \pm$, and $\bar\sigma = -\sigma$, denotes the Pauli basis spinors and Y_ℓ^m the spherical harmonics (superscipts $\sigma 1$, $\sigma 3$ mean $m = \sigma 1$, $\sigma 3$, i.e. ± 1, ± 3) (ℓ assumes the values 0,1,2,3 provided that $\ell \geq |m|$) (The functions belonging to Λ_5^3 are the time reversal transforms of the functions of Λ_4^3).

are seen to vanish for all four pairs $(\sigma,\bar\sigma)$; for others they are non-vanishing only for one particular pair $\sigma,\bar\sigma$, i.e. for only one spin direction σ of the final state (with $\sigma \to -\sigma$ for $Y_1^1 \to Y_1^{-1}$). Table 3 gives, for Y_1^1, these values σ (+,-) and indicates (by dots), which elements vanish altogether. For the example described by eqs.(93-95) as well as for the remaining cases, Table 3 shows that there are always transitions to both $\sigma = +$ and to $\sigma = -$. Consequently, the two diagonal ele -

(111)	final state Λ_6^1	Λ_6^2	Λ_6^3
Λ_6^1	..	..	-
Λ_6^2	..	..	-
Λ_6^3	+	+	..
$\Lambda_4^3\Lambda_5^3$	-	-	+

(initial state)

(001)	final state Δ_6^1	Δ_6^5
Δ_6^1	..	-
Δ_6^5	+	..
Δ_7^5	-	..
Δ_7^2	..	+
$\Delta_7^{2'}$	..	+

(initial state)

Table 3:
Electron spin polarization induced by circularly polarized light in dipole transitions between basis functions of irreducible double group representations (cf. table 2) of the point groups $C_{4v}(\Delta)$ and $C_{3v}(\Lambda)$. +(-) denotes + 100 % (-100 %) spin polarization (with respect to the positive z-axis pointing inward into the crystal), for positive helicity light propagating in the positive z direction (i.e. dipole operator $x + iy$ or Y_1^1). The results for negative helicity light ($x - iy$ or Y_1^{-1}) are obtained by interchanging + and -. Dots indicate that a transition is not dipole-allowed.

ments in the statistical operator ρ (eq.(75)) are in general non-vanishing and different from each other. It thus depends on the coefficients a_6^f and a_7^i, whether the resulting spin polarization P_z is positive or negative (or even vanishes). If, however, for example, $a_7^2 = a_7^{2'} = 0$ and $a_7^5 \neq 0$, only a transition to $\sigma = -$ is possible (for Y_1^1) and one obtains $P_z = -1$, i.e. -100 % spin polarization. The spin polarization values, which result for (001) and (111) surfaces, if the final and the initial state consist each of just a single basis function of the irreducible representation of the double group, are summarized in Table 3. Since in the special cases shown in Table 3 there is only one initial and one final state, the actual value of the non-vanishing matrix element, i.e. the strength of the transition, has no influence on the spin polarization. It is interesting to note that, for a fixed final state and a fixed photon helicity, P_z can be +1 or -1 depending on the initial state. This is not in contradiction to angular momentum conservation, since in both cases the photon angular momentum is added in the same way to the total angular momentum of the electron (e.g. $\Delta\mu = +1$). In the limit of vanishing spin-orbit interaction, table 3 remains formally valid (as no assumption was made about the strength of the spin-orbit interaction), but the actual final state will be purely of the type Λ_6^1 (or Λ_6^1) and the initial states of types Λ_6^5 and Λ_7^5 (or Λ_6^3 and $\Lambda_4^3\Lambda_5^3$) (in accordance with the "nonrelativistic" selection rules (cf. Table 1)); the latter will be degenerate and incoherent superposition will add up the spin polarization to zero, as it must be in the absence of spin-orbit coupling (cf. above). As can also be inferred from Table 3, a gradually increasing admixture of different basis functions (cf. eqs.(93,94)) ("hybridization") will reduce the magnitude of the spin polarization.

For emission normal to surfaces, which have an only two-fold rotation axis, analysis of the dipole matrix elements reveals a fundamentally different situation. As can be inferred from the selection rules (Table 1), the initial state in the case of a cubic (110) surface has, in analogy to eq.(94), the general form

$$|\hat{E},\hat{\sigma},\Sigma_5> = \sum_{i=1}^{4} a_5^i |\hat{E},\hat{\sigma},\Sigma_5^i> \quad . \tag{96}$$

The matrix elements of the operator $Y_1^{\pm 1} \hat{=} x \pm iy$ between the individual initial state basis functions and the principal final state basis function $|E,\sigma,\Sigma_5^1>$ are, from the selection rules (Table 1), easily seen to vanish for i = 1,2, while for i = 3 and 4 only the x or the y component, respectively, can produce a non-zero value. Circularly polarized

light thus acts on both the Σ_5^3 and the Σ_5^4 basis functions like linearly polarized light and consequently leads, for each of them alone, to
unpolarized photo-electrons. The analogue of Table 3 for the Σ direction would thus exhibit only zero spin polarization. If, however, a_5^3
$\neq 0$ and $a_5^4 \neq 0$, x and y can act simultaneously, and $Y_1^{\pm 1}$ may produce
spin polarization, the value of which depends on the values of a_5^3 and
a_5^4. For transitions to the other final state basis functions, the
situation is analogous. While for surfaces with an at least three-fold
rotation axis, "hybridization" between the basis function of the irreducible representations of the double group reduces the spin polarization (with respect to the "pure" cases), it thus turns out to be vital
for the very existence of spin polarization for surfaces with a two-
fold rotation axis. (A more formal group theoretical derivation may be
found in Wöhlecke and Borstel (1984) and references therein).

From the above results for the two special cases of linearly and
of circularly polarized light, some qualitative conclusions can be
drawn for the more general case of elliptically polarized light. Since
the latter state is intermediate between the limits of the two former
states, it will generally lead to a photo-electron spin polarization
reduced in magnitude compared to the circular polarization case. In
particular, Table 3 remains valid if one interprets "+" and "-" as a
positive and a negative P_z, respectively, with $|P_z| < 1$.

In summary, a number of general results can be obtained in the
framework of a one-step model: for normal emission from surfaces of
centrosymmetric crystals, simple spatial symmetry arguments imply zero
spin polarization for linearly polarized light (incident such that it
remains linearly polarized inside the crystal) and the possibility of
finite spin polarization for circularly or elliptically polarized
light. Some insight into the physical origin of this spin polarization
as well as qualitative predictions can be reached by means of a symmetry analysis of the dipole transitions elements. The latter also indicates that detailed information on symmetry types and the amount of
"hybridization" in final and initial states is contained in spin polar-

ization spectra. To fully extract this information from experimental data, it will in general be necessary to perform numerical calculations on the basis of eqs.(74,75 or 80) (as was recently done for Pt(111) by Ackermann and Feder (1985) and by Ginatempo et al. (1985)). For the special case of bulk interband transitions, some more insight may be obtained by considering the three stages of the "three-step model" (cf. eq.(84)) individually.

4.5.6.2 Bulk Interband Transitions

General features, which go beyond those derived above within the one-step model, emerge as a consequence of an additional symmetry of the bulk Hamiltonian H_b (compared to the Hamiltonian of the semi-infinite solid): lattice translational invariance also in the third dimension (the z direction), which implies k_z as a good quantum number (to be conserved in transitions if one disregards electron phonon interaction processes). For eigenstates of H_b, invariance under time reversal (operator T, expressed as $-i\sigma_y K$ in the Pauli and as βK in the Dirac formalism, where K denotes complex conjugation) then implies "Kramers degeneracy": given a state $|Eks\rangle$, where the spin label - in the presence of spin-orbit coupling - refers to the sign of the spin expectation value with respect to some direction, the time-reversed state $|E,-k,-s\rangle$ has the same energy but opposite k and opposite spin polarization. If the solid has an inversion centre, i.e. H_b is invariant under the inversion operation J, which reverses k but conserves the spin, the combined operation JT transforms the Bloch states as follows.

$$JT \ |Eks\rangle = |E,k,-s\rangle \quad . \tag{97}$$

The system consisting of the ground state and (for a general low-symmetry k) of a pair of final states $|Eks\rangle$ with s = +/- remains thus physically unchanged by JT. The same holds for linearly polarized light incident in an arbitrary direction q (JTq=q). Therefore, the bulk emission statistical operator ρ, which is defined as the s_i-sum of

the operators in eq.(83), must also be invariant under JT:
$\rho = JT \rho (JT)^+ = :\rho_{JT}$. The spin-averaged photoelectron intensity is then
$I = tr[\rho] = tr[\rho_{JT}]$, and the spin polarization vector

$$\underset{\sim}{P} = tr[\underset{\sim}{\Sigma}\rho]/\tilde{I} = tr[T^+J^+\underset{\sim}{\Sigma}JT\rho]/\tilde{I} = -tr[\underset{\sim}{\Sigma}\rho]/\tilde{I} \quad , \tag{98}$$

which implies $\underset{\sim}{P} = 0$. By incoherent superposition, this result is extended to unpolarized light. For circularly polarized light, JT reverses
(formally as a consequence of the complex conjugation K contained in T)
the helicity of the light (retaining the direction q). The complete
systems is therefore not invariant under JT, i.e. $\rho_{JT} \neq \rho$. Due to the
antiunitarity of JK, the spin-averaged intensity remains the same
$(tr[\rho_{JT}] = tr[\rho] = \tilde{I})$, but $\underset{\sim}{P}$ may now be different from zero, with
$\underset{\sim}{P}_{JT} = -\underset{\sim}{P}$. The above extends, on topological grounds, to special $\underset{\sim}{k}$
points of higher symmetry.

In summary, one has for any final-state bulk $\underset{\sim}{k}$: (a) linearly polarized and unpolarized light cannot produce electron spin polarization, (b) circularly (or elliptically) polarized light may lead to
non-zero spin polarization and (c) reversal of the photon helicity
reverses the spin polarization vector.

These bulk excitation results go beyond their analogues in the
one-step model, since they are valid for any direction of photon incidence and of Bloch electron propagation. Statement (a), while trivial at first sight, becomes intriguing if one compares it with its
counterpart in atomic physics: linearly polarized or unpolarized light
may induce spin polarization in (angle-resolved) photoemission from
single atoms, as has been predicted theoretically (Cherepkov 1978,1979;
Huang 1980, Klar 1980) and verified experimentally (Heinzmann et al.
1980). Borstel and Wöhlecke (1982), who obtained the above results by
explicit evaluation of ρ (cf. eq.(83)) with the aid of the JT symmetry
properties of final and initial states characterized by the same $\underset{\sim}{k}$,
ascribe this striking difference to $\underset{\sim}{k}$ conservation (direct transition
model) in the case of the bulk solid and its absence in atomic photoionization. Consequently, they expect the possibility of spin polariza-

tion by linearly polarized light if $\underset{\sim}{k}$ conservation is relaxed e.g. by phonon scattering (cf. e.g. Shevchik 1977) ("indirect transitions"). This should apply in particular for photoemission from highly localized core levels. From the above derivation (due to Feder (unpublished)) it appears, however, that the set of initial one-electron states is only required to be physically unchanged by the JT transformation, but that it need not have the same $\underset{\sim}{k}$ as the final states. Consequently, result (a) should still hold for non-$\underset{\sim}{k}$-conserving bulk transitions. The reason for the different spin polarization behaviour in "bulk photoemission" and atomic photoemission must therefore lie entirely in the different nature of the final states. In the atomic case this is a time-reversed scattering state consisting asymptotically of incoming spherical waves and an outgoing plane wave. Since the JT operation transforms the incoming spherical waves into outgoing ones, i.e. does not only reverse the spin but changes the spatial character of the state, the above JT-based argument,which is valid for Bloch states, is no longer applicable. It should hold, however, if one approximates the atomic final state by a plane wave (i.e. scattering phase shifts $\delta_\ell = 0$), and indeed the explicit expression given e.g. by Cherepkow (1979) for $\underset{\sim}{P}$ produced by linearly polarized light vanishes in this limit. Also, no net spin polarization occurs in photoexcitation (by linear light) to atomic bound states. On the other hand, as will be discussed below, non-zero spin polarization becomes possible in the case of the semi-infinite solid, for which proper scattering states with outgoing plane wave asymptotics exist. In the frame-work of the three-step model, this will be interpreted in terms of an unpolarized bulk state mixture diffracted in a spin-dependent way at the surface.

More information on the spin polarization, which may occur in bulk transitions induced by circularly polarized light, can be derived from symmetry properties of the dipole matrix elements (cf. eq.(82)). In addition to the special symmetry lines, which have already been considered above in the context of one-step model normal emission (cf. Tables 1,2,3), the reciprocal lattice associated with the infinite solid contains a number of $\underset{\sim}{k}$ points of higher symmetry. Transitions at such points and the corresponding spin polarization features have been analysed in detail by means of group theoretical methods (cf. Wöhlecke and Borstel 1984, Chapters 10 and 11 of this Book and references therein).

To obtain quantitative theoretical predictions at the bulk excitation stage of the three-step model, equations (82,83) have to be evaluated by numerical calculations. Explicit expressions for electric dipole matrix elements between Relativistic Augmented Plane Wave (RAPW) states were given by Koyama (1975), but not calculations were performed by this method. Numerical results were obtained (Borstel and Wöhlecke 1982) on the basis of a "second-principles" band structure method. In this "combined interpolation scheme" for fcc d-band metals, earlier versions of which (proposed by Hodges et al. 1966 and Mueller 1967) were refined by Smith and Mattheiss (1974) in view of photoemission calculations, a Hamiltonian matrix containing adjustable parameters is set up from a basis, which consists of atomic d orbitals and plane waves. The parameters are determined by least-squares fit to a first principles band structure at high-symmetry $\underset{\sim}{k}$ points. Relativistic effects are, in a two-component approximation (cf. eq.(5)), taken into account by explicitly including spin-orbit coupling into the scheme and fitting to a relativistic band structure (cf. also Benbow and Smith 1983). Scalar relativistic effects are thereby automatically taken care of. The matrix elements H_{mn}, with m and n labelling the basis states, are then known as analytic functions of $\underset{\sim}{k}$, and energy values and (initial and final) eigenstates for arbitrary $\underset{\sim}{k}$ are readily obtained by diagonalizing H. As pointed out by Smith (1979), the calculation of electric dipole transition matrix elements becomes very simple, since the effective momentum operator $\tilde{\underset{\sim}{p}}$ (cf. eq.(86b)) is related to the Hamiltonian as (cf. Blount 1962)

$$\tilde{\underset{\sim}{p}} = (m/\hbar) \, \nabla_k H(\underset{\sim}{k}) \quad . \tag{99}$$

Its matrix elements $P_{\sim mn}$ with respect to the basis states are thus obtained by differentation of the (analytically known) elements $H_{mn}(\underset{\sim}{k})$.

Denoting by u^f_{sm} and $u^i_{\sigma n}$ the elements of the representations of the final state $|E\underline{k}s>$ and the inital state $|E_i\underline{k}\sigma>$, respectively, the transition matrix element in eq.(82) is then

$$-(e/2mc)\ \underline{A}_o \cdot (\sum_{mn} u^{f*}_{sm}\ \underline{P}_{mn}\ u^i_{\sigma n})\ . \tag{100}$$

From the elements u^f_{sm} and the basis functions of the interpolation scheme, the $|E\underline{k}s>$ are obtained as two-component spinors, from which the (2x2) spin density matrix follows (cf. eqs. (82-84)).

The resulting bulk spin polarization is in general still different from the actual photo-electron spin polarization, since transport to the surface and in particular transmission through the surface can change the polarization vector. The spin dependence of these two stages of the three-step model is discussed in the following.

4.5.6.3 Transport Effects

For non-magnetic materials, the escape depth factor D in the spin-dependent three-step model expression eq.(84) is the same scalar as in the spin-less model. If the attenuation of the radiation field by photoexcitation processes is neglected, D is given (cf. Feibelman and Eastman 1974) by $(\hbar\ Imk_z(E))^{-1}$ or, equivalently

$$\lambda_n(E,\underline{k}) = -\hbar^{-1}\hat{\underline{n}}\cdot\nabla E(\underline{k})\tau(e)\ , \tag{101}$$

where $\underline{n}$ is the (inward-directed) surface normal and $\tau(E)$ is the electron life-time due to inelastic collision processes (especially electron-hole and plasmon excitation), which is related to the spatially uniform imaginary part of the effective potential: $\tau = (2|V_i|)^{-1}$ (cf. Section 4.4.4). The normal mean free path $\lambda_n(E,\underline{k})$ gives the average depth, from which an electron of energy E and crystal momentum $\hbar\underline{k}$ may reach the surface without being inelastically scattered. The optical absorption length α^{-1} has to be taken into account if it is of the same order of magnitude as λ_n or smaller. The transport factor D can then be expressed as (cf. Smith in Cardona and Ley 1978)

$$D(E,\underline{k}) = \lambda_n \alpha / (1 + \lambda_n \alpha) \quad . \qquad (102)$$

Obviously, a scalar D has no effect on the photo-electron spin polarization. If there is a paramagnetic overlayer on the surface, (elastic) spin exchange scattering of the outgoing photo-electrons in this layer may reduce the polarization (cf. Feuchtwang et al. 1978, Siegmann et al. 1984 and refs. therein).

Electrons, which have been inelastically scattered, may still reach the surface and contribute to the photo-current. (For a simple semi-classical treatment cf. e.g. Eastman 1972 and, including spin, Bringer et al. 1979). If the bulk-excited photo-electrons are unpolarized, the inelastically scattered ones are - for non-magnetic materials (as considered in this Section) - also unpolarized. If the former are polarized (with polarization vector $\underline{P}_{e\ell}$ parallel or antiparallel to the direction of propagation of the circularly polarized photons), the polarization vector $\underline{P}_{in}$ of the latter remains parallel to $\underline{P}_{e\ell}$ but is generally reduced in magnitude as a consequence of electron-hole pair excitation with spin-flip (cf. Section 4.4.2). The amount of this reduction depends on the ratio of the spin flip rate to the non-flip rates (of all relevant loss mechanisms) and therefore decrease with increasing final-bulk-state energy.

For a special class of non-magnetic cases, additional physical mechanisms associated with the transport stage in the three-step model are essential. Since near-threshold photoemission from GaAs, which provides a very efficient source of polarized electrons (cf. Chapter 5 and refs. therein), belongs to this class, a survey of these mechanisms seems pertinent. Let us consider emission with final state energy at the conduction band minimum Γ_6 of a III-V compound semiconductor, the surface of which has been treated such that Γ_6 is above the vacuum level (negative electron affinity). Since $\nabla E(\underline{k})=0$, equations (101,102) imply D=0, and there is no photcurrent in the framework of the above model. In reality, however, thermal diffusion does provide an efficient transport mechanism, with the diffusion length L taking over the role of λ_n in eq.(102). This expression was derived (James and Moll 1969), for spin-averaged photoemission from GaAs, by means of a one-dimensional diffusion model (cf. monograph by Bell 1973). In spin-polarized photoemission, one further has to take into account the effect of spin-flip scattering during transport. Such scattering is characterized by a "spin-flip" time τ_f or, more commonly, by a "spin-relaxation" time $\tau_s = \tau_f/2$, after which- for an initially fully polarized ensemble - half of the spins have flipped, i.e. the spin polarization is reduced to zero. Spin-relaxation of conduction electrons in III-V semiconductors may be due to a variety of physical reasons, the most important ones of which are: (a) spin splitting of the conduction band in the absence of an inversion centre (Dyakonov and Perel 1971), (b) exchange interaction between electrons and holes (Bir et al. 1976), (c) spin hybridization of conduction band states due to spin-orbit interaction (Elliot 1954 and Yafet 1963), and (d) hyperfine interaction between electron spin and nuclear spin. (An extensive survey and more references to the original literature may be found in a recent monograph by Meier and Zakharchenya (1984)).

Given a (fully or partially) polarized ensemble produced by bulk excitation, it is plausible that spin-flip scattering (with equal rates τ_s^{-1} for spin-up and spin-down electrons) leads to a depolarization if the spin relaxation time τ_s is comparable to or smaller than the electron life-time τ (determined by recombination processes). A quantitative estimate for this depolarization may be obtained from a spin-dependent one-dimensional diffusion model (cf. Bell 1973, Pierce et al. 1980). Denoting by $n_\sigma(z)$ (with $\sigma=+/-$) the spin-up/down electron densities, the steady-state diffusion equations are

$$D \frac{d^2 n_\sigma(z)}{dz^2} + G_\sigma(z) - \frac{n_\sigma(z)}{\tau} - \frac{n_\sigma(z) - n_{\bar\sigma}(z)}{2\tau_s} = 0 \quad , \tag{103}$$

where $\sigma = +/-$, $\bar\sigma = -\sigma$, D is the electron diffusion coefficient and $G_\sigma(z) = C_\sigma \exp(-\alpha z)$ is the generation rate by bulk excitation. The last term on the right-hand side expresses the two-fold action of spin-flip scattering (recall $\tau_f = 2\tau_s$): removal proportional to n_σ and generation proportional to $n_{\bar{\sigma}}$. The two equations are easily decoupled: adding and subtracting yields, with the definitions $N_\sigma(z) = n_+(z) + \sigma n_-(z)$, $T_+ = \tau$ and $T_- = \tau_s \tau/(\tau_s + \tau)$:

$$D\, d^2 N_\sigma(z)/dz^2 + G_\sigma(z) - N_\sigma(z)/T_\sigma = 0 \quad . \tag{104}$$

Each of these two equations (for $\sigma = +$ and $-$) has the form of the (spinless) diffusion equation considered by James and Moll (1969) and by Bell (1973). The outgoing solution $N_\sigma(z)$ at the surface (strictly speaking: at the beginning of the band bending region) (located at $z=0$) can be explicitly obtained from the Green function part $\exp[(z-z')/\sqrt{DT_\sigma}]$ and the source function $G_\sigma(z') = C_\sigma \exp(-\alpha z')$ by straightforward integration. The outgoing current density $J_\sigma = -eD(d/dz)N_\sigma(z)|z=0$ at the surface is then

$$J_\sigma = eC_\sigma/[1 + (\alpha\sqrt{DT_\sigma})^{-1}] \quad . \tag{105}$$

Since $C_-/C_+ = :P_0$ is the spin polarization produced by bulk excitation, the resulting spin polarization $P = J_-/J_+$ (note that $J_\sigma = J_\uparrow + \sigma J_\downarrow$) is conveniently expressed as

$$P = P_0 \, [\alpha\sqrt{D\tau} + 1] \, / \, [\alpha\sqrt{D\tau} + \sqrt{(\tau_s + \tau)/\tau_s}] \quad . \tag{106}$$

Spin relaxation is thus seen to reduce P_0 by a factor that depends in particular on the diffusion coefficient D and on the ratio τ/τ_s. In the limit of $\tau_s \gg \tau$ ($\tau_s \ll \tau$), this factor goes to 1 (0), as one would intuitively expect. If light absorption α is strong and/or the diffusion length $L = \sqrt{D\tau}$ is large, $P \approx P_0$; in the opposite case, the reduction factor becomes $[\tau_s/(\tau_s + \tau)]^{1/2}$, i.e. just the square root of the factor found for bulk recombination radiation (luminescence). Unless a significant further reduction occurs in the surface transmission step, the spin polarization in photoemission is therefore larger than the lu-

minescence polarization; but generally less than the polarization produced in the bulk excitation step.

A particularly interesting phenomenon was recently discovered
(Riechert et al. 1984) to occur in the band-bending region of non-centrosymmetric semiconductors. The spin-orbit-induced spin splitting
(lifting of the Kramers degeneracy) of the conduction band acts like
an effective magnetic field and causes a precession of the spin polarization vector away from its initial direction. This effect promises
to be useful for studying band-bending regions. (See also Chapter
10.3.2).

4.5.6.4 Surface Transmission

From the three-step model expression (eq.(84) in Section 4.5.4)
it is evident that the spin polarization of the photo-current is
affected by the transmission from the solid into the vacuum, if the
(2x2) matrices $T(E,\underline{k})$, which may be written in the form $t_1 + \underline{t}_2 \underline{\sigma}$, have
$\underline{t}_2 \neq \underline{0}$, i.e. in particular, if they have non-vanishing off-diagonal
("spin-mixing") elements. This is generally the case in the presence of
spin-orbit coupling, as a consequence of matching - at the surface -
the vacuum wave field, which has Pauli spinor eigenstates $|s\rangle$, with
the wave field inside the solid, which is necessarily of the form
$a_s|s\rangle + b_s|\bar{s}\rangle$. The existence of off-diagonal elements in $T(E,\underline{k})$ is
apparent from eq.(81), where the vacuum state $|E\underline{k}_\parallel s\rangle$ is a direct pro-
duct $|E\underline{k}_\parallel\rangle|s\rangle$, whilest the Bloch state $|E\underline{k}s\rangle$ contains an admixture of
$|\bar{s}\rangle$. If a surface potential barrier of finite thickness is taken into
account (cf. Fig. 2 in Section 4.2.2), the negligible role of spin-
orbit coupling in this region implies pure-spin states $|s\rangle$, and the
spin-dependence arises from matching these states to the mixed-spin
states inside the solid. The effect is also plausible from spin-polarized LEED (cf. Section 4.3) by recalling that the final state in photoemission is a time-reversed LEED state.

Spin dependence of the transmission through the surface manifests
itself in several ways. If the electrons excited in the bulk are un-

polarized (due to linearly or unpolarized light) (cf. Section 4.5.6.2), the photo-current can acquire a polarization vector $\underline{P}$. This is an inverse SPLEED effect and therefore subject to the same general symmetry properties (cf. Section 4.3.2). In particular, $\underline{P}=0$ at normal emission (for centrosymmetric systems); at off-normal emission, $\underline{P}$ is perpendicular to the emission plane if the latter is parallel to a mirror plane. Polarization (with values up to about 10 %) of this origin was found theoretically and experimentally for W(001) (Feder and Kirschner 1981b, Kirschner et al. 1981; see also Chapter 11 of this book). In that calculation, $T(E,\underline{k})$ was obtained by matching Bloch wave pairs propagating towards the surface with reflected ones and a set of plane waves emitted into the vacuum, and the bulk density matrix was assumed as a (2x2) unit matrix.

If the bulk photo-electrons are polarized (with $\underline{P}_b$) (due to circularly polarized light), $\underline{P}$ of the emitted current generally differs from $\underline{P}_b$. The process may be regarded as transmission SPLEED with a polarized-electron source inside the solid. For off-normal emission in a mirror plane (with normal incidence of the light), reversal of $\underline{P}_b$ with reversal of the photon helicity and the mirror symmetry dictate "$\underline{P}_b$ parallel to the emission plane"; transmission may firstly change direction and magnitude of $\underline{P}_b$ in the plane and secondly generate a component of $\underline{P}$ perpendicular to the plane. If the emission plane is not a mirror plane or if the light is not incident in this plane, $\underline{P}_b$ generally has a component $P_b^{\perp}$ normal to the plane. Reversal of $P_b^{\perp}$ by reversing the photon helicity changes the intensity of the emitted electrons. $P_b^{\perp}$ can thus be detected from an intensity asymmetry without polarization measurement. (For details and experimental results see Oepen 1985, and also Chapter 11 of this book).

As just shown, the experimentally accessible $\underline{P}$ in general originates (for circularly polarized light) from a combined bulk excitation and surface transmission effect. To extract $\underline{P}_b$ from $\underline{P}$-data is desirable for two reasons: (a) $\underline{P}_b$ contains information on the bulk electronic structure (in particular symmetry and hybridization features of the occupied quasi-particle states ("holes")); (b) knowing P_b and P, one knows (for bulk emission features) the surface transmission behaviour, which is not only interesting in its own right, but may provide (simultaneously) some information on the surface geometry. This extraction of P_b can generally be achieved with the aid of quantitative calculations, which reproduce the measured P and yield, for the same model assumptions and parameters, P_b. Since the three-step expression (eq. (84)), which first comes to mind for this purpose, is restricted to ideal bulk transitions and less adequate for quantitative purposes, it seems advisable to employ the one-step model theory (cf. Section 4.5.3). Designed to yield $\rho(E,k_\parallel)$ and thence P, this theory can, at practically no additional computational cost, also produce P_b: the calculated ρ is "fed back" into the solid to yield an internal density matrix and thence P_b:

$$\rho_b = \sum_{g\sigma} S_g^\sigma \, \rho \, S_g^{\sigma+} \text{ and } P_b = -\text{tr}[\underline{\sigma}\rho_b]/\text{tr}\rho_b \quad , \tag{107}$$

where the (2x2) matrices S_g^σ, with $= +/-$, transform a spinor amplitude of the reversed emitted beam into a spinor amplitude of the gth beam (in direction σ) sufficiently deep inside.

4.5.6.5 Emission from Adsorbates

The electronic structure of chemically "foreign" atoms adsorbed ("physi-" or "chemisorbed") on solid surfaces (cf. e.g. Chapter 1.4.2 and refs. therein) is a subject of great current interest, which has been investigated successfully by angle-resolved photoemission (cf. e.g. Williams et al. 1980, Westphal and Goldmann 1983, Ling et a. 1983 and refs. therein). Diffraction of the photo-electrons by the substrate may further reveal adsorption site geometry (cf. e.g. Barton et al. 1983 and refs. therein). As predicted theoretically (Feder 1977b and 1978) and verified experimentally (Schönhense et al. 1985), spin analysis is, also for nonmagnetic materials, capable of providing a wealth of additonal information. With reference to an extensive presentation of this topic (with an emphasis on recent experimental results and

their implications) in Chapter 11 of this book, it may suffice here
to outline a few key aspects.

Although a quantitative theoretical treatment should generally
be based on a "one-step model" (cf. above) including the adsorbate,
some more insight is, in the case of emission from states localized
within the adsorbate, reached by decomposing the photoemission process
into two successive steps: (a) excitation within the adsorbate, and
(b) diffraction by the substrate. Like in bulk excitation and surface
transmission (cf. Sections 4.5.6.2 and 4.5.6.4), spin-orbit coupling
can produce spin polarization effects in each step. Model calculations
for a localized adatom level on W(001) predict, for unpolarized light
and hence no spin polarization in step (a), substantial polarization
effects from the diffraction step (b), which are very sensitive to the
assumed position of the adatom relative to the substrate (Feder 1977).
For circularly polarized light, the excitation step (a) produces polarized electrons, and diffraction (b) gives rise to an intensity asymmetry (upon reversal of the photon helicity), i.e. the adsorbate
system has a "built-in" spin detector (which can replace external
spin analysis of the photo-current).

While the diffraction step promises to yield mainly geometrical
information, the excitation step reveals the electronic structure of
the adsorbate atoms as modified by the bindung to the substrate and
inter-adsorbate-atom interaction. A prototype example is physisorbed
Xe. The spin polarization of photoelectrons (due to circularly polarized light) from the energy-split $m_j = \pm 1/2$ and $m_j = \pm 3/2$ sub-levels of
$5p_{3/2}$ has opposite sign and (depending on the photon energy) maximal
values of 100 % (cf. Feder 1978, Schönhense et al. 1985 and Chapter 11
of this book). Finally a word of warning: the intuitive appeal of the
"two-step model" should not mislead one to overrate its applicability.

4.5.7 Transition Metal Ferromagnets

The concepts and theoretical aspects of photoemission, which were
presented in Sections 4.5.1 - 4.5.6, apply in particular to systems

with magnetic ordering in the bulk and/or at the surface. Since spin-resolved photoemission has recently been most fruitfully used for studying the elemental 3d transition metal ferromagnets Fe, Co and Ni (see Chapters 12 and 14 of this book and references therein), it seems pertinent to address, in addition to the above general treatment, a few points specific to photoemission from these materials, firstly at temperatures wll below the ferromagnetic transition temperature (Curie temperatur T_c) and secondly in the vicinity of T_c.

4.5.7.1 At Low Temperature

Well below T_c, the itinerant electron model (Stoner model) is valid (cf. e.g. Wohlfarth 1984 and refs. therein), in which the effective magnetic field $\underline{B}(E,\underline{r})$ entering in eq.(3) is parallel to a global magnetization axis and has the crystal-lattice periodicity.

It is instructive and, on the grounds of the fairly small atomic numbers ($Z = 26,27$ and 28 for Fe, Co and Ni), also useful for realistic calculations, to consider the Pauli approximation, i.e. the first three terms in eq.(5), in which spin-oribt coupling is neglected and the exchange term $\underline{\sigma}\cdot\underline{B}$ is the only spin-dependent interaction. One thus has two decoupled Schrödinger equations, with a spin-up and a spin-down potential $V^+(E,\underline{r})$ and $V^-(E,\underline{r})$ (cf. eq.(6)). Consequently, the lower and upper state Green functions G_1 and G_2, which are 2x2 matrices in the non-relativistic two-component approximation, also become diagonal with elements $G^s_{1,2}$ determined by effective spin-dependent potentials $V^s_{1,2}$, where $s = +/-$, and the subscripts have been added to indicate the energy dependence of the potentials. The interaction Hamiltonian H' is approximated by the standard spin-independent nonrelativistic form $\underline{A}p + p\underline{A}$, i.e. eq.(86a) with the magnetic dipole coupling $\underline{\sigma}\cdot\underline{B}_{\underline{r}}$ neglected. The matrix elements $I_{ss'}$ of the one-step model expression eq.(74) then also become diagonal, with elements $I_s: = I_{ss'}\delta_{ss'}$, and the photocurrent spin density matrix (eq.(75)) assumes the form

$$(E,\underline{k}_{\parallel}) = \begin{pmatrix} I_+ & 0 \\ 0 & I_- \end{pmatrix} \tag{108}$$

Spin is thus conserved in the photoemission process. According to eq.(1), the photoelectron spin polarization vector $\underset{\sim}{P}$ is then parallel to the z axis (i.e. the magnetic orientation direction) with magnitude

$$P = (I_+ - I_-) / (I_+ + I_-) . \tag{109}$$

Comparison with the conventional 'spin-less' one-step model expression eq.(70) shows that the diagonal elements I_s in eq.(74) now have exactly the same form (since $<s/s'> = \delta_{ss'}$). In the absence of spin-orbit interaction and dipole coupling to the radiation field, the calculation of spin-polarized photoemission from ferromagnets thus reduces to a succession of two spin-less calculations each based on an Schrödinger Hamiltonian with effective potentials $V_{1,2}^+$ and $V_{1,2}^-$, to yield a majority and a minority spin current I_+ and I_-. Specialising it to the three-step model (cf. eq.(84) in Section 4.5.4), one obtains the form

$$I_s(E,\underset{\sim}{k}_\parallel) = \sum_{i k g} t_s(E,\underset{\sim}{k}) \, d_s(E,\underset{\sim}{k}) \, M_s(E\underset{\sim}{k}; \, \underset{\sim}{k}_i s_i) \delta(E - \hbar\omega - E_i) \delta(\underset{\sim}{k}_\parallel + g_\parallel - k_\parallel^B), \tag{110}$$

where $s = +/-$, M_s denotes the square of the bulk transition matrix element for spin s, d_s the photo-electron escape depth (i.e. the mean free path with respect to inelastic collisions) and $t_s = |T_{ss}|^2$ the transmission probability across the surface.

The above non-relativistic one-step model has been successfully used for interpreting and analyzing experimental data from several Fe and Ni surfaces (cf. Chapters 12 and 14 of this book). In Chapter 14, there is also some discussion on how to approximate the spin-dependent potentials $V_{1,2}^s$ and in particular their complex self-energy parts.

Whether spin-orbit coupling, which is always present in nature, may be neglected in photoemission calculations for ferromagnets, depends on the features one wishes to study and on the level of accuracy reached in the corresponding experiments. As was demonstrated for non-magnetic cases (in particular Cu, with $Z = 29$) (cf. Goldmann et al. 1982, Przbylski et al. 1983, Ackermann 1985, and refs. therein), spin-orbit coupling effects, which manifest themselves in violations of non-relativistic dipole selection rules, are actually observable, although difficult to disentangle from finite-acceptance-angle effects. For ferromagnets, spin-orbit coupling produces fairly extended hybridization regions (along high-symmetry $\underset{\sim}{k}$-lines) between majority and minority spin states with a substantial reduction of the spin expectation value (cf. relativistic band structure calculations by Ackermann et al.(1984) and Eckardt et al.(1985), and refs. therein). Relativistic photoemission calculations (using the one-step theory outlined in Section 4.5.3) for Ni, for which spin-orbit and magnetic exchange interaktion are of comparable magnitude, reveal several interesting features (Ackermann 1985, Ackermann and Feder 1986): (a) extra peaks due to spin-orbit-induced initial-state band splittings, (b) a marked dependence on the orientation of the light polarization vector $\underset{\sim}{A}$ with respect to the magnetization direction for normal and even more so for off-normal emission, (c) opposite-spin satellites and (d) intensity asymmetries for circularly polarized light. Further, the spin-orbit-induced spin dependence of the surface transmission step is expected to lead, for linearly polarized or unpolarized light and off-normal emission in a plane perpendicular to the magnetization direction, to a left-right asymmetry, which acts as a built-in spin detector and could replace external spin analysis.

4.5.7.2 Near the Curie Temperature

A particularly interesting subject, to which spin-resolved photoemission provides a most powerful approach, is the magnetic and electronic structure of the 3d transition metal ferromagnets Fe, Co and Ni at finite temperatures, especially in the vicinity of the ferromagnetic to paramagnetic phase transition temperature (Curie temperature T_c) (cf. reviews by Gyorffy et al. (1984) and Moriya and Takahashi (1984), Chapter 2 of this book, and references therein). In view of the itinerant nature of the 3d electrons and the existence of fluctuating local moments (spin density fluctuations) even well above T_c, a theory of photoemission may proceed as shown above for low temperature, but with the effective magnetic field $\underline{B}(E,\underline{r})$ now varying in magnitude and direction from lattice site to lattice site. $\underline{B}$ thus consists of a sum of local magnetic fields $\underline{B}_i$ (in atomic cells around sites i). In the spirit of a magnetic Born-Oppenheimer approximation, the set $\{\underline{B}_i\}$ is taken as static, and time dependence is accounted for by averaging over the spin-dependent photocurrents obtained for a sequence of configurations $\{\underline{B}_i\}$ (for details see You and Heine 1982). Ferromagnetic ordering can be described by the correlation function

$$\rho_{ii'} = <\underline{B}_i \cdot \underline{B}_{i'}>/(<|B_i|^2><|B_{i'}|^2>)^{1/2} , \tag{111}$$

which may be decomposed into a long-range part corresponding to an average magnetization (non-zero for $T < T_c$) and a near-neighbour part characterizing short-range magnetic order. Information on the latter can be deduced from photoemission data by comparison with theoretical results calculated for configurations $\{\underline{B}_i\}$, in which the $\underline{B}_i$ vary randomly from site to site subject to assumed correlation function $\rho_{ii'}$ (cf. Haines et al. 1985 a,b). One can thus continuously go from the "disordered-local-moment" limit (no short-range order) to the opposite "local-band" limit (massive short-range order).

Since the disorder in $\underline{B}_i$ destroys the translation invariance (lattice periodicity) parallel to the surface, the layer-KKR one-step formalism of photoemission (cf. above) is no longer applicable. Instead, one could employ an alloy-type theory. This was recently done by Durham et al. (1984) for the special case of the disordered-local-moment limit. The more complicated case of an arbitrarily assumed amount of short-range magnetic order was treated by a simpler theory, a bulk-interband transition model with a plane-wave final state (Haines et al. 1985a,b and Clauberg et al.1985), which is adequate to describe the temperature behaviour of features identified as due to bulk transitions by one-step-model low temperature calculations. In this theory, the initial state is described in a tight-binding approximation including s and d bands, with Hamiltonian matrix elements

$$H^{ss'}_{i\ell m,i'\ell'm'} = h_{i\ell m,i'\ell'm'} \, \delta_{ss'} - \frac{1}{2}(\underline{\sigma}\cdot\underline{B}_{i\ell})_{ss'} \, \delta_{ii'} \, \delta_{\ell\ell'} \, \delta_{mm'} \quad (112)$$

between orbitals with angular momentum quantum numbers ℓm and $\ell'm'$, and spin indices s and s', on sites i and i'; δ is the Kronecker symbol and h the nonmagnetic part. The local magnetic fields $\underline{B}_{i\ell}$ are allowed to depend on ℓ (i.e. to be larger for d-orbitals than for s-orbitals) and could easily be generalized to account for anisotropic exchange splitting within the d-band. In choosing the tight-binding parameters, it is important to recall that the above H is acually the real part of a quasi-particle Hamiltonian and that it has to be supplemented by an imaginary self-energy part V^s_i (cf. Section 4.2.2). Consequently, the parameters must be fitted to reproduce quasi-particle bands, which are for example determined by low-temperature photoemission. The spin-dependent photoemission intensity $I^s(\underline{k}_f)$, for final states $|\underline{k}_{f,s}>$, is calculated (cf. McLean and Haydock 1977) as the initial-state Green function element with respect to the state $\underline{A}\cdot\underline{p}|\underline{k}_{f,s}>$ using Haydock's (1980) recursion scheme. While this is adequate for angle-resolved ultraviolet photoemission, the method is also easily transferred to X-ray photoemission and appearance potential spectroscopy by calculating the density of states as the site-projection of the same initial-state Green function (cf. Haines et al. 1985c).

4.6. Bremsstrahlung (Inverse Photoemission)

A very recently matured surface spectroscopy technique, which is very akin to photoemission conceptually and complementary to it with regard to the information it yields, is angle- and energy-resolved bremsstrahlung spectroscopy (more commonly referred to as "inverse photoemission"). (For details and references cf. review articles by Dose (1983, 1984)).

4.6.1 Principles

In scattering theoretical terms, bremsstrahlung is characterized by electrons of well-defined energy, momentum and spin in the input channel and photons in the output channel, which is just the opposite as in photoemission. The basic process is schematically indicated in Fig. 12 (Section 4.5.1). The incident electron (of energy E_2') makes a transition to an unoccupied state of energy $E_1'<E_2'$, generating a photon of energy $h\nu = E_2'-E_1'$, which is detected when emerging from the surface. The process is, loosely speaking, inverse to photoemission. While photoemission involves quasi-hole states with $E_1<E_F$, bremsstrahlung, however, is associated with quasi-particle states with $E_1'>E_F$. It thus "probes" the unoccupied states and provides information complementary to that from photoemission. In particular, it covers the energy range between E_F and E_{vac}, which is not accessible in such detail by other electron spectroscopies. As in photoemission, one distinguishes two important regimes according to the photon energy involved: the X-ray and the (vacuum-) ultraviolet regime. The latter (with $h\nu$ typically below about 100 eV) offers the advantage of greater momentum resolution and has therefore been pursued more widely in the past few years.

Spin-orbit coupling and ferromagnetism lead, in analogy to the situation in photoemission, to a dependence of the emitted photon intensity on the spin of the incident electrons, i.e. to an intensity asymmetry upon reversal of the electron spin polarization. For ferromagnets, one can thus study the unoccupied majority and minority spin (bulk and surface) states separately (cf. Chapters 13 and 14 of this book, and references therein).

4.6.2 Theory

From the above, it is plausible that the theoretical treatment of bremsstrahlung spectroscopy very closely parallels that of photoemission (cf. Section 4.5). It may therefore suffice here to focus on a few specific points.

Consider an electron of energy E and surface-parallel momentum $\underset{\sim}{k}_{\parallel}$ incident, and radiation with ω and $\underset{\sim}{A}$ emitted from the system. If spin is neglected, the photon flux I_p is (cf. Pendry 1981), except for a factor of geometrical origin, given by the photoemission expression eq.(70) as $I(E,-\underset{\sim}{k}_{\parallel})$. The equivalence with Pendry's (1981) form is easily seen from eq.(70), containing $-\underset{\sim}{k}_{\parallel}$, by using the time reversal relationships

$$G_-(E) = T\, G_+\, T^+ \quad \text{and} \quad T|E,-\underset{\sim}{k}_{\parallel}> = |E\underset{\sim}{k}_{\parallel}> . \tag{113}$$

If spin is taken into account, the incident electron beam is further characterized by a prescribed (i.e. experimentally set) spin polarization vector $\underset{\sim}{P}_0$, or by the corresponding spin density matrix $\rho_0 = (1+\underset{\sim}{\sigma}\underset{\sim}{P}_0)/2$. The photon flux is then related to the relativistic-one-step-model photoemission density matrix ρ (cf. eqs.(74,75)) as

$$I_p(E,\underset{\sim}{k}_{\parallel},\underset{\sim}{P}_0,\omega,\underset{\sim}{q},\underset{\sim}{A}) = tr[T\rho(E,-\underset{\sim}{k}_{\parallel})T^+ \rho_0 (\underset{\sim}{P}_0)] =$$

$$= (\widetilde{I}/2)\,(1 - \underset{\sim}{P} \cdot \underset{\sim}{P}_0) , \tag{114}$$

where $T = i\sigma_y K$, $\underset{\sim}{P}$ is the photo-electron polarization vector, $\widetilde{I} = tr\rho$, and $\rho = (\widetilde{I}/2)\,(1+\underset{\sim}{\sigma}\cdot\underset{\sim}{P})$ has been used. If the incident electron is in a pure spin state $|s>$, i.e. $\underset{\sim}{P}_0 = (0,0,\pm1)^T$, I_p reduces to the matrix element $I_{\bar{s}\bar{s}}(E,-\underset{\sim}{k}_{\parallel})$ of eq.(75), with $\bar{s} = -s$.

For ferromagnetic systems, it is important to note that the time reversal operation on G_+ (cf. eq.(113)) also reverses the effective magnetic field $\underset{\sim}{B}$. If spin-orbit coupling and the (spin-flipping) magnetic dipole term $\underset{\sim}{\sigma}\cdot\underset{\sim}{B}_r$ in H' are neglected, the photon flux from ferromagnets due to electrons fully polarized with $\pm\underset{\sim}{P}_0 \parallel \underset{\sim}{B} \parallel \hat{\underset{\sim}{z}}$ is, from eq.(11), seen to be given by the diagonal elements I_{++} and I_{--} of eq.(75), which can be calculated by using a Schrödinger Hamiltonian with majority and minority spin potential V^+ and V^- (cf. the discussion of photoemission in Section 4.5.7.1). Also, the high-temperature method indicated in Section 4.5.7.2 can then be used for inverse photoemission.

The bremsstrahlung intensity expression eq.(114) holds for any given radiation field $\underset{\sim}{A}$, which enters via H' (cf. eqs.(71,77)) into ρ (eq.(74,75)). Experimentally, $\underset{\sim}{A}$ can be set by passing the emitted photons through a polarization filter before detecting them. A convenient general characterization of the radiation field is given by means of a photon density matrix (cf. Blum 1978 and references therein) in a representation with respect to the two circularly polarized basis states of helicity $\lambda = +/-$ (with $\underset{\sim}{A} = -\lambda 2^{-1/2} A_0 (1, \lambda i, 0)^T$ in a coordinate system with $\hat{\underset{\sim}{z}}$ parallel to direction of propagation). Denoting by H'(λ) the electron-photon interaction (eq.(77)), there are four possible combinations of H'(λ) and H'$^+$(λ') in the electron spin density matrix elements in eq.(75). Consequently, eq.(114) gives four photon intensities $I_p(...,\lambda,\lambda')$, forming a (2x2) photon density matrix $\underset{\sim}{\tau}$, which can be written in the form

$$\underset{\sim}{\tau}/tr[\underset{\sim}{\tau}] = \frac{1}{2} \begin{pmatrix} 1+n_2 & -n_3+in_1 \\ -n_3-in_1 & 1-n_2 \end{pmatrix} \tag{115}$$

in terms of the Stokes parameters n_k (cf. e.g. Jackson 1962, 1975, and Born and Wolf 1980). The diagonal elements give the relative probabilities of observing a photon of positive and of negative helicity (in particular $n_2 = \lambda$ for fully circularly polarized light), $(1+n_3)/2$ and $(1-n_3)/2$ of observing linear polarization along the x and y axes, respectively.

In eq.(114), the bremsstrahlung intensity has been expressed in terms of the "elastic" part of the photoemission density matrix. Like the photoemission current comprises electrons that have undergone inelastic scattering after excitation (cf. Section 4.5.2, Fig. 14), this "primary" bremsstrahlung is augmented by a "secondary" contribution, which consists of photons produced by electrons that have already lost some energy w. For initial (upper) state energy E, their final (lower) state energy is E-w-ℏω. Since it must be above E_F, the secondary part increases, for constant ℏω ("isochromat spectroscopy"), as E increases from E_F + ℏω upwards. While general expressions involving flip and non-flip rates (cf. Section 4.4) are easily written down, realistic calculations have not yet been done, since their essential ingredient are the inelastic scattering rates, which are themselves poorly known.

Numerical results of spin-dependent inverse photoemission will be presented, for ferromagnetic Fe, in Chapter 14 and employed to interpret experimental data.

4.7   Concluding Remarks

In this Chapter, a unified theoretical picture of spin-dependent electron scattering and emission from ferromagnetic and non-magnetic solid surfaces has been presented on the basis of an effective one-electron Dirac Hamiltonian H, which incorporates many-body effects by means of an energy-dependent complex potential and magnetic field (in non-relativistic approximation equivalent to a spin- and energy-dependent complex potential V^S). The construction of V^S is a fundamental problem, which to date has been approximately solved (with empirical ingredients), and requires further study. Given V^S, the elastic scattering problem (or, more generally, the solution of the effective Dirac equation for a given energy E and prescribed boundary conditions) is the theoretically simplest, and has successfully been dealt with by multiple-scattering techniques (for both ordered and disordered systems). Photoemission and its inverse involve the solution of the scattering problem at two energies (E and E-ℏω) and the calculation of matrix elements. Realistic one-step model calculations have been performed even for ferromagnets with spin-orbit coupling. Electron-electron scattering, which is the dominant spin-dependent loss mechanism, is again more difficult theoretically, since it involves four energies and more complicated matrix elements. It currently presents a challenging problem still to be solved, both for its own sake and as a prerequisite for a first-principles self-energy.

The above theoretical framework has been very successful in spin-polarized low-energy electron diffraction (SPLEED) and ultraviolet photoemission and its inverse, leading to numerical results in quantitative agreement with experimental data. Although it has limitations in describing certain "many-body" phenomena like e.g. "satellites" and "core-hole screening" in photoemission at higher energies or Auger emission (cf. e.g. Gadzuk 1978, Gunnarsson and Schönhammer 1982 and references therein), its one-electron state (fully taking into account the discrete lattice structure) appears as an important ingredient for realistic calculations of such phenomena.

A final remark - on the threshold to the more experimental part of this book - concerns the relation between theory and experiment in spin-dependent electron scattering and emission. Quantitative agreement of numerical results - obtained by calculations, which are demanding for present-day computers - with their experimental counterparts is firstly gratifying as evidence of the adequacy of the theoretical model and formalism. Secondly, it permits the extraction of physical information from experimental data if one focuses on spectral features, which are selectively sensitive to changes in the physical model assumptions. Thirdly, the predictive power of theory thus established may be even practically used for the design of new experimental apparatus; an example is the construction of the first SPLEED spin detector (cf. Chapter 5 of this book), which was guided by theoretical SPLEED results.

References

Ackermann B 1985 Ph. D. Thesis (Dissertation), Duisburg/Jülich
Ackermann B and Feder R 1984 Solid State Sommun. 49 489 -
Ackermann B, Feder R and Tamura E 1984 J. Phys. F 14 L173
Ackermann B and Feder R 1985a J. Phys. C 18 1093
Ackermann B and Feder R 1985b Solid State Commun. 54 1077
Ackermann B and Feder R 1986 J. Phys. C
Akhiezer I A and Chudnowskii E M 1978 Physic Letters 65 A 433
Almbladh C O and Barth U V 1985 Phys. Rev. B 31 3231
Alvarado S F 1982 private communication
Alvarado S F, Feder R, Hopster H, Ciccacci F and Pleyer H 1982
    Z. Physik B 49 129
Alvarado S F and Weller D 1984 Verhandl. DPG Frühjahrstagung,
    and to be published
Awe B, Kemper F, Rosicky F and Feder R 1983 J. Phys. B 16 603
Baribeau J M, Lopez J and Le Bossé J C 1985 J. Phys. C 18 3083
Barton J J, Bahr C C, Hussain Z, Robey S W, Tobin J G,
    Klebanoff L E and Shirley D A 1983 Phys. Rev. Lett. 51 272
Bell R L 1973 Negative Electron Affinity Devices, Clarendon, Oxford
Benbow R L and Smith N V 1983 Phys. Rev. B 27 3144
Beni G, Lee P A and Platzmann P M 1976 Phys. Rev. B 12 5170
Bethe H A and Salpeter E E 1957 in ed. S. Flügge "Handbook of
    Physics" vol. 35 (Springer, Heidelberg)
Bir G L, Aronov A G and Pikus G E 1976 Sov. Phys. - JETP 42 705
Blount E I 1962 in ed. F. Seitz and D . Turnbull "Solid State Physics"
    (Academic, New York) 13 305
Blum K 1978 in ed. W. Hanle and H. Kleinpoppen "Progress in
    Atomic Spectroscopy" (Plenum, New York)
Born M and Wolf E 1980 Principles of Optics, Pergamon, Oxford
Borstel G, Neumann M and Wöhlecke M 1981 Phys. Rev. B 23 3121
Borstel G and Wöhlecke M 1982 Phys. Rev. B 26 1148
Breit G and Bethe H A 1954 Phys. Rev. 93 888
Bringer A, Campagna M, Feder R, Gudat W, Kisker E and Kulhmann E
    1979 Phys. Rev. Lett 42 1705
Callaway J 1976 Quantum Theory of the Solid State,
    Academic Press, New York
Callaway J and March N H 1984 in "Solid State Physics" ed.
    D Turnbull and H. Ehrenreich (Academic, New York) vol. 38 135
Cardona M and Ley L (eds.) 1978 Photoemission in Solids
    (Springer, New York)
Caroli C, Roulet B and Saint-James D 1978 in ed. L. Dobrzynski
    "Handbook of Surfaces and Interfaces", Gartland, New York
    & London
Caroli C, Lederer-Rozenblatt D, Roulet B and Saint-James D 1973
    Phys. Rev. B 8 4552
Castiel D 1976 Surface Sci. 60 24
Cherepkow N A 1978 J. Phys. B 11 L435
Cherepkow N A 1979 J. Phys. B 12 1279
Chung M S and Everhart T E 1974 J. Appl. Phys. 45 707
Clauberg R, Haines E and Feder R 1985 Z. Physik

Condon E U and Shortley G H 1977 The Theory of Atomic Spectra,
    Cambridge University Press, London
Cooke J F, Lynn J W and Davis H L 1980 Phys. Rev. B 21 4118
Cooke J F, Blackman J A and Morgan T 1985 Phys. Rev. Lett. 54 718
Cornwell J F 1969 Group Theory and Electronic Energy Bands in Solids,
    North Holland, Amsterdam
Courths R and Hüfner S 1984 Physics Reports 112 53
Dahl J P and Avery J 1984 (eds) "Local Density Approximations in
    Quantum Chemistry and Solid State Physics"
    (Plenum Press, New York)
Davis H L and Noonan J R 1983 Surface Sci. 126 245
Demangeat C, Mills D L and Trullinger S E 1977 Phys. Rev. B 16 522
De Wames R E and Vredevoe L A 1967 Phys. Rev. Lett. 18 853
De Wames R E and Vredevoe 1969 Phys. Rev. Lett.23 (1969) 123
Dirac P A M 1930 Proc. Camb. Phil. 26 376
Dose V 1983 Progress in Surface Sci. 13 225
Dose V 1984 J. Phys. Chem. 88 1681
Duke C B 1974 Adv. Chem. Phys. 27 1
Dunlap B I 1980 Solid State Commun. 35 141
Durham P J, Staunton J and Gyorffy B L 1984 J. Magn. Magn. Mater. 45 38
Dyakonov M L and Perel V I 1971 Soviet Physics JETP 33 1053
Eastman D E 1972 in ed. R.F. Bunshah "Metals", Wiley
Echenique P M and Titterington D J 1977 J. Phys. C 10 625
Eckardt H, Fritsche L and Noffke J 1984 J. Phys. F 14 97
Eckardt H, Fritsche L and Noffke J 1985 J. Phys. F
Elliot R J 1954 Phys. Rev. 96 266
Fadley C S 1984 Progr. Surface Sci.
Fano U 1969 Phys. Rev. 178 131
Feder R 1977 Surface Sci. 63 283
Feder R 1977b Solid State Commun. 21 1091
Feder R 1978 Solid State Commun. 28 27
Feder R and Pendry J B 1978 Solid State Commun. 26 519.
Feder R 1979 Solid State Commun. 31 821
Feder R 1980 Phys. Letters 78 A 103
Feder R 1981 J. Phys. C 14 2049
Feder R and Kirschner J 1981 Surface Sci. 103 75
Feder R and Kirschner J 1981b Sol. State Commun. 40 547
Feder R, Pleyer H, Bauer P and Müller N 1981 Surface Sci. 109 419
Feder R and Pleyer H 1982 Surface Sci. 117 285
Feder R, Alvarado S F, Tamura E and Kisker E 1983a Surface Sci. 127 83
Feder R, Rosicky F and Ackermann B 1983b Z. Phys. B 52 31 and B 53 144
Feder R, Gudat W, Kisker E, Rodriguez A and Schröder K 1983c
    Solid State Commun. 46 619
Feder R, Rodriguez A, Baier U and Kisker E 1984 Solid State Commun.
    52 57
Feder R, Awe B and Tamura E 1985 Surface Sci. 157 183
Feibelman P J and Eastman D E 1974 Phys. Rev. B 10 4932
Feibelman P J 1982 Prog. in Surf. Sci. 12 287
Fetter A L and Walecka J D 1971 Quantum Theory of Many-Particle
    Systems, McGraw-Hill New York

238

Feuchtwang T E, Cutler P H and Schmit J 1978 Surf. Sci. $\underline{75}$ 401

Feuerbacher B, Fitton B and Willis R F (eds.) 1978: Photoemission and the Electronic Properties of Surfaces (Wiley, New York)

Ford W K, Duke C B and Paton A 1982 Surf. Sci. $\underline{115}$ 195

Gadzuk J W 1978 in ed. B. Feuerbacher, B Fitton and R.F. Willis "Photoemission and the Electronic Properties of Surfaces" (Wiley, New York)

Gaspar R 1954 Acta Phys. Hung. $\underline{3}$ 263

Ginatempo B, Durham P J, Gyorffy B L and Temmerman W M 1985 Phys. Rev. Lett. $\underline{54}$ 1581

Gauthier Y. Baudoing R, Joly Y, Gaubert C and Rundgren J 1984 J. Phys. C $\underline{17}$ 4547

Glazer J and Tosatti E 1984 Solid State Commun. $\underline{52}$ 905

Glazer J and Tosatti E, Hopster H, Kurzawa R, Schmitt W, Walker K H and Güntherodt G 1985 to be published

Goldmann A, Westphal D and Courths R 1982 Phys. Rev. B 25 2000

Goldmann A, Rodriguez A and Feder R 1983 Solid State Commun. $\underline{45}$ 449

Gunnarsson O and Lundqvist B I 1976 Phys. Rev. B $\underline{13}$ 4274

Gunnarsson O and Schönhammer K 1982 Phys. Rev. B $\underline{26}$ 2765

Gyorffy B L and Stocks G M 1977 in "Electrons in Finite and Infinite Structures" eds. P. Phariseau and L.Scheire(Plenum, New York)

Gyorffy B L, Kollar J, Pindor A J, Stocks G M, Staunton J and Winter H 1984 in "The Electronic Structure of Complex Systems" ed. W. Temmerman and P. Phariseau (Plenum, New York)

Haines E M, Clauberg R and Feder R 1985a Phys. Rev. Lett. $\underline{54}$ 932

Haines E M, Heine V and Ziegler A 1985b J. Phys. F $\underline{15}$ 661

Haines E M, Clauberg R, Tamura E and Feder R 1985c Solid State Commun.

Hedin L and Lundqvist S 1969 Solid State Physics $\underline{23}$ 1

Hedin L and Lundqvist B I 1971 J. Phys. C $\underline{4}$ 2064

Heinz K and Müller K 1982 in "Structural Studies of Surfaces" (Springer, Heidelberg)

Heinzmann U, Schönhense G and Kessler J 1980 Phys. Rev. Lett. $\underline{42}$ 1603

Helmann J S and Baltensperger W 1980 Phys. Rev. B $\underline{22}$ 1300

Hermanson 1977 Solid State Commun. $\underline{22}$ 9

Hodges L, Ehrenreich H and Lang N D 1966 Phys. Rev. $\underline{152}$ 505

Hopkinson J F L, Pendry J B and Titterington D J 1980 Comp. Phys. Commun. $\underline{19}$ 69

Huang K N 1980 Phys. Lett. $\underline{77}$A 133

Ibach H and Mills D L 1982 "Electron Energy Loss Spectroscopy and Surface Vibrations (Academic Press, New York)

Ing B S and Pendry J B 1975 J. Phys. C $\underline{8}$ 1087

Inkson J C 1984 "Many-Body Theory of Solids" (Plenum Press, New York)

Jackson J D 1962, 1975 "Classical Electrodynamics" (John Wiley, New York)

James L W and Moll J L 1969 Phys. Rev. $\underline{183}$ 740

Jennings P J and Thurgate S M 1981 Surface Sci. $\underline{104}$ L210

Jepsen D W 1980 Phys. Rev. B $\underline{22}$ 5701

Jona F, Strozier J A and Yang W S 1982 Rep. Prog. Phys. $\underline{45}$ 527

Jones R O, Jennings P J and Jepsen O 1984 Phys. Rev. B $\underline{29}$ 6474

Keldysh L V 1965 Sov. Phys. JETP $\underline{20}$ 1018

Kemper F, Awe B, Rosicky F and Feder R 1983 J. Phys. B 16 1819
Kirschner J 1984a Phys. Rev. B 30 415
Kirschner J 1984b Solid State Commun. 49 39
Kirschner J and Feder R 1981 Surface Sci. 104 448
Kirschner J, Feder R and Wendelken J K 1981 Phys. Rev. Lett. 47 614
Klar H 1980 J. Phys. B 13 3117
Kleinherbers K K, Goldmann A, Tamura E and Feder R 1984
        Solid State Commun. 49 735
Kleinman L 1978 Phys. Rev. B 17 3666
Koelling D D 1969 Phys. Rev. 188 1049
Kohn W and Sham L J 1965 Phys. Rev. 140 A1133
Koyama K 1975 Z. Physik B 22 337
Lee P A and Beni G 1977 Phys. Rev. B 15 2862
Levy J C S 1981 Surface Sci. Reports 1 39
Liebsch A 1979 Festkörperprobleme XIX 209
Liebsch A 1981 Phys. Rev. B 23 5203
Lindgren S A, Walldén L, Rundgren J and Westrin P 1984
        Phys. Rev. B 29 576
Ling D T, Miller J N, Weissman D L, Pianetta P, Stepan P M,
        Lindau I and Spicer W E 1983 Surface Sci. 124 175
Lundqvist S and March NH 1983 (eds.) "Theory of the Inhomogeneous
        Electron Gas"(Plenum, New York)
Mahan G D 1981 "Many-Particle Physics" (Plenum Press, New York)
Malli G L 1983 (ed.) "Relativistic Effects in Atoms, Molecules and
        Solids" (Plenum Press, New York)
Marcus P M and Jona F 1984 (eds.) "Determination of Surface
        Structure by LEED" (Plenum Press, New York)
Mathon J 1981 Phys. Rev. B 24 6588
Matthew J A D 1982 Phys. Rev. B 25 3326
McLean R and Haydock R 1977 J. Phys. C 10 1929
Meier F and Zakharchenya B P 1984 (eds.) "Optical Orientation"
        (North Holland, Amsterdam)
Meister H J and Weiss H F 1968 Z. Physik 216 165
Messiah A 1965 "Quantum Mechanics" (North Holland, Amsterdam)
Mills D L 1967 J. Phys. Chem. Solids 28 2245
Moriya T and Takahashi Y 1984 Ann. Rev. Materials Sci. 14 1
Mueller F 1967 Phys. Rev. 153 659
Nesbet R K 1985 Int. J. of Quantum Chemistry, Symposium
Neve J, Westrin P and Rundgren J 1983 J. Phys. C 16 1291
Oepen H P 1985 Ph. D. Thesis, Jül-Report Nr. 1970, ISSN 0336-0885
Oles A M and Stollhoff G 1984 Phys. Rev. B 29 314
Pendry J B 1974 "Low Energy Electron Diffraction" (Academic,
        New York)
Pendry J B 1976 Surf. Sci. 57 679
Pendry J B 1981 J. Phys. C 14 1381
Pendry J B 1985 private communication
Penn D R 1980 Phys. Rev. B 22 2677
Persson B N J and Zaremba E 1985 Phys. Rev. B 31 1863
Petrillo C and Sacchetti F 1984 Solid State Commun. 50 521
Pickett W E and Wang C S 1984 Phys. Rev. B 30 4719

Pierce DT, Celotta R J, Wang G C, Unertl W N, Galejs A, Kuyatt C E
    and Mielczarek S R 1980 Rev. Sci. Instrum. 51 478
Pierce D T and Celotta R J 1981 Adv. El. and El.-Physics 56 219
Plummer E W and Eberhardt W 1982 Adv. Chem. Phys. 49 533
Przybylski H, Baalmann A, Borstel G and Neumann M 1983
    Phys. Rev. B 27 6669
Quinn J J and Ferrell R A 1958 Phys. Rev. 112 812
Raether H 1980 "Excitation of Plasmons and Interband Transitions
    by Electrons" (Springer Berlin)
Raith W 1984 in "Electronic and Atomic Collisions" eds. J. Eichler,
    I.V. Hertel and N. Stolterfoht (North Holland, Amsterdam)
Ramana M V and Rajagopal A K 1983 Adv. Chem. Phys. 54 231
Rasolt M and Davis H L 1979 Phys. Rev. B 20 5059
Rasolt M and Davis H L 1980 Phys. Rev. B 21 1445
Ravano G and Erbudak M 1982 Solid State Commun. 44 547
Rendell R W and Penn D R 1980 Phys. Rev. Lett. 45 2057
Riechert H, Alvarado S F, Titkov A N and Safarow V I 1984
    Phys. Rev. Lett. 52 2297
Ritchie R H and Ashley J C 1965 J. Phys. Chem. Solids 26 1689
Ritter A L 1985 Bull. Am. Phys. Soc. 30 219
Roman P 1965 "Advanced Quantum Theory" (Addison-Wesley, New York)
Rose M E 1961 "Relativistic Electron Theory" (John Wiley, New York)
Sacchetti F 1982 J. Phys. F 12 281
Sacchetti F 1983 J. Phys. F 13 1801
Saldin D K, Pendry J B, Van Hove M A and Somorjai G A 1985
    Phys. Rev. B 31 1216
Schäfer J, Schoppe R, Hölzl J and Feder R 1981 Surface Sci.107 290
Schaich W and Ashcroft N W 1971 Phys. Rev. B 3 2452
Schilling J S and Webb M B 1970 Phys. Rev. B 2 1665
Schönhense G, Eyers A, Friess U, Schäfers F and Heinzmann U 1985
    Phys. Rev. Lett 54 547
Selzer S and Majlis N 1982 Phys. Rev. B 26 404
Sham L J and Kohn W 1966 Phys. Rev. 145 561
Shevchik N J 1977 Phys. Rev. B 16 3428
Siegmann H C, Pierce D T and Celotta R J 1981 Phys. Rev. Lett. 46 452
Siegmann H C, Meier F, Erbudak M and Landolt M 1984
    Adv. El. El. Phys. 62 1
Slater J C, Wilson T M and Wood J W 1969 Phys. Rev. 179 28
Smith N V and Mattheiss L F 1974 Phys. Rev. B 9 1341
Smith N V 1979 Phys. Rev. B 19 5019
Strange P, Staunton J and Gyorffy B L 1984 J. Phys. C 17 3355
Sturm K 1982 Adv. Phys. 31 1
Tamura E and Feder R 1982 Solid State Commun. 44 1101
Tamura E and Feder R 1983 Vacuum 33 864
Tamura E, Ackermann B and Feder R 1984 J. Phys. C 17 5455
Tamura E and Feder R 1985 Solid State Commun.
Tamura E, Feder R, Krewer J, Kirby R E, Kisker E, Garwin E L and
    King F K 1985 Solid State Commun 55 543
Tran Thoai D B and Ekardt W 1981 Solid State Commun. 40 269
Treglia G, Ducastelle F and Spanjaard D 1982 J. Physique 43 341

Unguris J, Pierce D T and Celotta R J 1984 Phys. Rev. B 29 1381
Van Hove M A and Tong S Y 1979 "Surface Crystallography by LEED"
    (Springer, Berlin-Heidelberg-New York)
Wandelt K 1982 Surface Sci. Reports 2 1
Wang C S, Klein B M and Krakauer H 1985 Phys. Rev. Lett. 54 1852
Westphal D and Goldmann A 1983 Surface Sci. 131 113
White R M 1970 "Quantum Theory of Magnetism"(McGraw Hill, New York)
Williams R H, Srivasta G P and McGovern I T 1980
    Rep. Prog. Phys. 43 1357
William A R and Barth U V 1983 in "Theory of the Inhomogeneous
    Electron Gas" ed. by S. Lundqvist and N.H. March
    (Plenum, New York)
Wöhlecke M and Borstel G 1981a Phys. Rev. B 23 980
Wöhlecke M and Borstel G 1981b Phys. Rev. B 24 2857
Wöhlecke M and Borstel G 1984 in "Optical Orientation" ed. by
    Meier F and Zakharchenya B P (North Holland, Amsterdam)
Wohlfarth E P 1984 J. Magn. Magn. Mater. 45 1
Wolff P A 1954 Phys. Rev. 95 56
Woolfson M S, Gurman S J and Holland B W 1982 Surface Sci. 117 450
Yafet Y 1963 Solid State Physics 14 1 (ed. F. Seitz and D. Turnbull)
Yin S and Tosatti E 1981 International Centre for Theoretical Physics,
    Trieste, Report IC/81/129
You M V and Heine V 1982 J. Phys. F 12 177

Part II
Experiments and Results

Chapter 5

Sources and Detectors
for Polarized Electrons

J. Kirschner

Institut für Grenzflächenforschung und Vakuumphysik,
KFA Jülich, Postfach 1913, D-5170 Jülich, FRG

5.1   Sources of Polarized Electrons

5.1.1 Introduction

Much of the progress in surface and solid state physics with spin-polarized electrons over the past decade is due to the development of sources and detectors of free spin-polarized electrons. In this chapter on sources and the subsequent one on detectors some of the more recent developments will be discussed in detail. According to the general topic of the book the emphasis will be on devices which are easily compatible with standard Ultra High Vacuum (UHV) conditions since this is the preferred medium for surface studies. A considerable variety of sources has been developed in atomic physics which shall not be treated in detail here. We describe briefly their operating principles and give some references for the reader interested in more detail. A method used very early is to produce a beam of polarized alkali atoms (e.g. in a Stern-Gerlach type), to ionize the atoms by photons, and to collect the photoelectrons. The light may be unpolarized but the degree of polarization may be as high as 85 % with Li (Alguard et al. 1979). Since the discovery of the Fano effect it became apparent that the alkali atoms need not be polarized if instead the ionizing light is circularly polarized. The spin-polarization stems from spin-orbit coupling in the continuum states of the photoelectrons. Thus, the effect works best with heavy alkalis, such as Rb (Drachenfels et al. 1977) and Cs (Wainwright et al. 1978). The polarization is around 65 % in practical sources. The reference axis is the photon momentum and the polarization can be reversed easily by reversing the helicity of the light beam. An approach different from photoionization is chemiionization of aligned metastable atoms in a collision with other molecules. A microwave discharge is used to generate metastable He atoms in a flowing discharge afterglow. The He atoms are optically pumped by circularly polarized resonance radiation in order to populate the upper level preferentially with electrons of one spin orientation. In a collision with an injected target gas, such as CO_2, the excited electrons may be released while preserving their spin orienta-

tion and may be collected into a beam (Hodge et al. 1979). Spin polarization values of 40 % at 50 μA beam current have been reported (Gray
et al. 1983). Though polarization and current are two important criteria, others may be equally important for a particular application:
e.g. pulsed or continuous mode, initial energy spread of the electrons
and size of the source volume. For a discussion of these parameters
the reader is referred to articles by Wainrwright et al. 1978, Kessler
1979, Pierce et al. 1980, and Celotta and Pierce 1980. All the above
sources have in common that they operate in high vacuum. This may be
an advantage or a disadvantage depending on the type of experiment to
be done with the source. Since atomic physics experiments normally do
not require UHV these sources are well adapted, while for surface
physics experiments UHV is mandatory and the solid state sources to be
discussed below are suitable. Differential pumping is always possible,
though, and there are a few notable exceptions from the rule (see
Pierce and Celotta 1984, and references therein). In the following two
solid state sources are discussed. The one based on photoemission from
GaAs and its derivates has already found widespread application. The
other one based on field emission from ferromagnetic layers is less
frequently used but offers some unique features which might become important in the future.

5.1.2 Photoemission Sources

Photocathodes have encountered growing interest as electron
sources, also for non-polarized electron beams (Lee 1984, Lee et al.
1985). In connection with laser light sources they offer particular
advantages such as high current density, arbitrary beam shape, and
high electron-optical brightness, which may even exceed that of LaB_6
cathodes (Lee et al. 1985). As an additional bonus the electron beam
may be spin-polarized for a particular class of materials if the incident light is circularly polarized. The operating principle relies on
the action of selection rules for the 'magnetic' quantum number m_1 in
dipole transitions between electronic bands in materials with significant spin-orbit coupling. Because of the particular coupling of spa-

tial and spin parts in the wavefunctions of either the initial or the final state of the photoexcitation process the absorption of a photon may lead to selective excitation of the one or the other spin state in the (spin-degenerate) occupied bands. In GaAs this effect has been known for quite some time from luminescence studies with bulk material, observing the recombination radiation after exciting electrons across the direct band gap with circular polarized light. It was suggested by Garwin et al. 1974 and by Lampel and Weisbuch 1975, that this effect could be used as a polarized electron source if the electrons excited into the conduction band could be made to leave the crystal. Fortunately with GaAs (and many other substances, see Bell 1973) the work function may be lowered by a special surface treatment to such an extent that the vacuum level may lie below the minimum of the conduction band ("negative electron affinity", NEA). Then conduction band electrons may tunnel through the surface barrier into vacuum without any further rise in energy. This concept has been shown to work by Pierce et al. 1975 and the basic studies have been made by Pierce and Meier 1976 who showed that the electrons maintain their spin polarization to a large extent even when they travel many hundred nanometers towards the crystal surface. At present there are a few dozens of sources in operation, in a wide variety of applications (Pierce and Celotta 1984), as is also evidenced by this book.

The principle of the spin-selective excitation is visualized in Fig. 1 in simplified form. A section of the bandstructure of GaAs around the Γ point is shown in Fig. 1a, which is also approximately valid for GaAsP and GaAlAs. The conduction band edge is at Γ_6, separated from the valence band edge at Γ_8 by the gap energy E_g. The topmost valence band is doubly degenerate only since the spin-orbit coupling leads to an energy splitting Δ between Γ_8 and Γ_7 (in GaAs E_g = 1.42 eV at 300 K and Δ = 0.34 eV). For an estimate of the spinpolarization and of the relative transition probabilities into the conduction band a detailed knowledge of the wavefunction symmetries is necessary in principle. This is treated in detail by F. Meier in this book. It turns out that at the Γ point, with its high spatial symme-

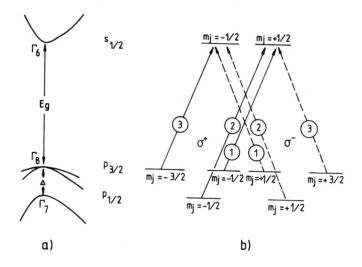

Fig. 1: Section of the bandstructure of GaAs around the Γ point (a)
and transitions induced by right-circular (σ^+) and left-circu-
lar light (σ^-) across the band gap. The calculated transition
rates lead to a polarization of ± 50 % for light with $h\nu = E_g$.

try, the wavefunctions can approximately be described by analogy to
the atomic physics nomenclature. Thus, the valence band edge at Γ_8
corresponds to a $p_{3/2}$ level while the conduction band edge corresponds
to an atomic level of s symmetry. The possible transitions with right
hand circular polarized light (σ^+) and left hand polarized light (σ^-)
are listed in Fig. 1b) together with relative transition probabilities. Each of the $s_{1/2}$ and $p_{1/2}$ levels is doubly degenerate, while the
$p_{3/2}$ level is fourfold degenerate. If we now apply the selection rules
for dipole transitions with circular polarized light ($\Delta\ell = \pm 1$, $\Delta m_j =$
± 1) we see that the parity selection rule is fulfilled by $\Delta\ell = -1$ for
excitations into the conduction band. For σ^+ light ($\Delta m = +1$) only
transitions $m_j = -3/2 \rightarrow m_j = -1/2$ or $m_j = -1/2 \rightarrow m_j = +1/2$ are allowed. For $h\nu = E_g$, i.e. band gap excitation, the evaluation of the angu-

lar parts of the wavefunctions yields a relative transition strength of 3 in the former case and of 1 in the latter case. Thus the upper $s_{1/2}$ level is populated preferentially with electrons of negative spin (relative to the photon momentum) but the spin polarization is incomplete:

$$P = \frac{3-1}{3+1} = 50 \ \%$$

This is due to the degeneracy of the $p_{3/2}$ level around the Γ point. If we had used light of higher energy, such that also transitions from the split-off band at Γ_7 become possible, the polarization of the total electron assembly in the conduction band would be close to zero, since the transitions $\Gamma_7 \rightarrow \Gamma_6$ occur with relative intensity 2, contributing electrons of the 'wrong' spin orientation. Thus we see that it is important to use a light source matched to the band gap and sufficiently monochromatic not to excite the split-off band. A feature of great practical importance is that the relative transition rates are exactly the same for light of opposite helicity (dashed arrows in Fig. 1b). This means that the spinpolarization of the source is reversed by reversing the light helicity, while the total beam current remains exactly the same (in principle). The light helicity can be inverted easily by electro-optical or electro-mechanical bi-refringent modulators outside the vacuum chamber. Since this can be done repeatedly with a fixed frequency lock-in techniques become applicable which greatly enhances the sensitivity. Thus spin-dependent intensity measurements with a relative sensitivity of 10^{-4} or less are feasible. There is a certain 'apparatus asymmetry' though since the optical modulation might lead to minute changes of the light intensity or of the position of the light spot on the photocathode which in turn might affect the position of the electron beam on the sample. It is left to the experimenter's art to suppress these effects as much as necessary. The feature of easy spin reversal without grossly affecting the electron optics of the polarized beam is one of the major advantages of the photoemission sources.

252

As mentioned above, the work function has to be lowered in order
to have the excited electrons available in vacuum. For this purpose
the crystal first is heavily p-doped, which pulls the Fermi level down
towards the valence band edge and produces a downward band bending in
the vicinity of the surface (to a depth of the order of 10 nm). This
alone is not sufficient to pull the vacuum level below the conduction
band, but the condensation of small quantities of alkali atoms on the
surface does give the desired effect. Mostly Cs is used, but others do
as well. The outermost electron is transferred in part to the substrate and the resultant surface dipole helps electrons to overcome
the surface barrier. It was found empirically that small amounts of
oxygen enhance this effect and also provide temporal stability of the
activated layer. The precise nature of the activation layer and of the
activation mechanism are still not very well known, and are the subject of current research. For example Su et al. 1983 pointed out that
a layer of oxygen bonded to GaAs might explain a yield improvement
upon a particular two-step activation procedure. The thickness of the
activation layer is estimated to be around 0.5 - 1 nm, with an interfacial potential barrier close to the GaAs surface. In Fig. 2 we

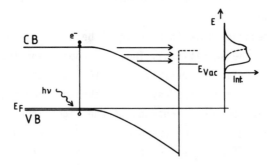

Fig. 2: Schematic potential diagram of strongly p-doped GaAs near the
surface. On the right hand side the energy distribution of
emitted electrons is shown for different amounts of negative
electron affinity.

sketch a strongly simplified potential model for the activated surface where we neglect the thickness of the activation layer relative to the bent-band region. Electrons diffusing from the bulk may directly escape into vacuum when NEA conditions are established, but since the transmission through the interface barrier is incomplete, they might be reflected back towards the bulk. On their way they may loose energy to optical phonons (of the order of 30 meV) and will be turned around in the bent-band region. Thus, an electron may travel back and forth many times until it escapes if its energy is still high enough. The energy distribution of the electrons in vacuum therefore shows a large inelastic contribution, which is cut off at low energies by the work function level (see Fig. 3 below).

The activation generally is preceded by a cleaning procedure, except for a few cases where the cathode is fabricated in situ by molecular beam epitaxy (Ciccacci et al. 1982, Alvarado et al. 1981). Before mounting in the vacuum chamber chemical etching is applied (a recipe may be found in Pierce et al. 1980). The final cleaning must be carried out under UHV conditions by heating close to the melting point, when Ga and As just start to evaporate. This requires a precise temperature control and is the most critical step in the activation procedure. If this fails because of too strong carbon or oxygen contamination a mild ion sputtering may be applied. This removes the contamination, but introduces damage wich reduces the photoemission sensitivity. The damage can be removed by annealing at $450^{\circ}C$ to $500^{\circ}C$ to a large extent, though not completely. Optimum sensitivity cathodes will perhaps not be obtained in this way, but the drawback of changing the wafer and repeating the whole vacuum preparation and cleaning procedure may be avoided. Details on the sputtering procedure may be found in Kirschner et al. 1983 and references therein. The activation is done first by applying Cs, most conveniently from a Cs-dispenser, until a certain saturation is reached. Then oxygen gas is added in intervals until maximum emission is reached. After switching off a well prepared cathode is observed to remain stable for a long time in good vacuum conditions. Decay time constants of more than 10 h (Pierce et

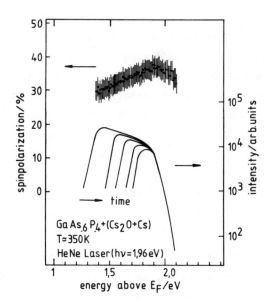

Fig. 3: Intensity distributions on a logarithmic scale as a function
of time after activation (at T ≈ 350 K; HeNe laser). The
shoulder at the right-hand side corresponds to emission from
the conduction band, the left-hand edge represents the low energy cut-off due to the work function. The spin polarization
distribution (linear scale) corresponds to the freshly activated cathode. The intensity distributions have been taken at
different time intervals.

al. 1980, Ciccacci et al. 1982) and even 100 h (Stocker 1975) have
been obtained. Reactivation can be carried out many times by repeating
the cesiation/oxygenation procedure.

From a practical point of view GaAs is not ideal as a source
material since the photon energy corresponding to the band gap lies in
the infra red. Though suitable solid state laser diodes are available,
the handling of the invisible and not well collimated light is not

easy. Alternatively the band gap may be enlarged by partly replacing As by P in $GaAs_{1-x}P_x$ compounds (Conrath et al. 1979, Reichert and Zähringer 1982). Up to $x \sim 0.45$ the compound remains a direct gap semiconductor and the behaviour as a source is very similar to that of GaAs, but the band gap is well matched to the HeNe laser light ($h\nu$ = 1.96 eV). These lasers are easy to handle, are relatively cheap, and yield a well collimated and highly polarized light beam. It should be noted that the same effect can be obtained by adding Al in $Al_xGa_{1-x}As$ compounds, though the Al is fairly reactive and requires special precautions (Alvarado et al. 1981).

The energy- and spinpolarization distributions of photoemitted electrons from $GaAs_{0.6}P_{0.4}$ (100) have been studied by Kirschner et al. 1983 under conditions of typical source operation. Results are shown in Fig. 3 for intensity (log. scale) and spin polarization (linear scale). The energy distribution is characterized by two edges, as mentioned above, of which the one stays near the conduction band edge while the low energy edge corresponds to the vacuum level and is seen to move towards higher energy with time. This corresponds to a reduction of the amount of negative electron affinity (by contamination of the activation layer in this case). The polarization distribution has a peak near the conduction band edge (1.86 ± 0.02 eV) and decays slightly towards both sides. The polarization is lower than the theoretical limit of 50 % for two reasons. First, at the operating temperature the laser energy is somewhat higher than the bandgap which causes less-polarized electrons to be excited off the Γ point and even into the X-valley. This is also obvious from the slight decay of polarization at the high energy edge of the intensity distribution. If the matching of band gap energy to photon energy is better (e.g. by enlarging the band gap upon cooling) there is no such decay and the peak polarization is around 44 %. Secondly, the electrons undergo depolarization on their way from the bulk to the surface mainly by two competing mechanisms. These are treated in detail at some other place in this book and shall only briefly be mentioned. The one is elastic electron-hole exchange interaction (Bir et al. 1975), the other one is due to the spin splitting of the conduction band outside Γ. As proposed by D'yakonov and Perel' 1971 (DP) the effect of this splitting on kinetic electrons can be viewed as the action of an effective magnetic field around which the electron spin processes. The first mechanism operates on thermalized electrons in the Γ valley and is thought to be the main reason for the reduction of the polarization of most emitted electrons. A further depolarization may occur in the escape process at or near the surface (Allenspach et al. 1984). The slight decay of the polarization towards low energies is not yet completely understood, but there seems to be agreement that it occurs while the electrons are temporarily trapped in the band bending region near the surface (Kirschner et al. 1983, Drouhin et al. 1985). Recently it was found by Riechert et al. 1984 with GaAs(110) that the polarization vector of the total yield was rotated away from the surface normal by values of the order of 10^0, which was successfully explained in terms of the DP mechanism. The rotation effect is absent for (100) and (111) surfaces, while a reduction of the length of the P vector might well occur (Drouhin et al. 1985). The above polarization measurements refer to GaAsP, but rather similar results have been found by Drouhin et al. 1985 for GaAs at near band gap excitation, taking the smaller amount of NEA into account.

Returning to the intensity distribution in Fig. 3 we note that the width may be rather large for strong NEA, and rather small for near zero electron affinity. The intensity loss in the latter case is substantial but still rather high emission currents may be obtained with a suitable laser source. This opens a new way to produce highly monochromatic intense electron beams (which need not be polarized for many applications). The conventional way is to monochromatize electrons from a thermal source in an electrostatic energy filter, which, however, is severely limited by space charge effects. It was demonstrated by Feigerle et al. 1984 that from GaAs a current of 1 µA at a full width at half maximum of 31 meV may be obtained. As seen in Fig. 4 this is at least one order of magnitude higher than what could be obtained by conventional means. It is certainly not allowed to extra-

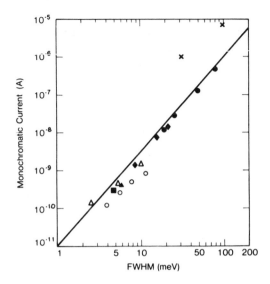

<u>Fig. 4:</u> Cathode current and distribution width of the EDC's obtained
from photoemission from NEA GaAs(x) are shown compared to that
available by coupling thermionic cathodes with electron monochromators. The data points which are shown as solid symbols
are from systems using spherical deflector monochromators,
while those shown as open symbols come from systems using cylindrical deflector monochromators. The solid sline follows
$I\alpha(\Delta E)^{5/2}$. For references on the data points see Feigerle et
al. 1984.

polate these data down to very small width, but if one used a photocathode in conjunction with a monochromator the space charge problems
could be reduced dramatically and much higher output currents might be
obtained.

At the end of this chapter an actual example for a source of
polarized electrons shall be given (see Fig. 5). Unlike other designs
published in the literature this source is particularly simple and

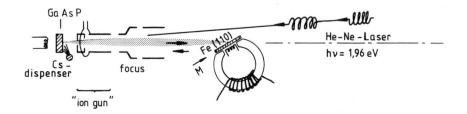

Fig. 5: A simple high intensity polarized electron source based on
photoemission from GaAsP. Up to 20 µA target current at 10 eV
have been obtained with a 15 mW laser. The electron beam is
polarized longitudinally. The distance from cathode to target
is about 15 cm.

compact and yields a longitudinally polarized electron beam (Kirschner
1984b). The cathode (GaAsP(100)) is heated by radiation from a filament and activated by means of a Cs-dispenser and oxygen gas. The circularly polarized light beam from a HeNe laser passes by the target
(rotated out of the paper plane) and hits the cathode at a few degrees
off normal. The emitted electrons are accelerated to the first anode
and focused onto the target by an electrostatic einzel lens. A part of
the electrode structure, together with a circular filament may be
electrically connected such as to form a simple ion gun which is used
to clean the photocathode if necessary. The electron beam polarization
vector is aligned with the direction of the laser beam and is nearly
coincident with the electron momentum. With 15 mW laser power emission
currents of more than 100 µA are routinely obtained. At low energies
not all of this can be brought to the target because of space charge
limitations. At 10 eV kinetic energy about 20 µA at the target have
been achieved. This source has been used in spin-polarized inverse
photoemission (Scheidt et al. 1983, Kirschner et al. 1984), spin-polarized LEED (Kirschner 1984b), inelastic electron scattering for the

investigation of Stoner excitations (Kirschner 1985c), and spin-polarized appearance potential spectroscopy (Kirschner 1984c).

5.1.3 The Field Emission Source

We found in the above chapter that photoemitters do have a good electron-optical brightness, which may even exceed that of conventional thermal electron sources. A source with high brightness emits a large number of electrons with small initial energy from a small spot size into a small solid angle. For spectroscopic purposes a small energy spread is also very often desirable. As we have seen GaAs may yield highly monochromatic electrons with low starting energy. The emitting area is ultimately limited by diffraction in the light optics (~ 1 μm $\emptyset$) but the limits to the actual source area have not yet been explored. The conventional electron source of the highest brightness of all is the field emission tip, primarily because of its extremely small virtual source size of the order of 10 nm. In this chapter we discuss polarized sources based on field emission from ferromagnetic materials, which offer high brightness together with high polarization, though they are not without drawbacks in practical use.

Already in 1930 it was suggested by Fues and Hellmann 1930 to extract polarized electrons directly from ferromagnetic materials by applying a strong electric field. While early experiments with iron tips were unsuccessful (von Issendorf and Fleischmann 1962, Pimbley and Mueller 1962) field emitted electrons from Gd were found to be polarized (Hofmann et al. 1967), though only in the 10 % range. Later experiments with Ni (Landolt and Campagna 1977) showed that Ni is not a suitable material since the polarization is too low (< 5 %). For selected low-index planes of Fe tips polarization values up to 25 % were observed by Landolt and Yafet 1978, which are, however, at variance with theoretical estimates predicting almost complete polarization. In other experiments by Chrobok et al. 1977 polarization up to 80 % was found. In these studies the field emission tip could not be characterized very well and the solid angle of emission was not well defined for electron-optical reasons. Thus the measured polarization data showed a large scatter (sometimes even sign reversals were found) depending on the pre-treatment of the tip. Producing clean iron field emitters is not an easy matter (Landolt and Yafet 1978) since iron wires are generally not sufficiently pure and rather fragile. Though the feasibility of this approach has been shown in principle, no actual source of polarized electrons has yet been built.

A different approach was successfully used by Müller et al. 1972 who covered a W tip by a thin layer of EuS. Below about 20 K this layer is ferromagnetic and spin polarization up to about 90 % was observed. It was pointed out that the ferromagnetic layer essentially acts as a spin filter for electrons tunneling from the Fermi level of the underlying W tip into the layer and from there into vacuum. The same system was subsequently studied by Baum et al. 1977 and by Kisker et al. 1978. From their work a simplified potential diagram of the tungsten tip with a ferromagnetic EuS layer of a few ten nm thickness and with an applied external field is reproduced in Fig. 6. The external field penetrates into the semiconducting layer and pulls down occupied and empty bands. The band bending at the interfaces is neglected here. Below the Curie temperature (16.6 K for bulk EuS) the conduction band is spin-split due to a local exchange interaction of the conduction electrons with the localized magnetic $4f^7$ states. Since the majority states are lower in energy than the minority states the internal potential barrier at the W-EuS interface is smaller for majority type electrons in W than for minority type electrons. Since the tunneling current depends exponentially on the barrier area the tunneling of majority electrons is strongly favoured over that of minority electrons and the internal barrier acts as a very efficient spin-filter. The external EuS-vacuum barrier is lowered by the electrostatic image potential so that conduction electrons inside the EuS layer can escape into vacuum if they do not experience strong energy losses. From this simple picture a spin polarization of 100 % should be expected and about 90 % at T ~ 10 K have been observed, indeed. As pointed out by Müller et al. 1972 and veryfied by model calculations

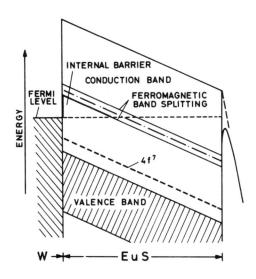

Fig. 6: Simplified band model for the W-EuS emitter. The applied external electric field penetrates into the insulating EuS layer
and causes. The emission current is governed by the internal
barrier which is of different heights for the two spin states
due to the ferromagnetic splitting of the EuS conduction band
(Baum et al. 1977).

(Nolting and Reihl 1979) the conduction band states are not pure spin
states due to spin-orbit coupling which lowers the degree of spin polarization of emitted electrons. When the layer thickness is much
smaller than assumed in Fig. 6, say only a few nm, the conduction
bands do no more cross the Fermi level. Even in this case high polarization (89 ± 7 %) was observed by Müller et al. 1972 which was attributed to tunneling assisted by the $4f^7$ states or direct tunneling from
these states. Since these are highly polarized below T_c a strong polarization should be expected.

The energy distribution of the emitted electrons was found to depend strongly on the annealing procedure applied to an evaporated film (Kisker et al. 1978). Besides a peak at the Fermi energy often a broad structure at lower energies is found which was attributed to energy losses in the film. For a rather narrow range of annealing temperatures around 840 K very narrow energy distributions of less than 100 meV FWHM can reproducibly be obtained. Since the polarization is high (~ 85 %) field emission through EuS under these conditions represents a source of highly polarized monochromatic electrons with very high brightness. The volume in phase space, expressed as the product of starting energy E, source area A and solid angle Ω, is $EA\Omega \sim 2 \cdot 10^{-9}$ eV·mm^2·sr compared to $EA\Omega \sim 4 \cdot 10^{-2}$ eV·mm^2·sr for the GaAs source (Pierce et al. 1980). One might wonder whether this gain of roughly 6 orders of magnitude in brightness should not be accompanied by a few disadvantages, and this is indeed the case. A schematic diagram of a source assembly is shown in Fig. 7 (from Kisker et al.

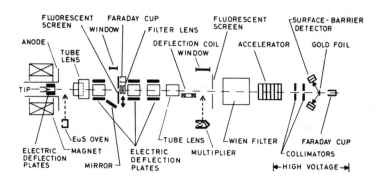

Fig. 7: Schematic view of a polarized electron source based on field emission from EuS, together with a Mott detector for spin polarization analysis. For explanation see text.

1978). The right hand side may be disregarded since this part of the apparatus refers to the polarization analysis by means of a Mott detector. On the left hand side we see the tip, surrounded by electric deflection plates and immersed into the magnetic field of an external solenoid. The tip has to be cooled to below 20 K which requires a liquid Helium supply. In front of the tip a retractable evaporation source of EuS may be inserted. After evaporation the tip is annealed which requires a rather precise temperature control to obtain optimum conditions. Subsequently the emission pattern may be imaged onto a fluorescent screen in combination with the electrostatic lenses. A particular part of the emission pattern may be steered through the probe hole and analyzed with respect to energy by means of a filter lens and with respect to polarization by Mott scattering. A certain technical complexity is apparent from the above and a good deal of experience is required to reproduce the tip and the operating conditions. A certain disadvantage of the source is that an external magnetic field is required in operation to stabilize the magnetization of the EuS layer. Polarization reversal of the source beam requires reversal of the magnetic field. Because of inevitable stray fields and alignment errors the electron beam path may change upon polarization reversal. It may require considerable efforts to prevent the beam from changing in intensity or to move on the sample which might introduce unwanted apparatus asymmetries. Finally, the obtainable current is limited to about 10^{-8} A, though about half a microampere has been obtained with some compromise on polarization and energy width (Kisker et al. 1978).

The limits of the field emission approach to polarized electron sources have not yet been fully explored and substantial improvements may be within reach. However, because of the current limitations and its relative technical complexity it will probably be most useful in applications where extreme brightness and high polarization are of primary concern.

5.2   Detectors for Polarized Electrons

5.2.1 Introduction

An ideal detector for polarized electrons should work like an optical polarization filter: the electrons with the desired spin orientation pass through while the others do not. For polarized atoms such a device does exist: the famous Stern-Gerlach experiment with inhomogeneous magnetic fields. Unfortunately for electrons this does not work since electrons have too small a mass-to-charge ratio. A particularly lucid description of why the Stern-Gerlach device does not work for electrons has been given by Kessler 1977. Thus, other means have to be found, and have been found, but all of them suffer from low efficiency, orders of magnitude worse than an ideal spin filter. In principle, most of the mechanisms which have been used for the production of polarized electrons could also be used for their detection by reversing the experiment. For example, if polarized electrons were injected into the conduction band of GaAs the recombination radiation would be circularly polarized. It would be an easy matter to detect the corresponding photons and their polarization by electron-optical means, but unfortunately the radiative recombination is of extremely low efficiency and no such device is known. The same problem of low intensity would be encountered if one were to use spin-polarized inverse photoemission (see the chapter by Dose in this book).

All detectors developed so far rely on electron scattering in one form or the other. The main spin-dependent scattering mechanisms are exchange interaction and spin-orbit coupling. The exchange interaction has been used in the so-called Møller scattering (Kessler 1977). It is based on the fact that in electron-electron scattering the ratio of the triplet to singlet scattering cross section has a pronounced minimum near a scattering angle of $45°$ (in the non-relativistic limit). Thus, in scattering of electrons from magnetized iron foils at the critical angle an intensity asymmetry is observed for polarized primary electrons when the magnetization is reversed. Since this effect occurs also at relativistic energies this scheme has been used in beta decay studies. Møller scattering suffers, however, from low spin polarization sensitivity since the surplus of majority electrons in ferromagnets is relatively small. Thus, in the vast majority of cases the operating principle of spin polarization detectors has been the spin-orbit coupling in scattering electrons from atoms with large atomic number. This is true for the traditional "Mott-detector" as well as for the modern "LEED detector" and the "adsorbed current detector" to be discussed below. We will give a more detailed discussion of the two latter devices, while for the former a brief account only is given of more recent developments since there exist a number of excellent reviews on this subject.

5.2.2 Mott Detectors

The basic phenomenon upon which the Mott detectors rely is demonstrated in Fig. 8, showing elastic scattering of 300 eV electrons from free Hg atoms. In the upper panel we see on a quasi-logarithmic scale the differential cross section as a function of scattering angle Θ for electrons with spin parallel to the normal to the scattering plane $(e^{\uparrow})$ or antiparallel to the normal $(e^{\downarrow})$. Both partial cross sections follow roughly the same oscillatory behaviour with angle but the one or the other may be slightly larger or smaller at certain angles. This is more clearly seen in the lower panel where the normalized difference $A(\Theta)$ of the partial cross sections is plotted:

$$A(\Theta) = \frac{\sigma^{\uparrow}(\Theta) - \sigma^{\downarrow}(\Theta)}{\sigma^{\uparrow}(\Theta) + \sigma^{\downarrow}(\Theta)}$$

This 'asymmetry' $A(\Theta)$ reaches values up to 80 % which means that the differential cross section for the one spin orientation is an order of magnitude larger than that for the other orientation. This effect is due to the coupling of the angular momentum of an electron with its spin during the scattering process. This gives rise to an additional scattering potential which may be added to or subtracted from the (screened) Coulomb potential. Thus we have a spin-dependent scattering potential and hence a spin-dependent differential cross section. A

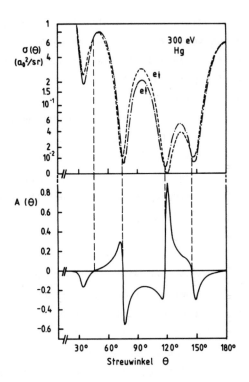

σ(θ)
(a₀²/sr)

300 eV
Hg

A(θ)

Streuwinkel Θ

Fig. 8: Differential cross section $\sigma(\theta)$ (upper panel) and asymmetry $A(\theta)$ (lower panel) for completely polarized electrons scattered elastically from free Hg atoms at 300 eV.

characteristic feature of the asymmetry function is its point symmetry with respect to zero scattering angle:

$$A(-\theta) = -A(\theta)$$

This means that up-electrons are preferentially scattered into an angle of, say, $\theta = +130°$ while down-electrons are preferentially scattered into $\theta = -130°$. Now we can easily design a spin-polarization detector by placing two identical electron detectors at $\theta = \pm130°$: If the incident beam is up-polarized the detector at $\Theta = +130^0$ will count more electrons than that at $\Theta = -130^0$ and vice-versa for down-polarized incident electrons. We may form the normalized count rate difference of the two counters and will find that it is proportional to the degree of spin polarization of the incident electrons with a proportionality constant just equal to A. Thus we have reduced the polarization measurement to a relative intensity measurement. This is the basic operating principle of the spin-polarization detector.

One of its charateristic features is that the electrons have to be polarized normal to the scattering plane, defined by the incident beam and the direction of observation. Otherwise only the component normal to the plane will be detected. Longitudinal polarization (i.e. spin along the electron momentum) cannot be detected. In this case the electron beam may be deflected by an electrostatic field, which does not affect the spin since the macroscopic fields involved are much too weak.

A real spin polarization detector could be built directly along the experiment of Fig. 8. We could scatter a transversely polarized electron beam from a mercury vapor beam, select the elastically scattered electrons by a simple retarding field from the inelastic ones and count them by two channeltrons placed at conjugate angles. Such devices have in fact been built and found to operate satisfactorily if the intensity of the electrons to be analyzed is reasonably high (Deichsel and Reichert 1965). The detection probability can be increased by increasing the density of scattering atoms, i.e. by using a solid target. Unfortunately the electrons tend to be scattered several times in the solid which washes out their angular definition and as a result decreases the polarization sensitivity. To keep plural scattering at a tolerable level thin foils (mostly of Au) and high energies have been used (typically 100 to 200 keV). A further gain in intensity is possible by increasing the solid angle seen by the detectors, but this cannot be carried too far. The asymmetry function, also called Sherman function, has a pronounced angular variation (see Fig. 8), in-

cluding sign reversals, which makes the evaluation of the effective
Sherman function S_{eff} difficult for large solid angles. Mott detectors
frequently are not calibrated experimentally, since it is difficult
(though not impossible) to have a source of precisely known spin polarization. Instead, measured asymmetry data are extrapolated to zero
foil thickness and set equal to the calculated effective Sherman function. In this way estimated absolute accuracies around 5 % have been
achieved (Müller 1979). In many cases absolute accuracy is of minor
importance and may be sacrificed for intensity, which is mostly of
primary concern.

In order to design a polarization detector of high sensitivity
one might be tempted to place the electron counters at angles where
the asymmetry is large. This is, however, in general not to be recom-
mended since the differential cross section is mostly small at these
angles. It is an empircal rule of thumb in atomic physics that maxima
of polarization are correlated to minima of intensity. Thus a suitable
compromise has to be found between reasonable polarization sensitivity
and high intensity. A criterion of general importance is the statistical accuracy of a polarization measurement for a given number of
electrons sent into the detector. For the Mott detector it has been
shown (Kessler 1977) that the quantity

$$F_m = S^2_{eff} \cdot I/I_0$$

should be as large as possible. S_{eff} is the effective Sherman function
and I is the number of detected electrons per number I_0 of incident
electrons. This 'figure of merit' may also be useful in comparing different detectors (though it is by no means the only criterion). Typical effective Sherman function values for high energy Mott detectors
lie between 0.2 and 0.3 and typical detection probabilities are in the
range 10^{-4} to 10^{-3}. Thus a typical figure of merit is of the order of
a few times 10^{-5}. The detection probability can be increased by opening up the solid angle subtended by the detectors as far as possible.
The effective Sherman function decays considerably, but though it enters quadratically this may be overcompensated by the gain in intensi-

ty. A figure of merit of $2 \cdot 10^{-4}$ has been reported (Raue et al. 1984). Though efficiency is of major concern, other considerations deserve similar attention. For example the apparatus asymmetry, since its experimental determination and its stability determine the precision of the polarization zero (Müller 1979). Also unwanted scattered electrons have to be suppressed since they diminish the polarization sensitivity. The thickness and quality of the scattering foils also become important for high precision measurements. All these points have been studied and documented in great detail and shall not be repeated here. The reader is referred to a number of excellent articles and the references therein (Eckstein 1970, Kessler 1979, Van Klinken 1966, Jost et al. 1981).

The traditional Mott detector generally is a rather bulky apparatus, of the order of several m^3, which is mainly due to the safety precautions imposed by the high voltage. By suitable high voltage engineering this size has been reduced considerably (Hodge et al. 1979, Koike and Hayakawa 1984), but further size reduction requires lower operating voltages. It has been demonstrated recently (Gray et al. 1984) that a very compact design can be achieved at operating voltages of 20-40 kV with a reasonable efficiency. This design is shown in Fig. 9. The electrons to be analyzed enter from the right, are accelerated by the field between the outer and inner hemispheres and hit the Au-foil at the center of the two spheres. Backscattered electrons leave the inner hemisphere by two holes, are decelerated in the radial field and are detected by two channeltrons on the outer hemisphere. A bias voltage at the channeltron entrance apertures provides some additional energy discrimination against inelastically scattered electrons. The efficiency was found to be rather independent of the operating voltage and a figure of merit of $F_m \sim 2 \cdot 10^{-5}$ was reported, determined by calibration with a high energy Mott detector.

High and medium energy Mott detectors offer particular advantages in certain applications. First, the high accelerating voltage allows electrons with a large phase space product to be analyzed,

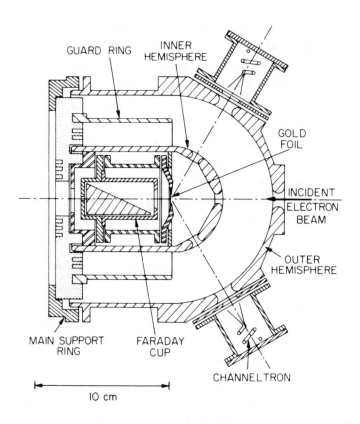

Fig. 9: Schematic diagram of a medium energy Mott analyzer. The analyzer has rotational symmetry about the horizontal axis.

given by their energy spread, emittance area and solid angle. This is useful when a total electron yield is to be analyzed and/or a broad energy spectrum (Koike and Hayakawa 1984). Secondly, the high scattering energies at the Au foil make low-Z surface contamination (mostly water and hydrocarbon) negligible. This is advantageous in many atomic physics experiments where ultra-high-vacuum conditions are not required normally. By contrast, the modern low energy detectors discussed in the following are best suited for UHV environments.

5.2.3 The Absorbed Current Detector

The absorbed current detector is a fairly recent device, which is small and relatively simple and operates at low energies. Its operating principle is the following:

It is well known that the net current to a solid target bombarded by energetic electrons may be zero. This is the case if the incident electron flux equals the flux of electrons leaving the sample. Besides elastically and inelastically scattered primary electrons this flux mainly consists of secondary electrons. For conductors there are in general two such zero crossings at two largely different energies. As a function of energy the first one (at several 100 eV) occurs because with increasing energy there are more and more secondaries generated per primary electron. The second zero crossing appears at several keV to some 10 keV because the range of the primaries increases with increasing energy, the secondaries are produced at larger depth, and fewer electrons arrive at the surface with sufficient energy to overcome the surface barrier. The precise location of these zero crossings depends on a number of parameters, e.g. the material, the angle of incidence, the surface roughness, or the work function. It was first observed by Siegmann et al. 1981 for a ferromagnetic glass that the low energy zero crossing may split into two crossings, depending on the orientation of the electron spin of the primaries relative to the magnetization. This effect was attributed to the spin dependence of the elastic scattering from ferromagnets via exchange interaction. The same phenomenon was shortly afterwards observed on non-magnetic single crystals of Au (Erbudak and Müller 1981) and W (Celotta et al. 1981). In these cases it is the spin-orbit interaction which causes the splitting. Subsequently the effect was observed also for polycrystalline materials (Erbudak and Ravano 1981, Pierce et al. 1981). The origin of the phenomenon has not yet been investigated in much detail, though it is very likely that inelastic scattering contributes in addition to elastic scattering. Nevertheless, the effect can be exploited immediately for a polarization analysis of the incident beam. The spin-dependent absorption has very clearly been demonstrated on Au(110) by Erbudak and Müller 1981. Fig. 10 shows a typical result for the plane of incidence normal to the $[1\bar{1}0]$ direction. The upper part of the figure shows the energy value E_0 at which the zero crossing of the sample current occurs for unpolarized primary electrons as a function of the angle of incidence Θ. The insert shows at $\Theta = 56.5^0$ the absorbed current as a function of energy for up-spin ($I_\uparrow$), down-spin ($I_\downarrow$), and unpolarized primary electrons (dashed). Evidently the zero-crossing for unpolarized electrons ($E_0 = 118.4$ eV) splits up into two crossings displaced by about ± 1 eV with respect to E_0. The corresponding curves are not necessarily straight lines and they are not necessarily displaced parallel to each other, though they may appear as such over a small energy interval. The lower part of Fig. 10 shows the current difference $I = I_\uparrow - I_\downarrow$ relative to the primary current I_0 at the energy E_0 as a function of the angle of incidence. For the parameters chosen I/I_0 is seen to be of the order of 10^{-3} to 10^{-2}. After a calibration, either by a primary beam of known polarization or by comparison to another polarization detector, this intensity difference can be used as a measure for the polarization of the incoming beam. For the absorbed current the statistical uncertainty of a polarization measurement has to be evaluated in a different way than for the Mott detector. The fluctuations of the measured current I are not determined by the current I itself but rather by the fluctuations of the incident beam and by the statistics of the secondary electron generation. This is evident from a measurement with unpolarized electrons at E_0: The average net current to the sample is zero, but the fluctuations of this current are far from zero. As shown by Pierce et al. 1981 the figure of merit of this detector is

$$F_m^! = \frac{1}{2\delta^2} \left(\frac{I}{I_0}\right)^2$$

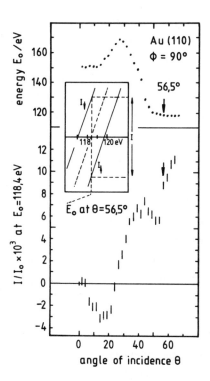

Fig. 10: Spin dependence of the net current to a clean Au(110) surface in the vicinity of the zero crossing. Explanation see text.

where δ^2 is the variance of the statistics of secondary electron generation. Assuming Poisson statistics to be valid, the variance equals the mean value and we have $\delta^2 \approx 1$ since the secondary electron yield is close to unity. The largest I/I_0 reported so far in an actual application of the detector is $I/I_0 = 1.5 \cdot 10^{-2}$ (Erbudak and Ravano 1983) for a polacrystalline Au surface. With these data the figure of merit is $F_m' = 1.1 \cdot 10^{-4}$, which is less than but comparable to that of the LEED detector described in the next section.

Absorbed current detectors exploiting the spin-orbit interaction are sensitive to the polarization component normal to the plane of incidence. Therefore two transversal components of the polarization vector can be obtained by rotating the detector assembly about the incident beam. These components can, however, only be measured consecutively, not simultaneously as with the Mott detector and the LEED detector.

An actual example of an absorbed current detector together with an electron energy analyzer is shown in Fig. 11 (Erbudak and Müller 1984). Note the small size and simple design of this polarization detector. The absorbed current is measured by I_A while the total incident current is nearly equal to I_S (by definition since I_A is small, see Fig. 10). When the spin of the electrons to be analyzed can be rapidly inverted, as for example by using primary electrons from a GaAs source, lock-in techniques are applicable. Under these conditions a detection limit of $0.5 \cdot 10^{-15}$ A for I_A was reported. With a current $I_S \sim 5 \cdot 10^{-12}$ A a polarization measurement within 1 % absolute accuracy can be made in about 500 sec (Erbudak and Müller 1984). This corresponds to a total number of about 10^{10} electrons fed into the detector.

The working point E_0 and the energy splitting of the two zero crossings was observed to be subject to drifts, most likely caused by contamination from the residual gas. Therefore UHV conditions are mandatory and re-calibrations are necessary, at least occasionally. The detector is sensitive to the angle of incidence, the energy, and the energy spread of the incident beam. As pointed out by Pierce et al. 1981 each of these factors must be handled the same way at the time of calibration and at the time of measurement. An advantage of this detector over the LEED detector is that it is not bound to single crystal surfaces. A contaminated surface may be renewed by evaporating a fresh layer of gold, followed by annealing and recalibration. A major drawback of the absorbed current detector is that relatively large primary currents are required and that single electron counting is not

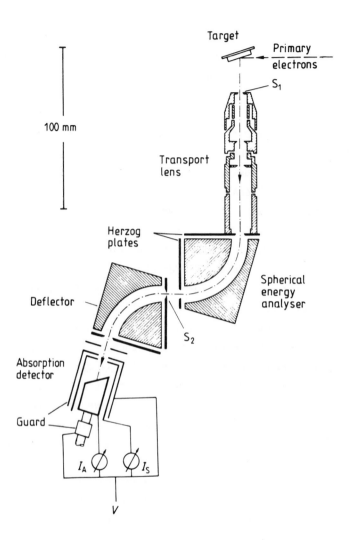

Fig. 11: Scale drawing of the adsorbed current polarization detector
together with an electron energy analyzer system (Erbudak and
Müller 1984).

possible, a feature that is most desirable at low intensity levels. Therefore this detector appears most useful in applications where a high-intensity source of known polarization is available.

5.2.4 The LEED Detector

The LEED detector relies on spin-dependent effects in low energy electron diffraction. When an unpolarized primary beam impinges normally onto a low-index crystal surface diffracted beams involving equivalent two-dimensional reciprocal lattice vectors are in general of equal intensity. When the scattering potential is spin-dependent, either via exchange interaction or via spin-orbit coupling, this intensity degeneracy may be removed if the primary beam is spin-polarized. The basic idea of the LEED detector is to use intensity differences of normally equivalent beams as a measure of the polarization of the incident electron beam. This principle works with ferromagnetic crystals as well, but for practical reasons spin-orbit coupling with high-Z materials is preferred. The existence of spin-dependent diffraction was first demonstrated experimentally by O'Neill et al. 1975 by measuring the spin polarization of electrons diffracted from W(001). Intensity asymmetries as mentioned above were first observed by Kirschner and Feder 1979 in a double diffraction experiment, and the importance of multiple scattering for the spin polarization was pointed out. A prototype LEED spin polarization detector was described in this work, which has since been investigated in detail and improved technically. As a detector crystal W(100) was used. Tungsten was used because of its high atomic number and the relative ease to clean the surface (see also the discussion below). The (100) plane is used because it has a fourfold rotational symmetry and is not reconstructed above room temperature (except for a rather large contraction). The fourfold symmetry with two orthogonal mirror planes facilitates the simultaneous detection of two orthogonal components of the polarization vector: each two equivalent beams in a mirror plane measure the polarization component in the other mirror plane. Several of the diffracted beams have been investigated, and the < 2,0 > beams were found

to be most suitable at a particular scattering energy. Results for polarization and intensity of the (2,0) beam as a function of energy are
shown in Fig. 12. Fig. 12 b) shows the polarization which equals ex-

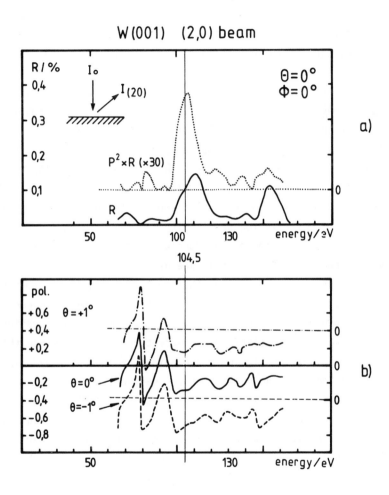

Fig. 12: Experimental data for the reflectivity R, the figure of merit
$P^2 \cdot R$ (a), and the polarization of the detector-beam and its
sensitivity with respect to the angle of incidence θ (b). The
"working point" at 104.5 eV is indicated.

actly the asymmetry since the scattering plane coincides with one of
the mirror planes (Kirschner 1985a). The polarization was measured for
normal incidence ($\Theta = 0^0$) and for $\Theta = \pm 1^0$ off normal. In Fig. 12 a) we
see the elastic reflectivity $R = I_0/I_{(2,0)}$ of this beam and the quantity $P^2 \cdot R$ which is formed by analogy to the figure of merit expression
for the Mott detector. The reflectivity R has a maximum near 110 eV
with a shoulder at 105 eV, and a second maximum around 155 eV. The polarization curve for $\Theta = 0^0$ exhibits a strong plus/minus feature near
80 eV and a relatively flat behaviour above 100 eV. The optimum working point may be read from the efficiency curve $P^2 \cdot R$ in Fig. 12 a). It
is evident that the efficiency is not maximum for the largest polarization values, but has a pronounced peak around 105 eV, in spite of
substantially smaller polarization. The second reflectivity maximum,
though of similar magnitude, is obviously not useful because of lack
of polarization. From these results the scattering energy was fixed at
104.5 eV. The polarization sensitivity was found to be A = -0.27 ±
0.02 by self-calibration of the double diffraction experiment quoted
above. If two identical crystals are used and if the scattering conditions are exactly the same then the measured intensity asymmetry
equals just the square of the polarization or asymmetry in each of the
diffraction processes. This is a purely experimental calibration,
without involving calculated data, except for the sign since this is
lost by squaring. The sign may be obtained by theory or by comparison
to other detectors. The quality of the self-calibration may be judged
from the fact that two independent groups, one working with a high energy Mott detector, the other with the LEED detector, quoted exactly
the same polarization value of -78 % for a particular feature in the
specular beam from W(001) (Wendelken and Kirschner 1981, Kalisvaart et
al. 1978, Riddle et al. 1979). Though this precise agreement might be
accidental it shows that polarization measurements can be carried out
with good accuracy by either technique.

The statistical uncertainty of a polarization measurement by
means of the LEED detector is given by the counting statistics since
single electrons are detected like in the Mott detector. Therefore the

same figure of merit applies. From the polarization sensitivity and
the reflectivity data above we obtain for the figure of merit of this
LEED detector

$$F_m = 1.6 \cdot 10^{-4}$$

Though the figure of merit is an important criterion, it is by no
means the only one. This becomes strikingly apparent by comparing with
the absorbed current detector which has a similar size and figure of
merit. For the same accuracy of a polarization measurement the absorbed current detector needs about two orders of magnitude more electrons
than the LEED detector! This is due to the technical difficulty of
measuring extremely low absorbed currents compared to counting single
free electrons. Another important consideration is the sensitivity of
the detector with respect to the solid angle and the energy spread of
the incident electron beam. We see from Fig. 12 b) that at the working
point the variations of the polarization are only of the order of a
few percent. Thus a divergence of $\pm 2^0$ of the incoming beam is acceptable. Even higher divergence may be tolerable but a re-calibration of
the detector would be required. We also learn from Fig. 12 a) that the
energy width of the beam is not critical within a few eV around the
nominal energy. A width of about 3 eV FWHM may be easily accommodated.

A particular advantage of the LEED detector is its small size
and weight combined with good efficiency. These features allow to use
it as an "add-on" to existing electron spectrometers. One of the more
advanced momentum- and spin-resolving electron spectrometer systems to
date, which is capable of measuring all three components of the polarization vector, is shown schematically in Fig. 13 (Oepen 1985, Oepen
et al. 1985). Electrons from the sample are accepted by a zoom lens,
accelerated or decelerated to the pass energy chosen and focused into
the spectrometer. The electrostatic spectrometer of the cylindrical
mirror type, with nearly second order focusing, operates with a virtual entrance slit. This allows the exciting radiation to pass through
the whole electron optical system onto the sample. After energy analysis the electrons are accelerated or decelerated to the working point

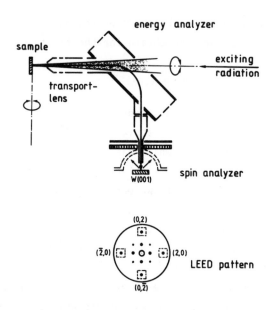

Fig. 13: A momentum- and spin-resolving electron spectrometer system as for example used in photoemission with circularly polarized synchrotron radiation. The spin-analyzer is attached to the electrostatic energy analyzer and is moved with it about two perpendicular axes. The LEED pattern is observed on an oscilloscope with single electron counting.

energy of the spin analyzer and focused onto the W(001) crystal. The inelastically scattered electrons are suppressed by hemispherical retarding grids. Behind these a two-stage channelplate is mounted, followed by a collector of the resistive anode type. In this way an x-y readout of single electron diffraction events is obtained (Stair 1980) and a LEED pattern as shown schematically in Fig. 13 may be observed on an oscilloscope screen. Electronic windows are set around the diffraction spots and the beam intensities are measured in a quasi-parallel way. In the geometry shown in Fig. 13 the $(2,0/(\bar{2},0)$ beam pair

detects the transverse polarization component normal to the axis of
the transport lens and normal to the paper plane. The $(0,2)/(0,\bar{2})$ beam
pair detects the longitudinal polarization component along the transport lens axis. Of the remaining beams, the <1,1> beams measure independently the polarization projections at 45°. The <1,0> beams, which
have small polarization sensitivity at the working point, serve to
control the apparatus asymmetry. The complete spectrometer system is
rotatable about the vertical axis lying in the sample surface. The
second transverse component of the spin polarization vector, lying in
the paper plane, can also be measured. For this purpose the spectrometer including the spin analyzer is rotated by 90° out of the paper
plane about the axis of the tranport lens. The transverse component is
measured by the $(2,0)/(\bar{2},0)$ beam pair while the longitudinal component
is measured redundantly. In this way the complete polarization vector
of energy- and angle-selected electrons is measured. The drawing is
not to scale but shows approximately correct proportions. (The resistive anode of the spin analyzer has a linear dimension of 10 cm). This
system or a similar one with discrete channeltrons instead of the
channelplate has been used in a variety of applications: spin-polarized LEED (Kirschner and Feder 1979, Feder and Kirschner 1981, Kirschner and Feder 1981), spin-polarized photoemission (Kirschner et al.
1981, Oepen 1985, Oepen et al. 1985), secondary electron analysis
(Kirschner 1985 b) and inelastic electron scattering (Kirschner 1985
c): In some applications where high count rates are encountered, such
as in LEED or in a scanning electron microscope (Kirschner 1984,
Kirschner 1985 b), the channelplate/resistive anode scheme imposes severe count rate limitations. The total count rate should not exceed a
few times 10^4 s^{-1}, which means about 10^3 s^{-1} per beam. Thus, where
speed is the major concern discrete channeltrons are preferable.

While the LEED detector has a number of undisputable advantages
it also has one important limitation: a clean surface is needed during
operation. This precludes its use under non-ultra-high-vacuum conditions. In atomic physics experiments, for example, it will be applicable only if the detector crystal and its surrounding are kept in a

low pressure region. But even in UHV a clean surface becomes notably contaminated after some time which depends strongly on the sticking coefficient. Tungsten has a sticking coefficient near 1 for all common residual gases and appears the worst choice. At first glance noble metals like Au and Pt appear much better suited but the sticking coefficient is not the only criterion. Since 'inert' surfaces also become contaminated after some time, it is equally important to have simple cleaning procedures at hand which work fast and without further control of the surface cleanliness. From this point of view W is a good choice, since a flash to ~ 2500 K removes all adsorbed gases and restores the crystal to its previous state. In routine operation of the LEED detector with tungsten crystal in the 10^{-10} Torr vacuum range a flash is made every 15 to 30 min. This interrupts the measurement for about 1 min., including the cool-down, and works fully reproducibly.

A general remark shall be made at the end of this section on sources and detectors. The status of the sources of polarized electrons is not unsatisfactory. The GaAs source and its derivatives are relatively simple to operate. One would like to have a higher polarization but the beam currents are quite satisfactory. The status of the polarization detectors, however, is far from being satisfactory. Though some important improvements have been made, and some are foreseeable, in particular for the LEED detector (Kirschner, 1985a) the detectors still are a weak point in the field. The polarization sensitivity is not better than 30 % and we still loose many hundreds of electrons for each one detected. Some new ideas shall be welcome ...

References

Alguard M J, Clenendin J E, Ehrlich R D, Hughes V W, Ladish J S, Lubell M S, Schüler K P, Baum G, Raith W, Miller R H and Lysenko W 1979 Nucl. Instr. Meth. <u>163</u> 29

Allenspach R, Meier F and Pescia D 1984 Appl. Phys. Lett. <u>44</u> 1107

Alvarado S F, Ciccacci F. Valeri S, Campagna M, Feder R and Pleyer H 1981 Z. Phys. B <u>44</u> 259

Baum G, Kisker E, Mahan A H, Raith W and Reihl B 1977 Appl. Phys. <u>14</u> 149

Bell R.L. 1973 <u>Negative Electron Affinity Devices</u> (Oxford: Clarendon)

Bir G L, Aronov A G and Pikus G E 1975 Zh. Eksp. Teor. Fiz. <u>69</u> 1382 (Sov. Phys. - JETP <u>42</u> 705 (1976))

Celotta R J and Pierce D T 1980 Advances Atomic and Molec. Phys. <u>16</u> 101

Celotta R J, Pierce D T, Siegmann H C and Unguris J 1981 Appl. Phys. Lett. <u>38</u> 577

Chrobok G, Hofmann M, Regenfus G and Sizmannn R 1977 Phys. Rev. B <u>15</u> 429

Ciccacci F, Alvarado S F and Valeri S 1982 J. Appl. Phys. <u>53</u> 4395

Conrath C, Heindorff T, Hermanni A, Ludwig E and Reichert E 1979 Appl. Phys. <u>20</u> 155

Deichsel H and Reichert E 1965 Z. Phys. <u>185</u> 169

Drachenfels W von, Koch U T, Müller Th M, Paul W and Schaefer H R 1977 Nucl. Instr. Meth. <u>140</u> 47

Drouhin H-J, Hermann C and Lampel G 1985 Phys. Rev. B <u>31</u> 3872

D'yakonov M I and Perel' V I 1971 Zh. Eksp. Teor. Fiz. <u>60</u> 1954 (Sov. Phys. - JETP <u>33</u> 1053 (1971))

Eckstein W 1970 Report IPP 7/1 Max-Planck-Institut für Plasmaphysik, Garching, Germany

Erbudak M and Müller N 1981 Appl. Phys. Lett. <u>38</u> 575

Erbudak M and Müller N 1984 J. Phys. E <u>17</u> 951

Erbudak M and Ravano G 1981 J. Appl. Phys. <u>52</u> 5032

Erbudak M and Ravano G 1983 Surface Sci. <u>126</u> 120

Feder R and Kirschner J 1981 Surface Sci. 103 75

Feigerle C S, Pierce D T, Seiler A and Celotta R J 1984 Appl. Phys. Lett. 44 866

Fues E and Hellmann H 1930 Z. Phys. 31 465

Garwin E, Pierce D T and Siegmann H C 1974 Helv. Phys. Acta 47 393

Gray L G, Giberson K W, Cheng Ch, Keiffer R S, Dunning F B and Walters G K 1983 Rev. Sci. Instr. 54 271

Gray L G, Hart M W, Dunning F B and Walter G K 1984 Rev. Sci. Instr. 55 88

Hodge L A, Moravec T J, Dunning F B and Walters G K 1979 Rev. Sci. Instr. 50 5

Hofmann M, Regenfus G, Schärpf O and Kennedy P J 1967 Phys. Lett. 25 A 270

Issendorf H von and Fleischmann R 1962 Z. Phys. 167 11

Jost K, Kaussen F and Kessler J 1981 J. Phys. E 14 735

Kalisvaart M, O'Neill M R, Riddle T W, Dunning F B and Walters G K 1978 Phys. Rev. B 17 1570

Kessler J 1977 Polarized Electrons (Berlin, Heidelberg, New York: Springer)

Kirschner J 1984a Scanning Electron Microsc. 1984/III 1179

Kirschner J 1984b Surface Sci. 138 191

Kirschner J 1984c Solid State Commun. 49 39

Kirschner J 1985a Polarized Electrons at Surfaces, Springer Tracts in Modern Physics, Vol. 106 (Springer, Heidelberg, Berlin, New York)

Kirschner J 1985b J. Appl. Phys. A 36 121

Kirschner J 1985c to be published

Kirschner J and Feder R 1979 Phys. Rev. Lett. 42 1008

Kirschner J and Feder R 1981 Surface Sci. 104 448

Kirschner J, Feder R and Wendelken J F 1981 Phys. Rev. Lett. 47 614

Kirschner J, Glöbl M, Dose V and Scheidt H 1984 Phys. Rev. Lett. 53 612

Kirschner J, Oepen H P and Ibach H 1983 Appl. Phys. A 30 177

Kisker E, Baum G, Mahan A H, Raith W and Reihl B 1978 Phys. Rev. B 18 2256

Koike K and Hayakawa K 1984 Jap. J. Appl. Phys. 23 L85 and L187

Lampel G and Weisbuch C 1975 Sol. State Commun. 16 877

Landolt M and Campagna M 1978 Surface Sci. 70 197

Landolt M and Yafet Y 1978 Phys. Rev. Lett. 40 1401

Lee Ch 1984 Appl. Phys. Lett. 44 565

Lee Ch, Oettinger P E, Sliski A and Fishlein M 1985 Rev. Sci. Instr. 56 560

Müller N 1979 Report IPP 9/23 Max Planck Institut für Plasmaphysik, Garching, Germany

Müller N, Eckstein W, Heiland W and Zinn W 1972 Phys. Rev. Lett. 29 1651

Nolting W and Reihl B 1979 J. Magnetism Magnet. Mat. 10 1

Oepen H P 1985 Report Jül-1970, KFA Jülich, Jülich Germany

Oepen H P, Hünlich K, Kirschner J, Eyers A, Schäfers F, Schönhense G and Heinzmann U 1985 Phys. Rev. B 31 6846

O'Neill M R, Kalisvaart M, Dunning F B and Walters G K 1975 Phys. Rev. Lett. 34 1167

Pierce D T and Celotta R J 1984 in Optical Orientation ed F Meier, B P Zaharchenya (Amsterdam: Elsevier)

Pierce D T, Celotta R J, Wang G-C, Unertl W N, Galejs A, Kuyatt C E and Mielczarek S R 1980 Rev. Sci. Instr. 51 478

Pierce D T, Girvin S M, Unguris J and Celotta R J 1981 Rev. Sci. Instr. 52 1437

Pierce D T, Meier F and Zürcher P 1975 Appl. Phys. Lett. 26 670

Pierce D T and Meier F 1976 Phys. Rev. B 13 5484

Pimbley W T and Mueller E W 1962 J. Appl. Phys. 33, 238

Raue R, Hopster H and Kisker E 1984 Rev. Sci. Instr. 55 383

Reichert E and Zähringer K 1982 Appl. Phys. A 29 191

Riddle T W, Mahan A H, Dunning F B and Walters G K 1979 Surface Sci. 82 511 and 517

Scheidt H, Glöbl M, Dose V and Kirschner J 1983 Phys. Rev. Lett. 51 1688

Siegmann H C, Pierce D T and Celotta R J 1981 Phys. Rev. Lett. 46 452

Stair P C 1980 Rev. Sci. Instr. 51 132

Stocker B J 1975 Surface Sci. 47 501

Su C Y, Spicer W E and Lindau I 1983 J. Appl. Phys. 54 1413

Van Klinken J 1966 Nucl. Phys. 75 161

Wainwright P F, Alguard M J, Baum G and Lubell M S 1978 Rev. Sci. Instr. 49 571

Wendelken J F and Kirschner J 1981 Surface Sci. 110 1

Chapter 6

Elastic Spin-Polarized
Low Energy Electron Diffraction
from Non-Magnetic Surfaces

F.B. Dunning and G.K. Walters

Rice University, Houston, Texas, USA

6.1 INTRODUCTION

The reliability of low energy electron diffraction (LEED) intensity analysis as a means for determining the geometrical arrangements of atoms at the surfaces of crystalline solids has improved steadily as the precision and sophistication of both experimental and theoretical methods have advanced. In particular, the increasing power of modern computers, which offer unprecedented speed and economy, has made possible dynamical LEED calculations for ever more complex structures of both clean and adsorbate-covered surfaces (Marcus and Jona 1984). However, the experimentalists' inability to fully deconvolute instrumental response functions from measured intensity profiles, and the theorists' reliance on approximation methods in structural model calculations, may limit the reliability of surface parameters deduced solely from LEED studies.

In recent years, measurements of spin polarization effects in LEED have become feasible, and the question naturally arises as to whether or not spin-polarized low energy electron diffraction (SPLEED) offers promise of improved and more reliable surface structural information. Indeed, early SPLEED theoretical investigations were motivated by this possibility, and provided the original stimulus for the experimental developments that followed (Jennings 1970, 1971, 1974; Jennings and Sim 1972; Feder 1971, 1972, 1973, 1974, 1975, 1976).

Since the first reported SPLEED investigations of W(001) a decade ago (O'Neill et al 1975) technological advances in both spin-polarized electron sources and electron polarimeters have made SPLEED measurements little more complicated than those involved in conventional LEED. As will be discussed more fully in Section 6.3, measurements of diffracted beam intensities and their polarization dependences can now be undertaken simultaneously with only minor increase in measuring time and effort. Comprehensive reviews of the pertinent technology appear both in this volume and elsewhere.

After a brief discussion of the spin-orbit interaction, which is the origin of the spin dependence that is sensitive to surface structure, this Chapter focuses on experimental methods and a survey of the growing body of SPLEED experimental and theoretical results on both clean and adsorbate-covered surfaces. Aspects of SPLEED that would appear to offer advantages for surface structure determinations relative to conventional LEED, and/or complementary information, are also developed in the text, and are summarized in a final section on future prospects. Underlying theoretical principles appear in Chapter 4 of this volume. The interested reader is also referred to earlier excellent review articles by Feder (1981), and Pierce and Celotta (1981).

6.2 Spin-Dependent Elastic Scattering - Basic Concepts

Spin-orbit interactions and - in the case of magnetically ordered surfaces - exchange interactions, can give rise to spin-dependent effects in LEED. Here we are concerned with the spin-orbit interaction, which alone is sensitive to surface geometrical structure. Studies of surface magnetism based on exchange scattering are reviewed in Chapter 7 of this volume. As will be discussed, it is not difficult experimentally to distinguish between spin-orbit and exchange contributions in SPLEED.

The physical basis of spin-dependence in elastic scattering can most easily be visualized for the case of electron scattering from a free atom. For motion in a central potential V(r), the spin orbit term in the interaction Hamiltonian can be written (Kessler 1985)

$$V_{so} = \frac{1}{2m^2c^2} \frac{1}{r} \frac{dV}{dr} \vec{s} \cdot \vec{\ell} \tag{1}$$

where $\vec{s}$ is the spin of the incident electron and $\vec{\ell}$ its orbital angular momentum with respect to the scattering center. If we regard an unpolarized incident beam of energy E as consisting of equal numbers of electrons with "spin-up" (e↑) and "spin-down" (e↓) relative to the

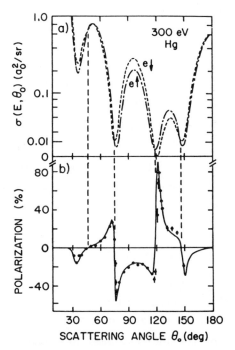

Fig. 1. (a) Calculated differential cross sections $\sigma(E,\theta_0)$ for scattering of spin-up and spin-down 300 eV electrons from atomic mercury (Kessler 1985). The pseudologarithmic cross-section ordinate is defined by $\log[1 + 50\sigma(E,\theta_0)/(a_0^2/sr)]$. (b) Corresponding electron polarizations $P(E,\theta_0)$. The experimental data points are from Jost and Kessler (1966).

scattering plane, these electrons experience different scattering potentials as a consequence of V_{so}. Thus, the numbers of spin-up and spin-down electrons scattered in a particular direction θ_0 relative to the incident beam, $N_\uparrow(E,\theta_0)$ and $N_\downarrow(E,\theta_0)$, will in general not be the same, i.e., the corresponding differential scattering cross-sections $\sigma_\uparrow(E,\theta_0)$ and $\sigma_\downarrow(E,\theta_0)$ are different. This is illustrated by the theoretical curves in Figure 1a for elastic scattering of 300 eV electrons from mercury (Kessler 1985). Since $N_\uparrow(E,\theta_0)$ and $N_\downarrow(E,\theta_0)$ are proportional to the corresponding cross-sections, unpolarized electrons elastically scattered through θ_0 are polarized to a degree

$$P(E,\theta_0) = \frac{N_\uparrow(E,\theta_0)-N_\downarrow(E,\theta_0)}{N_\uparrow(E,\theta_0)+N_\downarrow(E,\theta_0)} = \frac{\sigma_\uparrow(E,\theta_0)-\sigma_\downarrow(E,\theta_0)}{\sigma_\uparrow(E,\theta_0)+\sigma_\downarrow(E,\theta_0)} \ . \tag{2}$$

Theoretical differential scattering cross section curves can therefore be used to construct $P(E,\theta_0)$ vs θ_0 profiles. Such a profile is shown in Figure 1b together with the experimental results of Jost and Kessler (1966). The excellent agreement between theory and experiment is typical of that obtained in atomic scattering studies.

The differences in $\sigma_\uparrow(E,\theta_0)$ and $\sigma_\downarrow(E,\theta_0)$ also lead to spin dependences in the scattering of a polarized incident beam. This spin dependence is characterized by an asymmetry parameter

$$A(E,\theta_0) = \frac{1}{|P_b|} \frac{I_\uparrow(E,\theta_0)-I_\downarrow(E,\theta_0)}{I_\uparrow(E,\theta_0)+I_\downarrow(E,\theta_0)} = \frac{\sigma_\uparrow(E,\theta_0)-\sigma_\downarrow(E,\theta_0)}{\sigma_\uparrow(E,\theta_0)+\sigma_\downarrow(E,\theta_0)} \tag{3}$$

where $I_\uparrow(E,\theta_0)$ and $I_\downarrow(E,\theta_0)$ are the scattered electron currents at angle θ_0 for incident electrons with spin-up and spin-down, respectively. $1/|P_b|$ is a normalization factor accounting for the fact that the magnitude of the incident beam polarization P_b is in general less than unity. In the case of atomic scattering, it is apparent that $A(E,\theta_0) \equiv P(E,\theta_0)$, though, as discussed below, this is true in SPLEED only if certain symmetry conditions are satisfied.

It is important to note that, to a good approximation, measured values of A and P will be independent of the instrument response function. This results because these are absolute <u>ratios.</u> The instrument response function, which appears in both numerator and denominator, cancels out point-by-point. This is in marked contrast to LEED intensity measurements where the data must be corrected for a response function that is generally very difficult to determine.

An interesting aspect of P and A, apparent from the construction of Figure 1, is that large polarization features occur in atomic scattering only near cross-section minima, because only in such regions are there sizable differences in the relative magnitudes of $\sigma_\uparrow$ and $\sigma_\downarrow$. Though this is not strictly true in SPLEED because of multiple scattering effects, it is nevertheless observed that SPLEED

polarization features tend to be most pronounced in, and thus emphasize, energy regions near relative minima in the scattered intensities.

With the exceptions noted in the paragraphs above, the basic
concepts governing spin polarization effects in electron-atom elastic
scattering carry over to SPLEED. Of course, SPLEED calculations are
much more complex, primarily because of multiple scattering, but the
underlying physics is fully developed. Figure 2 illustrates some of
the differences to be expected in extending spin polarization
measurements from atoms to surfaces (Feder 1977a). Figure 2a shows
the angular dependence of $P(E,\theta_0)$ calculated for scattering from a

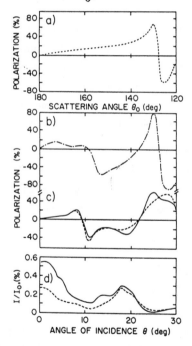

Fig. 2. Calculated polarizations for (a) scattering from a single
"muffin-tin" W atom, (b) the (00) beam from a muffin-tin monolayer, (c) the (00) beam from a W(001) surface: --- top layer
spacing as in bulk, —— top layer spacing contracted by 10%
(Feder 1977a). (d) The corresponding intensities for W(001).
The incident electron energy is 82 eV. The scattering and
incidence angles are related by $\theta = (180° - \theta_0)/2$.

single "muffin-tin" tungsten atom. Calculated polarizations and intensities for the specular (00) beam from a "muffin-tin" tungsten monolayer and from an unrelaxed, and a 10% surface-layer-contracted, W(001) surface are shown in Figures 2b and 2c, respectively, as a function of the angle of incidence θ. Pronounced differences between these curves are evident and the calculated polarizations do not simply reflect those for scattering from a single "muffin-tin" atom. This is a consequence of multiple scattering effects which, as evident from Figure 2, also relax the correlation between polarization maxima and intensity minima.

Time-reversal and spatial symmetries impose useful restrictions on spin-orbit induced effects in SPLEED (Feder 1981). In particular, if the scattering plane is a mirror symmetry plane of the crystal, it is easy to show that $A \equiv P$, both having non-zero components only normal to the scattering plane. Under these circumstances, it is particularly easy to discriminate against exchange effects in scattering from magnetized targets by ensuring that the magnetization vector lies in the scattering plane.

In the more general case that the scattering plane is not a symmetry plane, A and P need no longer be equal or normal to the scattering plane. Exchange scattering effects in this case are most easily eliminated by demagnetizing the sample, or working above the Curie temperature.

6.3 Experimental Considerations

In the earliest successful SPLEED experiments, undertaken on W(001) (O'Neill et al 1975, Kalisvaart et al 1978) and Au(110) (Müller 1977), an unpolarized incident beam was used and the spin polarizations $P(E,\theta,\phi)$ of the scattered LEED beams were measured. Targets of high atomic number Z were selected, primarily because the spin-orbit contribution to the scattering potential increases rapidly with Z. Large polarizations (~80%) were detected from both W(Z=74) and Au(Z=79). More recent studies of the lower-Z Ni(001) surface revealed smaller, but still easily measurable spin-orbit-induced spin polarizations (Alvarado et al 1981, Lang et al 1982).

The advent of a reliable UHV-compatible polarized electron source in the late 1970's (Pierce et al 1979) has greatly simplified SPLEED measurements. By simply replacing a conventional unpolarized LEED electron gun with a GaAs (or GaAsP) polarized electron source, the experimenter is equipped with the capability of simultaneously measuring both the intensities $I(E,\theta,\phi)$ and spin asymmetries $A(E,\theta,\phi)$ of the scattered beams without need for analyzing the beam polarizations $P(E,\theta,\phi)$.

Figure 3 shows a schematic diagram of the present Rice University SPLEED apparatus (Jamison et al 1985). The system incorporates both a polarized GaAs electron source (Pierce et al 1979) for measurements of A and a Mott spin-polarimeter which is used either for measurements of P or the polarization P_b of the electron beam from the GaAs source. Principles and performance characteristics of the polarized electron source and spin polarimeter are discussed in Chapter 5 of this volume.

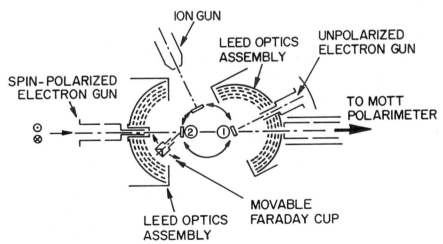

Fig. 3. Schematic diagram of the Rice University SPLEED aparatus (Jamison et al 1985).

For measurements of P, the target crystal is rotated to position (1) indicated in Figure 3. The crystal manipulator, in conjunction with the moveable LEED optics and (unpolarized) electron gun, allows any one of several LEED beams to be directed through a narrow horizontal slit cut in the phosphor screen and into the small fixed entrance aperture of a system of electron lenses. The area surrounding this entrance aperture is phosphor-coated to aid in centering the chosen beam on the aperture. Electrons entering the lenses through the aperture are accelerated to 100 keV, and their spin polarization determined by the conventional Mott scattering technique (Kalisvaart et al 1978). P measurements are tedious and very time consuming, both because of the low ($\sim 10^{-4}$) efficiency of Mott polarimeters, and because, for non-specular beams, the target crystal and LEED optics assembly must each be rotated when the incident beam energy is changed. Determination of a single polarization vs. energy (PV) profile requires several hours for the specular beam, and usually more than a day for a non-specular beam because of the angular adjustments noted above. Beam intensities are determined by positioning the beam immediately adjacent to the aperture and measuring the light output from the phosphor by use of a spot photometer; a correction is made for this small ($< 1°$) angular adjustment.

In contrast, asymmetry parameter vs. energy (AV) profiles are obtained simultaneously with the corresponding intensity vs. energy (IV) profiles, and with no extra effort. For these measurements the target crystal is rotated to position (2) in Figure 3. The polarization of the incident electron beam is modulated from up ($\uparrow$) to to down ($\downarrow$) at a frequency of 500 Hz. The diffracted beam currents are measured using a movable Faraday cup and are separated electronically into dc (spin-averaged) and ac (spin-dependent) components. These correspond, respectively, to $I = (I_{\uparrow} + I_{\downarrow})/2$ and $I_{\uparrow} - I_{\downarrow}$, where $I_{\uparrow}$ and $I_{\downarrow}$ are the diffracted beam currents for incident beam polarizations ($\uparrow$) and ($\downarrow$). The spin-averaged intensity I and the asymmetry parameter $A = (I_{\uparrow} - I_{\downarrow})/|P_b|(I_{\uparrow} + I_{\downarrow})$ are recorded by a dedicated computer which also controls and steps the electron energy. Specular-beam IV and AV profiles can be obtained in only a few minutes. Non-specular beams take longer because the position of the Faraday cup must be adjusted manually at each beam energy. The Mott polarimeter indicated in Figure 3 is used to measure the incident beam polarization P_b at regular intervals. Typically, $|P_b| \simeq .28$ and is quite stable.

Figure 4 shows SPLEED data recorded by Wang et al (1979) for specular scattering at an angle of incidence $\theta = 16°$ from W(001). The diffracted beam currents $I_\uparrow$ and $I_\downarrow$, normalized to $|P_b| = 1$, are plotted at the bottom of the figure, and the asymmetry parameter A at the top. The spin-dependent effects are quite sizeable. Note that the largest values of A tend to occur near relative minima in the intensities, though this correlation is not so strong as it is in scattering from atoms (see Figure 1) because of multiple scattering.

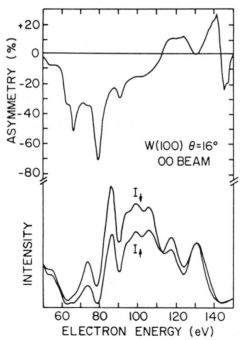

Fig. 4. Asymmetry parameter and spin-dependent scattered intensities I↑ and I↓, normalized to $|P_b| = 1$, for the specular beam from W(001) at an angle of incidence $\theta = 16°$ (Wang et al 1979).

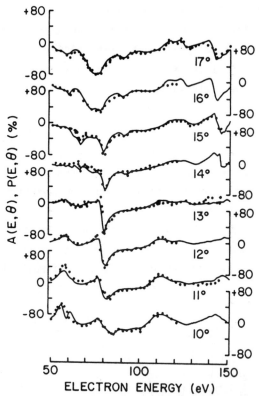

Fig. 5 Comparison between the polarization measurements (...) of Kalisvaart et al (1978) and the asymmetry parameter measurements (——) of Wang et al (1979), for the specular beam from W(001) at several angles of incidence.

The excellent laboratory-to-laboratory repoducibility of SPLEED measurements is illustrated in Figure 5, in which the filled circles represent the polarization data of Kalisvaart et al (1978) taken at Rice University, and the solid lines the asymmetry data of Wang et al (1979) taken at NBS. Since the scattering plane in each case is a (010) mirror symmetry plane of the target crystal, symmetry considerations require that the two data sets be equivalent, as is observed.

Another experimental configuration that has been used extensively in SPLEED is illustrated in Figure 6 (e.g., Bauer et al 1980a). In this method, both the longitudinal polarization component P_k and the transverse polarization component P_n normal to the scattering plane are measured for an unpolarized primary beam of fixed energy, incident at a fixed angle θ. The target crystal is rotated azimuthally about the surface normal to produce so-called rotation diagrams. By electrostatically deflecting the diffracted beam of interest through 90° about the axis normal to the scattering plane, both P_k and P_n are left mutually perpendicular and transverse to the beam velocity vector and can be separately measured by a Mott polarimeter.

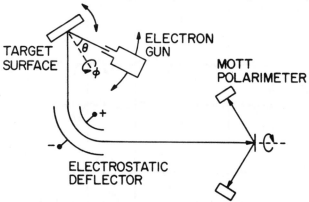

Fig. 6 Schematic diagram of the SPLEED apparatus designed for measuring both the longitudinal component P_k of the polarization and the component P_n normal to the scattering plane. The target crystal is rotated azimuthally about the surface normal to produce specular beam rotation diagrams (Bauer <u>et al</u> 1980a).

In this approach, multiple-scattering effects are experimentally separated from kinematic (single-scattering) effects, since $P_k=0$ in the strictly kinematic limit. Also, since $P_k \equiv 0$ when the original scattering plane is a mirror symmetry plane of the target crystal, regardless of multiple scattering contributions, the resulting P_k rotation diagram directly samples the symmetry of the crystal. This is illustrated in Figure 7, which shows both P_k and P_n as a function of azimuthal angle φ, for specular scattering from Pt(111) (Bauer <u>et al</u> 1980), at an angle of incidence θ=43.5⁰ and an energy 60 eV. Also shown are the results of relativistic LEED calculations of $P_k(\phi)$, $P_n(\phi)$, and the remaining (unmeasured) transverse polarization compo-

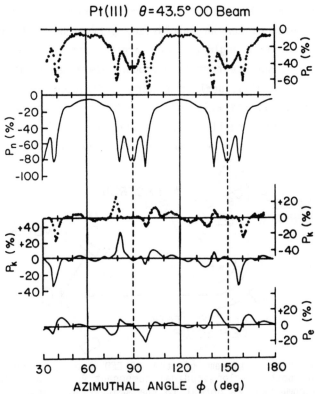

Fig. 7 Spin polarization rotation diagrams for the specular beam
from Pt(111) at E = 60 eV and θ = 43.5°. Measured values of the
polarization component P_n normal to the scattering plane, and the
longitudinal polarization component P_k are shown by dots. Calculated values (solid lines) are shown for P_n, P_k and the (unmeasured) in-plane transverse component P_e (Bauer et al 1980a).

nent $P_e(\phi)$. Both P_k and P_e vanish for ϕ = 60° and 120°, which are
mirror planes of the crystal. Note also the excellent agreement
between experiment and theory.

Finally, it should be noted that Waller and Gradmann (1982) have
demonstrated that spin-orbit and exchange effects in scattering from
ferro-magnetic surfaces can easily be separated by suitable choice of
geometry. The interested reader is referred to their publication for
discussion of the procedure.

6.4 Experimental Results and Comparisons with Theory

In the ten years since the first successful SPLEED experiment was reported, a growing number of laboratories around the world have entered the field. The development of the GaAs polarized electron source, the first use of which was reported only six years ago (Wang et al 1979), provided the necessary impetus by making SPLEED capabilities so easily accessible.

A summary of SPLEED experimental investigations on a variety of surfaces is presented here, and compared to available results of dynamical SPLEED calculations. Early work focussed on high-Z elements where spin-dependent effects are most pronounced. However, in recent years high-quality SPLEED data have also been obtained for low-Z surfaces, demonstrating that the method is generally applicable to any surface suitable for conventional LEED investigation.

6.4.1 Tungsten

The surface first and most extensively studied by SPLEED is the unreconstructed tungsten (001) surface. From an experimental viewpoint, this surface is attractive because of its high Z and well-documented cleaning procedures. However, the principal reason early experiments focussed on W(001) was that pioneering SPLEED theoretical calculations predicted large spin-dependent effects that were expected to be very sensitive to surface structural parameters such as top-layer spacing (Jennings 1970, 1971, 1974, Jennings and Sim 1972, Feder 1971, 1972, 1973, 1974, 1975, 1976). Conventional LEED investigations have revealed a rich and interesting complexity of reconstructed phases of W(001), induced by adsorbates and/or temperature. These phases also have been investigated by SPLEED.

6.4.1.1 W(001)(1x1)

At least five different laboratories have reported SPLEED investigations of the unreconstructed W(001)(1x1) surface, which is stable above approximately 370K (O'Neill et al 1975, Kalisvaart et al 1978,

Wang et al 1979 and 1981, Calvert et al 1977, Kirschner and Feder 1979, Feder and Kirschner 1981, Koike and Hayakawa 1983). Results of the two earliest studies of the specular beam, shown in Figure 5, demonstrate the excellent agreement expected between measurements from different laboratories, and illustrate the equivalence of measurements of P and A when the scattering plane is a mirror symmetry plane of the crystal. As had been predicted, the spin dependence is strikingly large and very sensitive to energy and incidence angle. Polarizations and asymmetries as high as 80% are observed. Comparison of Figure 5 with corresponding intensity profiles that appear in the original publications (Kalisvaart et al 1978, Wang et al 1981) shows that PV profiles are complementary to IV profiles, in that pronounced polarization features tend to occur at energies and angles near minima in the corresponding intensities.

The most recent extensive PV and IV data sets for W(001), which are in good overall agreement with the earlier results discussed above, were obtained by Feder and Kirschner (1981) as part of a comprehensive comparison of SPLEED theory and experiment. Their study demonstrated the sensitivity of calculated PV and IV profiles to variations in top-layer spacing, and to details of the scattering potential model. A comparison between the results of their model calculations and the measured specular beam polarization and intensity profiles for $\theta = 17°$ and $\phi = 0°$, obtained by three different groups, is shown in Figure 8. Theoretical profiles are included for assumed top-layer contractions of 0, 5% and 10%, and for two assumed potentials, a band structure potential V_B with constant exchange and an ion-core potential V_E incorporating both an exchange interaction that decreases with increasing energy and an imaginary term to account for localized absorption due to 4f excitation. The better agreement with experiment of calculations using V_E is evident, and the analysis suggests that the surface layer spacing is contracted by $7\pm1.5\%$ relative to the bulk spacing.

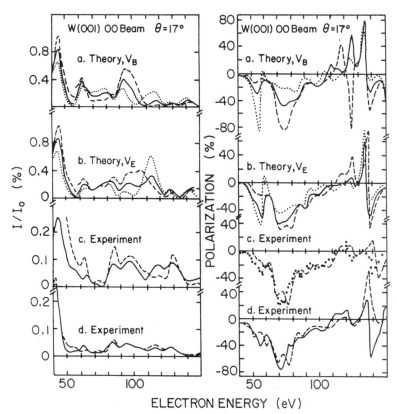

Fig. 8 IV and PV profiles for the specular beam from W(001) at θ =
17° and φ = 0. Theoretical profiles (Feder and Kirschner 1981)
are shown for top interlayer spacings contracted by 0% (···),
5% (——) and 10% (---). (a) Band structure potential (V_B).
(b) Potential with energy-dependent exchange (V_E). (c) Experimental results of Kalisvaart et al (1978) (——, ···) and Wang et al
(1981) (---) for nominal incidence angle θ = 17°. (d) Experimental results of Feder and Kirschner (1981) for θ = 17° (——)
and θ = 16° (---). I and I_0 are the scattered and incident
currents, respectively.

Figure 9 shows AV and IV profiles measured by Wang et al (1981)
for the (0$\bar{2}$) and (1$\bar{1}$) beams from W(001) together with the calculations of Feder and Kirschner (1981) using the potential V_E of Figure
8 and assuming 7% contraction of the surface layer spacing.

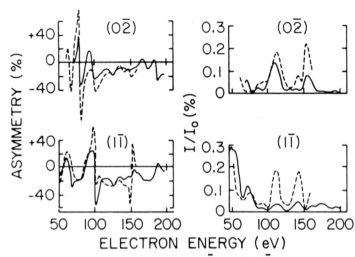

Fig. 9 AV and IV profiles for the $(0\bar{2})$ and $(1\bar{1})$ beams from W(001) at normal incidence; (——) measurements of Wang et al (1981), (---) calculations of Feder and Kirschner (1981) assuming a 7% contraction of the topmost layer spacing. I and I_0 are the scattered and incident currents, respectively.

Temperature effects in SPLEED from the unreconstructed W(001) surface have been investigated by Calvert et al (1977), Riddle et al (1978) and Kirschner and Feder (1981). For small angles of incidence, specular beam intensities are observed to decrease with increasing temperature, due largely to enhanced thermal lattice vibrations, while thermal lattice expansion causes shifts of peak positions to lower energy in both IV and PV profiles. Kirschner and Feder (1981) showed, however, that to account fully for changes in the PV profiles, which are particularly sensitive to top-layer spacing, the top-layer contraction must be taken to be temperature dependent with a thermal expansion coefficient some two-to-three times that of the linear bulk expansion coefficient.

6.4.1.2 Adsorbate- and Temperature-Induced Reconstruction

Riddle et al (1971a,b) and Mahan et al (1980) were the first to explore changes in PV profiles from W(001) caused by the deposition

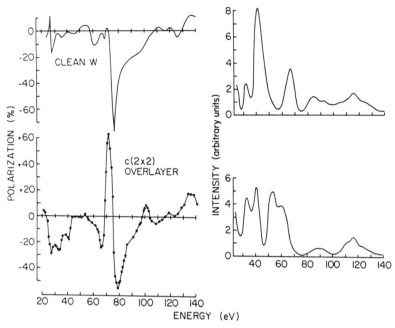

Fig. 10 Measured PV and IV profiles for the specular beam at θ = 13° and. φ = 0° from a clean W(001) surface and from W(001) with an ordered c(2x2) CO overlayer (Riddle _et al_ 1977b).

of adsorbates. As illustrated in Figure 10, deposition of an ordered c(2x2) CO overlayer leads to marked changes in the observed PV profiles. Indeed, PV profiles for the specular beam were found to be more sensitive to surface condition than the corresponding IV profiles. Further, as shown in Figure 11, striking similarities were noted between the PV and IV profiles obtained with c(2x2) overlayers of CO and N_2, suggesting the possibility that both adsorbates induce similar reconstruction of the W(001) surface (Mahan _et al_ 1980). This conjecture is consistent with a later study by Griffiths _et al_ (1981), which attributed LEED patterns observed from W(001) with a nitrogen coverage of ~0.4 monolayer to a structure composed of islands in which the surface-layer tungsten atoms are displaced to produce a contracted interatomic spacing.

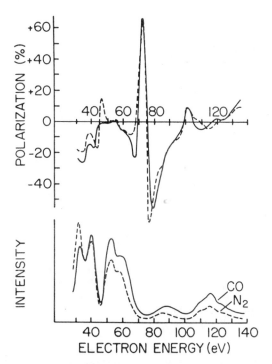

Fig. 11 Comparison of PV and IV profiles for the specular beam
at θ = 13° and φ = 0° from a W(001) surface following N₂ (---) or
CO (——) adsorption (Mahan et al 1980).

Wendelken and Kirschner (1981) have investigated ordered p(4x1)
and p(2x1) oxygen overlayers in W(001) with SPLEED, and suggest that
the similarity of their measurements to those of Mahan et al (1980)
implies that at least the early stages of reconstruction are similar
for oxygen and nitrogen adsorption.

In a very interesting comparative study, Wang et al (1982) were
able to demonstrate with SPLEED that the reconstruction of W(001)
induced by lowering the temperature produces a different structure
than does reconstruction induced by hydrogen adsorption. Their data
are shown in Figure 12. While the c(2x2) LEED patterns and IV profiles from the two reconstructed surfaces are the same within experimental uncertainty, large differences are found in the AV profiles.

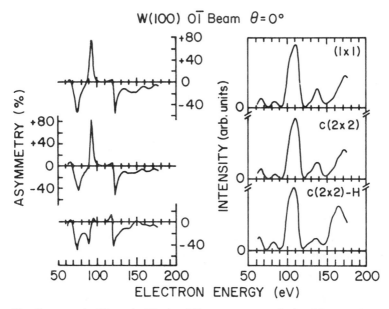

W(100) O̅1 Beam $\theta = 0°$

Fig._12 Measured AV and IV profiles at normal incidence for the (01) beam from W(001)(1x1) at 420K, W(001)c(2x2) at 110K and W(001)c(2x2)H at 370K (Wang et al 1982).

Because the spin-orbit effect in scattering from hydrogen is very small, these differences must result from different arrangements of the tungsten substrate atoms. However, dynamical SPLEED calculations will be required to determine the actual structures.

6.4.1.3 Surface Barrier Resonances

The shape of the surface potential barrier seen by an electron approaching a metal surface is an important component of models used for dynamical LEED (and SPLEED) calculations. Interference effects near a beam emergence threshold between the specularly reflected beam and non-specular beams diffracted back into the specular beam give rise to fine structure in specular beam intensities (McRae 1979). Calculations of this fine structure are very sensitive to details of the assumed surface barrier potential.

It was predicted some years ago, on the basis of SPLEED model calculations, that the mechanism responsible for such surface barrier "resonances" should also produce sizeable polarization effects in near threshold PV and AV profiles that are even more sensitive than corresponding IV profiles to both real and imaginary components of the assumed surface potential barrier. These predictions were confirmed in measurements on W(001) by Pierce et al (1981) that provided a deeper understanding of electron reflection at surfaces, especially of mechanisms important in the grazing emergence of diffracted beams near threshold (McRae et al 1981).

These fine structure data also were analyzed by Jones and Jennings (1983) using dynamical SPLEED theory to test surface potential barrier models. They showed that a saturated image barrier

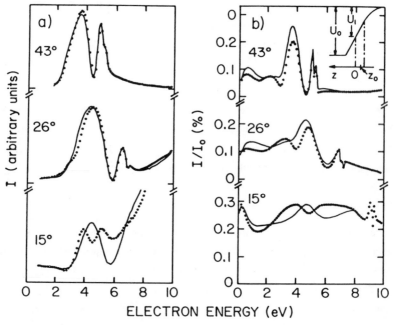

Fig. 13 (a) Specularly reflected intensities for beams of spin-up (•••) and spin down (——) electrons from W(001) for $\theta = 43°$, $26°$ and $15°$ (Pierce et al 1981). (b) Corresponding calculations of Jones and Jennings (1983) using the saturated image barrier model shown schematically in the inset.

model provides a satisfactory description, and determined the model parameters for W(001) that provided the best fit to the spin-dependent scattering data. They were unable to reproduce all the observed structure with simpler models such as those commonly used in LEED intensity calculations.

The spin-dependent scattering data of Pierce et al (1981) are compared with the calculations of Jones and Jennings (1983) in Figure 13. The barrier model is shown in the inset. Far from a jellium surface, the image potential has the form $U=1/2(z-z_0)$ (Lang and Kohn 1973) where z_0 denotes the center of mass of the image charge. This asymptotic form is extended linearly (Dietz et al 1980) to the point where it joins continuously to the crystal inner potential $(-U_0)$. The parameters used in defining the barrier are z_0 and U_1, the absolute value of the potential at the jellium edge $(z=0)$. The barrier that gave the best fit to the data $(z_0=3.3$ bohr, $U_1=0.7$ Ry) was used to derive the theoretical curves in Figure 13.

Tamura and Feder (1983) have recently proposed a more general scattering potential which takes account of spatially periodic surface potential barrier variations in the plane of the surface. Preliminary calculations based on this three-dimensional barrier show better agreement with the interesting spin-dependent measurements in Figure 13 near 4 eV for $\theta = 15°$ (Feder 1985).

6.4.2 Gold

The Au(110) surface was the second studied by SPLEED (Müller et al 1977, Feder et al 1977, Müller et al 1978), predating the development of the GaAs polarized electron source. As in the case of W(001), PV profiles show large maxima (~80%) and evidence of strong multiple scattering effects.

This surface has very interesting structural properties. At room temperature the clean surface exhibits a (1x2) superstructure that gives way to the (1x1) bulk structure at temperatures above about 750K. This transition is evident in the PV profiles shown in

Figure 14, which change markedly as the temperature is increased from 320K to 830K (Müller et al 1978). SPLEED calculations (Feder et al 1977) on the high temperature (1x1) phase show reasonable agreement with experiment, though not so good as in the cases of W(001) and Pt(111). This is not surprising since most of the data were taken in the 710-750K temperature range, where the real Au(110) surface is still in transition from the (1x2) to the (1x1) structure. No theoretical studies of the (1x2) room temperature superstructure, nor of the transition region, have yet been reported.

Discrepancies in earlier polarization experiments (Reihl 1981) on the $(0 \frac{1}{2})$ beam of the low temperature phase have been resolved by Müller et al (1981). The low temperature Au(110) surface was also the subject of the first experiments in which the polarization component P_n normal to the scattering plane, as well as the in-plane transverse component P_k, were both measured (Erbudak et al 1982).

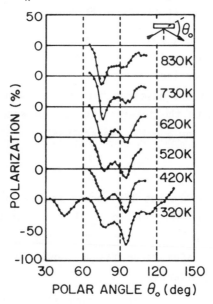

Fig. 14 Polarization vs. polar scattering angle θ_0 for 50 eV electrons specularly reflected from Au(110). Changes in the polarization profiles with temperature are due to surface reconstruction (Müller et al 1978).

6.4.3 Platinum

The Pt(111) surface structure is known to be very close to a simple termination of the bulk lattice from LEED intensity analysis (Kesmodel et al 1977, Adams et al 1979), and Rutherford ion scattering (Davies et al 1980). Thus this is an attractive surface for exploratory SPLEED investigations focusing on non-structural model features, without the need to disentangle structural effects. Comparison of experiment with theory (Bauer et al 1980a,b) clearly demonstrated the superiority of an ion-core scattering potential incorporating an energy-dependent exchange term over a simple band structure potential, as is also the case for W(001). This is illustrated in Figure 15, which shows the spin polarization of the (10) beam from Pt(111) at 95 eV as a function of the angle of incidence θ (Bauer et al 1980b). Very good agreement between experiment and theory is obtained using the potential with energy dependent exchange and an exponentially smooth surface barrier. The theoret-

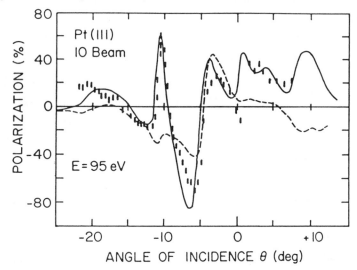

Fig. 15 Polarization of the (10) beam from Pt(111) at 95 eV vs. incidence angle θ. The scattering plane is normal to the surface. The experimental data (⌶ ⌶ ⌶) are compared to theory based on a band structure potential (---) and a potential with energy-dependent exchange (——) (Bauer et al 1980b).

ical rotation diagrams shown in Figure 7 are obtained using the identical potential (Bauer et al 1980b), and also show excellent agreement with experiment. A detailed investigation of the sensitivity of calculated polarization profiles to assumed variations in top layer spacing (Feder et al 1981) suggests the possibility of a slight outward relaxation of the topmost layer (0.5% ± 1.0% of the bulk interlayer spacing). The surface Debye temperature was found to be close to the bulk value of 230K.

More recently, measurements have been made of the dependence of longitudinal and transverse spin polarization components on angle of incidence θ and azimuthal angle ϕ, over the energy range 60 to 70 eV (Bauer et al 1983).

6.4.4 Nickel

Early SPLEED measurements focused on materials of relatively high Z, for which the spin-orbit coupling is strong. However, lower-Z surfaces have also been shown to be accessible to SPLEED investigation, despite the fact that the magnitudes of the features observed in PV and AV profiles are generally rather small.

Indeed, in a remarkable LEED experiment performed well over fifty years ago on the Ni(111) surface, Davisson and Germer (1928a, 1929) attempted the first test for polarization of the electron waves they had previously demonstrated (Davisson and Germer 1927a,b, 1928b). In the absence of any clear theory of spin polarization effects in electron scattering, they based their interpretation on an assumed analogy between electron and light polarization and concluded that electron waves are not polarized by reflection. Many years later, Kuyatt (1975) reinterpreted their measurements to show that in fact the data suggest a 10-15% polarization in the reflected beam. However, an effect of the magnitude observed could also have been caused by very small systematic alignment errors ($\lesssim$ 1%), and it is very doubtful that Davisson and Germer actually observed spin polarization in LEED (Feder 1977b).

Here we review recent work on spin-orbit induced polarization effects on the clean Ni(001) surface and on Ni(001) with an ordered c(2x2) tellurium adlayer.

6.4.4.1 Ni(001)

This surface was first investigated by Alvarado et al (1981) who measured spin-orbit induced asymmetries in the specularly scattered beam, using a polarized incident beam. Lang et al (1982) measured intensity and polarization profiles for the (00) and (10) beams,

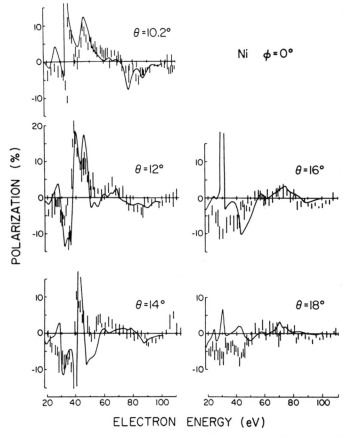

Fig. 16 Comparison of experimental and theoretical PV profiles for the specular beam from Ni(001) for an azimuthal angle $\phi = 0°$ and angles of incidence θ from 10.2° to 18° (Lang et al 1982).

using an unpolarized incident beam. The results of these studies are in good agreement.

A comparison of theoretical and experimental Ni(001) (00) beam polarizations is shown in Figure 16 (Lang et al 1982). The theoretical calculations are based on model characteristics derived from earlier LEED intensity data. The polarization profiles are very sensitive to the angle of incidence θ. Agreement between theory and experiment was improved by assuming that the measured angles of incidence were systematically 1° too small, and by application of a -2 eV energy shift to the theoretical calculations. These adjustments are incorporated in the profiles shown in Figure 16. Agreement between theory and experiment is seen to be generally satisfactory, especially for the smaller values of θ.

6.4.4.2 Ni(001)c(2x2)Te

This sytem was chosen for study to explore the possibility that spin-orbit induced polarization effects might be dominated by the higher-Z tellurium adlayer (Z = 52, as compared to Z = 28 for the Ni substrate), thus making PV profiles more sensitive than IV profiles to adlayer properties. Previous LEED studies (Demuth et al 1973a,b, 1975; Demuth and Rhodin 1974a,b) of the Ni-Te system had shown that the tellurium adlayer induced little or no surface reconstruction, suggesting that the Ni-Te system might also be suitable for the first SPLEED calculations for an adsorbate-substrate system.

Deposition of tellurium resulted in marked changes in both the intensity and polarization profiles, as compared to clean Ni(001). In general, the magnitudes of the polarization features are somewhat larger when the tellurium adlayer is present. As an example of the changes induced by tellurium deposition, Figure 17 compares PV and IV profiles for the (10) beam at normal incidence from Ni(001) and Ni(001)c(2x2)Te (Lang et al 1982).

Theoretical calculations were undertaken using a relativistic multiple scattering formalism (Feder 1981) generalized to allow for

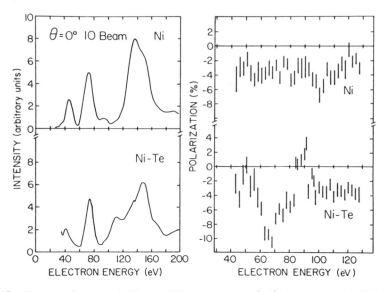

Fig. 17   Measured PV and IV profiles for the (10) beam from Ni(001) and Ni(001)c(2x2)Te at normal incidence (Lang <u>et al</u> 1982).

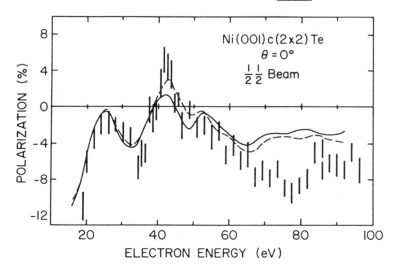

Fig. 18   Comparison of experimental and theoretical PV profiles for the (1/2,1/2) beam from Ni(001)c(2x2)Te at normal incidence (Lang <u>et al</u> 1982).   The solid and dashed theoretical curves are for assumed Te-Ni interlayer distances of 1.90 Å and 1.84Å respectively (Feder 1983).

an ordered c(2x2) overlayer. As illustrated in Figure 18 for a fractional order beam (Lang et al 1982, Feder 1983), the agreement of early calculations with experiment is encouraging. As is also evident from Figure 18, a 3% change in the spacing of the Te layer from the topmost Ni layer leads to significant changes in the calculated PV profiles. Discrepancies such as those in Figure 18 above 65 eV are not surprising since further calculations (Feder 1983) indicate that PV profiles are highly sensitive to small variations not only of the geometry, but also of the ion-core potential, the Debye temperature and the complex inner potential chosen for the Te adlayer. Spin polarization analysis thus promises a more accurate determination of these adsorbate properties than is possible by intensity analysis alone.

6.4.5 Cu_3Au: Order-Disorder Transformations

In recent years there has been renewed interest in the study of order/disorder transitions at Cu_3Au surfaces. Cu_3Au is a classic ordering alloy that undergoes a discontinuous bulk order/disorder transition at a critical temperature T_c of 390°C. The behavior of Cu_3Au surfaces in the vicinity of T_c has been investigated using a number of techniques, including low energy ion scattering (Buck et al 1983), and LEED (Sundaram et al. 1973, 1974; Sundaram and Robertson 1976; Potter and Blakely 1975; McRae and Malic 1984). These studies indicate a continuous surface order/disorder transition, with the possibility of a sudden component at T_c. Since the spin-orbit effect, which gives rise to polarization effects in LEED, is strongly dependent on atomic number, substantial changes in composition or geometrical arrangement at a Cu_3Au surface due to an order/disorder transition might be expected, in view of the marked difference in atomic number of copper and gold (Z = 29 and 79 respectively), to lead to significant changes in observed LEED polarization features.

The observed temperature dependence of the AV profile for the (00) beam from Cu_3Au (001) at an angle of incidence $\theta = 13°$ is shown in Figure 19 (Jamison <u>et al</u> 1985). Heating the crystal leads to a decrease in the magnitudes of the various asymmetry features. However, no pronounced changes in the form of the AV profiles, or new asymmetry features, are observed as the crystal is heated through T_c. Much, if not all, of the decrease in the magnitudes of the asymmetry features as the crystal is heated is attributable to the increase in the diffuse background, which is spin independent.

The SPLEED data are consistent with the results of earlier investigations and provide no evidence of sudden changes in surface composition or order at T_c; rather the data suggest that the surface disorders continuously and retains local order above T_c.

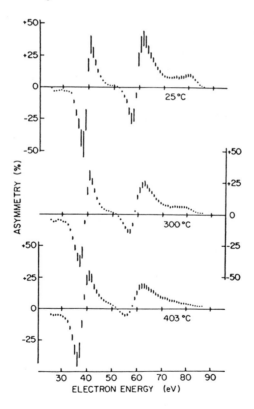

Fig. 19 Temperature dependence of the AV profiles for the specular beam from $Cu_3Au(001)$ at $\theta = 13°$ and $\phi = 0°$.

6.5   Conclusions and Future Prospects

In the ten years since the first SPLEED measurements on W(001) were reported, a variety of clean surfaces and adsorbate systems have been investigated in considerable detail.   Extensive comparisons between experimental data and dynamical SPLEED calculations have demonstrated the utility of spin-orbit SPLEED in surface geometrical structure determinations.

Polarization and intensity studies naturally complement one another in that they depend in different ways on the scattering potential, and polarization profiles tend to emphasize regions of (E,θ,ϕ) space where intensities are small.   Indeed, studies to date have shown that polarization features often are considerably more sensitive than intensities to surface relaxation and certain details of the scattering potential model.

From an experimental viewpoint, SPLEED measurements are now little more complex than conventional LEED investigations.   With the advent of the GaAs polarized electron source, simultaneous measurements of both intensities and asymmetries can be readily undertaken. The initial upgrade of a conventional LEED apparatus to incorporate a polarized electron gun and appropriate data acquisition capabilities can be accomplished with moderate expense of time and money.

As structural analyses are extended in the future to surfaces of ever greater complexity, and as the sophistication of dynamical SPLEED model calculations continues to advance, it seems evident that the additional information provided by SPLEED, as compared to conventional LEED, will be of ever increasing value.

Acknowledgment.   The research by the authors and their colleagues discussed here was supported by the U. S. Department of Energy and the Robert A. Welch Foundation.

REFERENCES

Adams D L, Nielsen H B and Van Hove M A 1979 Phys. Rev. B 20 4789
Alvarado S F, Hopster H, Feder R and Pleyer H 1981 Sol. State Comm. 39 1319
Bauer P, Feder R and Müller N 1980a Sol. State Comm. 36 249
Bauer P, Feder R and Müller N 1980b Surf. Sci. 99 L395
Bauer P, Eckstein W and Müller N 1983 Z. Phys. B 52 185
Buck T M, Wheatley G H and Marchut L 1983 Phys. Rev. Lett. 51 43
Calvert R L, Russell G J and Haneman D 1977 Phys. Rev. Lett. 39 1226
Davies J A, Jackson D P, Norton P R, Posner D E and Unertl W N 1980 Sol. State Comm. 34 41
Davisson C J and Germer L H 1927a Nature 119 558
Davisson C J and Germer L H 1927b Phys. Rev. 30 705
Davisson C J and Germer L H 1928a Nature 122 809
Davisson C J and Germer L H 1928b Proc. Natl. Acad. Sci. U.S.A. 122 809
Davisson C J and Germer L H 1929 Phys. Rev. 33 760
Demuth J E and Rhodin T N 1974a Surf. Sci. 42 261
Demuth J E and Rhodin T N 1974b Surf. Sci. 45 249
Demuth J E, Jepsen D W and Marcus P M 1973a Phys. Rev. Lett. 31 540
Demuth J E, Jepsen D W and Marcus P M 1973b J. Phys. C 6 L307
Demuth J E, Marcus P M and Jepsen D W 1975 Phys Rev. B 11 1460
Dietz R E, McRae E G and Campbell R L 1980 Phys. Rev. Lett. 45 1280
Erbudak M, Ravano G and Müller N 1982 Phys. Lett. 90A 62
Feder R 1971 Phys. Status Solidi B 46 K31
Feder R 1972 Phys. Status Solidi B 49 699
Feder R 1973 Phys. Status Solidi B 56 K43
Feder R 1974 Phys. Status Solidi B 62 135
Feder R 1975 Surf. Sci 51 297
Feder R 1976 Phys. Rev. Lett. 36 598
Feder R 1977a Surf. Sci. 68 229
Feder R 1977b Phys. Rev. B 15 1751
Feder R 1981 J. Phys. C 14 2049
Feder R 1983 Phys. Scripta T4 47
Feder R 1985 Private communication
Feder R and Kirschner J 1981 Surf. Sci. 103 75
Feder R, Müller N and Wolf D 1977 Z Phys. B 28 265
Feder R, Pleyer H, Bauer P and Muller N 1981 Surf. Sci. 109 419
Griffiths K, Kendon C, King D A and Pendry J B 1981 Phys. Rev. Lett. 46 1584
Jamison K D, Lind D M, Dunning F B and Walters G K 1985 Surf. Sci. in press
Jennings P J 1970 Surf. Sci. 20 18
Jennings P J 1971 Surf. Sci. 26 509
Jennings P J 1974 Jpn. J. Appl. Phys. Suppl. 2 661
Jones R O and Jennings P J 1983 Phys. Rev. B 27 4702
Jost K and Kessler J 1966 Z. Phys. 195 1
Kalisvaart M, O'Neill M R, Riddle T W, Dunning F B and Walters G K 1978 Phys. Rev. B 17 1570
Kesmodel L L, Stair P C and Somorjai G A 1977 Surf. Sci. 64 342

320

Kessler J 1985 Polarized Electrons 2nd ed (Berlin and New York: Springer-Verlag)

Kirschner J and Feder R 1979 Phys. Rev. Lett. 42 1008

Kirschner J and Feder R 1981 Surf. Sci. 104 448

Koike K and Hayakawa K 1983 Jpn. J. Appl. Phys. 22 1332

Kuyatt C E 1975 Phys. Rev. B 12 4581

Lang J K, Jamison K D, Dunning F B, Walters G K, Passler M A, Ignatiev A, Tamura E and Feder R 1982 Surf. Sci. 123 247

Lang N D and Kohn W 1973 Phys. Rev. B 7 3541

Mahan H, Riddle T W, Dunning F B and Walters G K 1980 Surf. Sci. 93 550

Marcus P M and Jona F (Ed.) 1984 Determination of Surface Structure by LEED (New York and London: Plenum Press)

McRae E G 1971 Rev. Mod. Phys. 51 541

McRae E G and Malic R A 1984 Surf. Sci. 148 551

McRae E G, Pierce D T, Wang G-C and Celotta R J 1981 Phys. Rev. B 24 4230

Müller N 1977 Dissertation (Münich)

Müller N, Erbudak M and Wolf D 1981 Sol. State Comm. 39 1247

O'Neill M R, Kalisvaart M, Dunning F B and Walters G K 1975 Phys. Rev. Lett. 34 1167

Pierce D T and Celotta R J 1981a Advances in Electronics and Electron Phys. 56 219

Pierce D T, Celotta R J, Wang G-C, Unertl W N, Galejs A, Kuyatt C E and Mielczarek S R 1979 Rev. Sci. Instrum. 51 478

Pierce D T, Celotta R G and Wang G-C 1981 Sol. State Comm. 39 1053

Potter H C and Blakely J 1975 J. Vac. Sci. Tech. 12 635

Reihl B 1981 Z. Phys. B 41 21

Riddle T W, Mahan A H, Dunning F B and Walters G K 1978 J. Vac. Sci. Technol. 15 1686

Riddle T W, Mahan A H, Dunning F B and Walters G K 1971a Surf. Sci. 82 511

Riddle T W, Mahan A H, Dunning F B and Walters G K 1979b Surf. Sci. 82 517

Sundaram V S and Robertson W D 1976 Surf. Sci. 55 324

Sundaram V S, Farrell B, Alben R S and Robertson W D 1973 Phys. Rev. Lett. 31 1136

Sundaram V S, Alben R S and Robertson W D 1974 Surf. Sci 46 653

Tamura E and Feder R 1983 Vacuum 33 864

Wang G-C, Dunlap B I, Celotta R J and Pierce D T 1979 Phys. Rev. Lett. 42 1349

Wang G-C, Celotta R J and Pierce D T 1981 Phys. Rev. B 23 1761

Wang G-C, Unguris J, Pierce D T and Celotta R J 1982 Surf. Sci. 114 L35

Wendelken J F and Kirschner J 1981 Surf. Sci. 110 1

Waller G and Gradman U 1982 Phys. Rev. B 26 6330

Chapter 7

Elastic Spin-Polarized Low-Energy Electron Scattering from Magnetic Surfaces

U. Gradmann

Physikalisches Institut, Technische Universität Clausthal,
D-3392 Clausthal-Zellerfeld, FRG

and

S.F. Alvarado

Institut für Festkörperforschung, KFA Jülich, D-5170 Jülich, FRG

7.1. Introduction: Spin Polarized Electron Scattering in Comparison with other Experimental Probes of Surface Magnetism.

The question how scattering of an electron from magnetic surfaces depends on the orientation of its spin with respect to the surface magnetization was raised first by Davisson and Germer (1929) who tried to observe spin polarized electron scattering in a double scattering experiment using Ni(111)-surfaces, soon after their pioneering discovery of electron diffraction. According to their analysis, they could not detect the effect. A reinterpretation by Kuyatt (1975) implied that Davisson and Germer's data contained exchange-induced magnetic scattering asymmetries of up to 27%. However, Feder (1977), on the basis of theoretical calculations, suggested that the data could be explained also by small misalignment of the crystals. In the light of recent experimental results on Ni(001) surfaces (Alvarado et al 1982c, Feder et al 1983), where the exchange-induced scattering asymmetries were at a few percent, it appears that Feder's suggestion and Davisson-Germer's original conclusion are right: Davisson and Germer did not detect magnetic effects in electron scattering, for which they searched for the first time.

Half a century after Davisson and Germer, it remains a challenging task to use spin-polarized low-energy electron diffraction (SPLEED) for the magnetic analysis of surfaces, just like conventional LEED for their structural analysis. The final aim is the quantitative analysis of space-dependent spin-densities from quantitative comparison of relativistic dynamical scattering calculations with SPLEED experiments. The first steps in this direction were done by theory, starting with a kinematical approach by

Vredevoe and de Wames (1968) and a first dynamical study by
Feder (1973); for a full exposition of the theoretical analysis compare chapter 4 of this book. Experiments in the
field became possible only after the introduction of the
GaAs-source for spin-polarized electrons (G.C. Wang et al
1979), starting with the pioneering SPLEED experiment of
Celotta et al (1979). It is the aim of the present paper to
give a review of physical principles and present results in
this field, emphasizing the experimental point of view, in
comparison with other experimental techniques of surface
magnetism. We can report encouraging first results of high
physical interest; quantitative refinement is still a challenging task for future research.

Surface magnetism, being of considerable scientific
interest since many years (Gradmann 1977), became a subject
of high actuality recently, both in fundamental research
and with respect to applications. For a theoretical review
compare Freeman (1983) and chapter 1 of this book, for experimental reviews emphasizing Mössbauer methods and electron capture spectroscopy, resp., compare Bayreuther
(1983a) and Rau (1982).

The merit of SPLEED, in comparison with other probes
of surface magnetism, is given by its general applicability
to any ferromagnetic surface, and its ability to give information on just the few topmost atomic layers, for which
deviation from the bulk magnetization must be considered in
the ground state. The difficulties of a quantitative dynamical analysis should be emphasized, however. Further, the
use of low energy electrons prevents the analysis in finite
magnetic fields. This problem is common with photoemission
of spin polarized electrons, compare chapter 12 of this
book. Finally, SPLEED cannot be applied to interfaces.

For comparison, local analysis both in surfaces and interfaces, and in finite magnetic fields, can be done by Mössbauer spectroscopy. It provides a unique probe mainly for the most convenient Mössbauer isotope ^{57}Fe and magnetic order in Fe, if it is possible to deposit the ^{57}Fe-Mössbauer isotope separately in an atomic probe layer. This approach, which has been introduced by Owens et al (1979), using probes of 2 to 6 layers, and Fe(110) films prepared on Ag(111) and coated by several materials, has been reviewed by G. Bayreuther (1983 a,b). Only recently, literally monolayer probe Mössbauer analysis has been performed by Korecki et al (1985) for clean and Ag-coated Fe(110)-surfaces. A fundamental problem consists in that one measures magnetic hyperfine fields, which provide a local, but only rough measure of the magnetization (Ohnishi et al 1984).

A further complement to SPLEED is given by the direct measurement of magnetic moments in oligatomic ferromagnetic films (Gradmann 1974, 1977, Bayreuther 1983b, Gradmann et al 1984). Measurements can be done in finite fields; torsion magnetometry results in magnetic anisotropies in addition to and independently of magnetic moments. As an integral method, magnetometry seems not appropriate for local analysis. However, it is extremely useful in the study of changes of magnetic moments at surfaces, caused by any reaction or structural change near the surface (Bergter et al 1985). The method is applicable to interfaces.

Electron capture spectroscopy (Rau 1982) has the unique merit to probe magnetic order in the topmost layer strictly separately. It can be applied in finite fields and in UHV, to study magnetic changes connected with surface processes. Interpretation of the experimental data in terms of surface magnetization is a very challenging problem for this method.

Spin-dependent band-structures, surface states and excitation spectra near magnetic surfaces can be analyzed by spin-resolved photoemission, spin-dependent bremsstrahlung and inelastic electron scattering, to be discussed in other chapters of this book.

It becomes clear from this concise discussion of the most important probes of surface magnetism, that SPLEED must be used in combination and in comparison with other experimental methods, and theoretical analysis, to get a complete picture of surface magnetism. Our discussion is guided by this point of view. The discussion is based on chapters 4.3.1 and 4.3.7 of this book where the main definitions and theoretical principles of SPLEED are given, and on chapter 5, where the GaAs-source of polarized electrons is discussed.

7.2. Surface Magnetization of 3d-Magnets at Low Temperatures

For comparison with extended modern theoretical work on the electronic and magnetic structure of solid surfaces in the ground state (chapter 1 of this book), we need low temperature experiments. For 3d-ferromagnets, however, because of their high Curie temperatures, room temperature experiments can be taken, for the present problems, as a reasonable approximation of low temperature experiments. Finite temperature corrections can be included. Low temperature experiments in this sense are discussed in this section for Ni(001) and Fe(110)-surfaces.

In the experimental analysis of magnetic surfaces using SPLEED, two general problems must be solved: First, the use of low energy electrons forbids the use of external or stray magnetic fields during the scattering experiments. Therefore, the samples must be used in a stray-field free geometry and a remanent state of magnetic saturation. Celotta et al (1979), in their first demonstration of magnetic SPLEED, used a Ni single crystal platelet forming part of a closed magnetic circuit; more elegantly, Alvarado et al (1982a-d), in their work on Ni(001), used a picture-frame Ni-single crystal forming the closed magnetic circuit in itself. The geometry of the experiment is shown in fig. 1. Finally, Waller et al (1982), in their work on Fe(110), used epitaxial Fe(110)-films on W(110), about 100 Å thick, with negligible stray fields, avoiding, at the same time, the difficult preparation of clean bulk Fe surfaces. A saturated remanent state could be established in both cases.

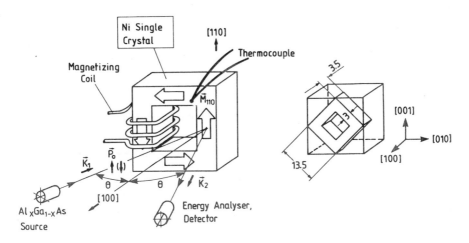

Fig. 1: Geometry of the picture-frame Ni-single-crystal, magnetic circuit and scattering experiment for SPLEED-experiment of Alvarado et al (1982a-d) and Feder et al (1983). The temperature is stabilized by indirect pulse heating. The crystal is magnetized after each heating pulse. Data collection is switched off during the heating-magnetizing pulses (from Alvarado et al 1982b).

Secondly, spin-orbit-coupling effects (compare chapters 4, 6 of this book) are superimposed on the magnetic exchange effects asked for in this chapter and must be separated. One way to separate exchange from spin-orbit effects was used by Celotta et al (1979) and by Kirschner (1984a,b), who used an electron polarization lying in the scattering plane, thus eliminating spin-orbit asymmetries by symmetry. The other way, which was used both by Alvarado et al (1982a-d) and by Waller et al (1982), uses a geometry where both the sample magnetization (magnetic easy axis) and the spin polarization coincide with the normal of the scattering plane, compare fig. 1. This requires transversal polarization of the electron beam. As both the AlGaAs source used by Alvarado et al (1982a-d) and the GaAsP source of Waller et al (1982) give longitudinally polarized electrons, the beam must be deflected in an electrostatic field (spherical or cylindrical condenser) to get the desired transversally polarized beam.

As discussed first by Alvarado et al (1982c), simultaneous probing of both exchange and spin-orbit polarizations can be performed as follows (for a theoretical foundation compare chapter 4.3): With wave vectors $\vec{k}_1$ and $\vec{k}_2$ of incident and of scattered electrons, the scattering normal $\vec{n}$ is given by $\vec{n} = \vec{k}_1 \times \vec{k}_2 / |[\vec{k}_1 \times \vec{k}_2]|$. If $\vec{k}_1$ and $\vec{k}_2$ are given, the reflected intensities I_σ^μ depend on whether the primary beam polarization vector $\vec{P}_o$ is parallel ($\sigma = +$) or antiparallel ($\sigma = -$) to $\vec{n}$ and on whether magnetization $\vec{M}$ is antiparallel ($\mu = +$) or parallel ($\mu = -$) to $\vec{n}$.

Experimentally, one determines the scattering asymmetries

$$A^\mu = (1/|P_o|)(I_+^\mu - I_-^\mu)/(I_+^\mu + I_-^\mu). \qquad (1)$$

This can easily be done using a GaAs-source by optical modulation of P_o and detection of $I_+^\mu - I_-^\mu$ using lock-in techniques (Celotta et al 1979, Waller et al 1982) or by syn-
chronous counting techniques (Alvarado et al 1982a-d).

The experimental quantities

$$A_{ex} = (A^+ - A^-)/2 \quad \text{and} \tag{2}$$

$$A_{so} = (A^+ + A^-)/2 \tag{3}$$

then represent, in a very good approximation for Ni and Fe,
the exchange-induced asymmetry in the absence of spin-orbit
coupling and the spin-orbit-induced asymmetry in the absence of magnetism.

The advantages of simultaneously probing A_{ex} and A_{so}
have been demonstrated by Feder et al (1983) in their
SPLEED-analysis of Ni(001). Experimentally, rocking curves
were taken, that means A (θ) was measured at fixed energies
and fixed scattering plane as a function of the angle of
incidence, θ; this was done both for the specular and one
non-specular beam. In the first stage of analysis, the measured $A_{so}(\theta)$ curves were compared with their theoretical
counterparts, which were calculated by a relativistic LEED
formalism for a nonmagnetic crystal (i.e. using a spin-averaged potential). This confirmed the adequacy of an energy-dependent exchange-correlation potential, suggested the
importance of localised loss processes (3p excitation) (as
phenomenologically described by imaginary phase shift parts
δ_{11}), and determined the topmost interlayer spacing as
bulk-like, with a possible outward relaxation by 1%. These
results were used in subsequent calculations of $A_{ex}(\theta)$ by
means of a non-relativistic scalar LEED theory to determine

the layer-dependent surface magnetization M_n (with M_1 belonging to the topmost layer). It is important to note that M_n can be obtained for any temperature, at which A_{ex} is measured, via comparison with calculations for assumed trial values. To reduce the computing effort and to make contact with first-principles theoretical results for T = 0K, Feder et al (1983) focussed, instead, on four surface magnetization models at T = 0 and extrapolated them to the experimental temperature with the aid of a semi-empirical Heisenberg model in the mean field approximation. Three of these models are illustrated in fig. 2 for T = 0, 300 and 520 K.

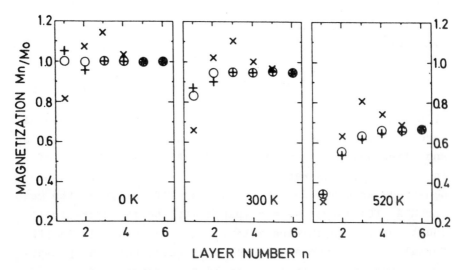

Fig. 2: Spin magnetic moments M_n (relative to the zero temperature bulk magnetic moment M_o) of the topmost six monoatomic layers of ferromagnetic Ni(001). At T = 0K: homogeneous model (o), predictions by Jepsen et al (1980) (+) and by Wang and Freeman (1980) (X), taking the central layer moment als the bulk moment. The corresponding results at T = 300 K and 520 K were therefrom obtained by a mean field treatment of an effective Heisenberg model.

The four models are characterized as follows:

(1) WF-model (Wang and Freeman 1980): M_1 reduced by 20%; oscillating enhancement of M_2 and M_3.

(2) JMA-model (Jepsen et al 1980): M_1 enhanced by 5%; M_2 slightly reduced.

(3) Homogeneous model: $M_1 = M_2 = \ldots = M_{bulk}$.

(4) Dead layer model: $M_1 = 0$, $M_2 = M_3 = \ldots = M_{bulk}$.

Theoretical $A_{ex}(\theta)$ results are given in fig. 3 in comparison with the experimental values, for 300 K.

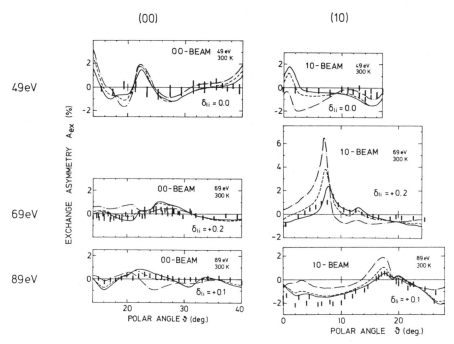

Fig. 3: Exchange asymmetry $A_{ex}(\theta)$ of Ni(001) for the (00) and (10) beam in a (110) scattering plane, for fixed energies (49 eV, 69 eV, 89 eV) as a function of polar angle (angle of incidence) θ. Experiments at 300 K (vertical bars) are compared with the WF model (- -), the JMA (——) and the "dead layer" model (—·—·—) extrapolated to 300 K. Calculations using relaxation $\delta_{12} = 0$ and imaginary phase shifts as indicated, taken from the fit of $A_{so}(\theta)$ (from Feder et al 1983).

A clear preference for the JMA-model can be seen, for which good agreement is achieved. The most decisive information comes from the non-specular beam. The asymmetries calculated for the homogeneous model are very close to those for the JMA-model, and the homogeneous model is within the uncertainty of the analysis. The analysis performed at 520 K confirmed these findings, and an enhancement of M_1 by 5% $\pm$ 5% at T = 0K could be concluded.

Meanwhile the enhancement of M_1 has also been found in refined calculations of Freeman's group (Freeman 1983). A slight enhancement of the surface magnetization M_1 in Ni(001) is therefore now well established.

A similar analysis has been performed for Fe(110)-surfaces by Tamura and Feder (1982), compare also Gradmann et al (1983), using experimental data of Waller and Gradmann (1982). Only the specular beam was measured, at fixed angle of incidence, θ, in a (001) scattering plane as a function of energy at room temperature. Experiments were compared with calculations for model crystals, in which the magnetization M_1 of the first layer was changed. As shown in fig. 4, more pronounced structures in $A_{ex}(E)$ were observed than for Ni and higher magnitudes up to 32% for A_{ex}; good qualitative agreement between theory and experiment could be achieved for all main features of $A_{ex}(E)$. Special emphasis was given to the theoretical peak for $\theta = 45°$, E = 74 eV, because it reacts very sensitively to a change of M_1, as shown in fig. 5. It is this sensitive feature which was used to estimate M_1. Clearly, the best fit to the experiment is given for $M_1 = 1.3\ M_B$ (bulk magnetization M_B). According to this finding, the surface magnetization is enhanced by about 30% in Fe(110). This is compatible with a 20% enhancement obtained in a subsequent tight-bind-

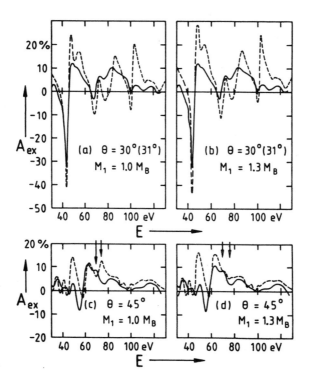

Fig. 4: Exchange asymmetry A_{ex} as a function of kinetic energy E for specular reflection of spin-polarized electrons on Fe(110). Experiments for incident angles θ = 31° and 45°, resp. (full curves) in the (001)-scattering-plane are compared with calculations for θ = 30° and 45°, resp. (dotted curves). For figs. 4a, 4c the calculations were done with a magnetization M_1 of the topmost layer given by M_1 = 1.0 M_B (bulk magnetization M_B). For figs. 4b, 4d, M_1 = 1.3 M_B is assumed for the calculations (from Gradmann et al 1983).

ing calculation by Victora and Falicov (1984). Surface enhanced magnetization in Fe was predicted previously for the case of a seven layers Fe(001) film (M_1 = 1.5 M_B) by Wang and Freeman (1981) and for a Fe(001)-monolayer (M_1 = 1.66 M_B) by Noffke and Fritsche (1981).

334

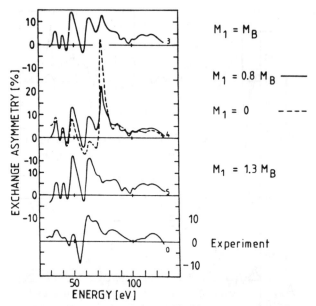

Fig. 5: Exchange asymmetry A_{ex} of the specular beam from $\overline{Fe(110)}$ in the (001) plane for $\theta = 45°$. Theoretical profiles for different values of M_1 show the sensitivity of the peak at 74 eV on M_1 (from Támura and Feder 1982).

In summary, a surface enhancement of the ground state magnetization, which was obtained in ab initio calculations for Ni(001), Fe(001) and Fe(110), and determined by analyzing experimental SPLEED data for Ni(001) and Fe(110), seems to be a general tendency for 3d-ferromagnetic surfaces. Magnetic "dead layers" are definitely absent in clean Ni and Fe surfaces (compare also chapter 1 of this book).

7.3. Surface Magnetism and Chemisorption at 3d-Magnet Surfaces

Magnetic phenomena connected with chemisorption, like changes of magnetic moments and magnetic surface anisotro-

pies (Gradmann 1984) must be included in a coherent understanding of chemisorption in general. Some steps in this
direction have been done by magnetometry of superparamag-
netic small particle systems of Fe, Co and Ni by Selwood
(1975) and by FMR of polycrystalline films by Göpel (1979).
In general, the magnetic moment is reduced by chemisorption; for example, one H atom apparently destroys the magnetic moment of one Ni-atom. For single crystal surfaces,
this could be qualitatively confirmed by spin-polarization
methods: In spin-polarized field emission from Ni(100),
H-adsorption extinguished spin polarization completely
(Landolt and Campagna 1977); in electron capture spectroscopy on Ni(110), spin polarization decreased from - 96% to
- 8% by adsorption of H (Rau 1982). A reduction of spin polarization by adsorption of O and CO on Ni(110) has been
detected recently by spin polarized photoemission (Schmitt
et al 1985) (compare also chapter 12 of this book).

These emission-type spin-polarization methods primarily give information on the spin-polarized surface band
structure; their interpretation is by no means easy. It is
of high interest to extend them and complement them by spin
polarized electron scattering along the lines of the last
section to determine changes of magnetic moments in the
surface. Finally, these spin polarization studies should be
combined with magnetometry of oligatomic films during chemisorption, to detect directly the change of the magnetic
moment, as has been done recently by Bergter et al (1985)
for the adsorption of Cu on Ni(111).

At present, this challenging program is only in the
beginning. S and O on Fe(110) are the only adsorption systems for which magnetic SPLEED data are available. Experiments have been performed by Kirschner (1984a) with longi-

tudinal polarization of the electrons, thus eliminating spin-orbit-asymmetries. Intensity and asymmetry were observed for the specular beam at 45° incidence as a function of electron energy. For oxygen, 1/4 monolayer changes the intensity only weakly, indicating that there is no reconstruction of the substrate. This is shown in fig. 6.

However, strong changes were found for the exchange asymmetry, mainly at energies below 30 eV, with minor changes only between 30 and 100 eV. Tentatively, the changes for low energies, i.e. large information depths, were explained from changes in the surface barrier, whereas the minor changes for high energies, i.e. low information depth, indicate that only minor changes in the magnetic moment took place.

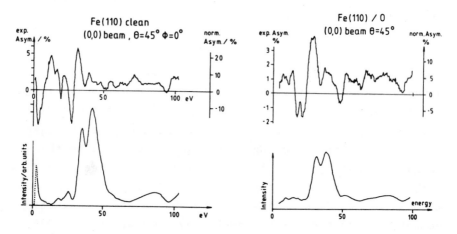

Fig. 6: Asymmetry (upper panels) and intensity (lower panels) of the specular beam as a function of kinetic electron energy for clean Fe(110) (left panels) and p(2x2)-0 (right panels). The nomenclature refers to the primitive lattice, oxygen coverage is about a quarter of a monolayer. The left-hand scale indicates the measured asymmetry, the right-hand scale is normalized by the primary beam polarization (from Kirschner 1984a).

For sulfur which was brought to the surface by segregation from the volume, both intensity and asymmetry showed strong changes for a p(2x2)S-coverage, indicating a substrate reconstruction, which was found previously by Shih et al (1981) using LEED intensities. Again, only minor changes of the asymmetry were observed at higher energies (low information depths), indicating again only minor changes of the magnetic moment.

In a theoretical study of Fe(110)-p(2x2)S, Tamura and Feder (1984) found that the specular beam is nearly independent both in intensity and in asymmetry of both top layer magnetization and reconstruction of the Fe(110)-substrate. Little similarity could be detected between these theoretical results and the data of Kirschner (1984a). It has been suggested, that the experimental data might have been influenced by subsurface S, as the p(2x2)S-structure was created by segregation. Therefore, further experiments with S adsorbed from the vacuum side are desired. In contrast to the specular beam, in the nonspecular (10) beam sharp polarization features were calculated which depend very sensitively on the surface magnetization, as shown in fig. 7.

Corresponding experiments are not available. However, these theoretical results are encouraging in that they suggest A_{ex} as a sensitive measure for the surface magnetization also in chemisorption systems. The outstanding importance of nonspecular beams, found for clean surfaces, was confirmed. As a whole, the available data form a strong indication that the study of magnetic phenomena connected with chemisorption by SPLEED, in combination with other spectroscopic methods and with magnetometry, forms a promising field of future research.

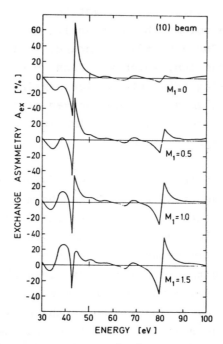

Fig. 7: Calculated A_{ex} of the (10) beam for normal incidence on Fe(110)p(2x2)-S with reconstructed top Fe layer
and with various values for the top layer magnetization M_1
as indicated (from Tamura and Feder 1984).

7.4. Critical Behavior of Ni-Surfaces

The critical behavior of magnetic phenomena near the
surface of magnetic crystals has been a field of great
theoretical interest since many years (compare chapter 3 of
this book). Experimental data on clean single-crystal surfaces, however, have been scarce. Pioneering work was done
by Palmberg et al (1968) on antiferromagnetic NiO-surfaces,
using the temperature-dependence of half-order LEED spots
which are caused by antiferromagnetic order and therefore
disappear at the Néel-temperature. They first found that

the critical exponent of the surface magnetization β_1 is close to 1, in comparison with $\beta \approx 1/3$ for the bulk.

Spin polarized electron scattering can be used as a convenient experimental tool to analyse critical behavior, if A_{ex} is proportional to the surface magnetization m_1, as a function of temperature. This appears reasonable near T_c, as the magnetic coherence length $\xi(t)$ diverges with $t = (T_c - T)/T_c$ going to zero, and so ξ becomes large in comparison with the information depth of the low energy electrons, which therefore probe only a region which scales with the surface layer magnetization. The proportionality of A_{ex} and surface magnetization of a ferromagnet in the critical region was confirmed in an extended theoretical analysis by Feder and Pleyer (1982). This analysis further showed that this proportionality could not be expected in general outside the critical region and deviations were indeed found experimentally for Fe(110) for $t \geq 0.1$ (Kirschner 1984b).

Considering this problem, an experimental analysis must be restricted to $t \lesssim 0.1$, and the results must be checked by comparing $A_{ex}(t)$ for various energies. This has been done in an experimental study of the critical behavior of Ni(001) and Ni(110) surfaces given by Alvarado et al (1982a,b). The experimental setup has been discussed in section 7.2, compare fig. 1. Specular beams were used only, but both kinetic energy E_K and angle of incidence θ were varied. As examples, log-log plots of $A_{ex}(t)$ are shown in fig. 8 for Ni(001) at $E_K = 13$ eV, $\theta = 60°$ and for Ni(110) at $E_K = 49$ eV, $\theta = 15°$.

A critical power law $A_{ex} \sim t^{\beta_1}$ is clearly confirmed and the surface critical temperature is the same as the bulk one. The critical exponent β_1 of A_{ex}, interpreted as

exponent of the first monolayer, is determined as
$\beta_1 = 0.80\pm0.02$ (mean).

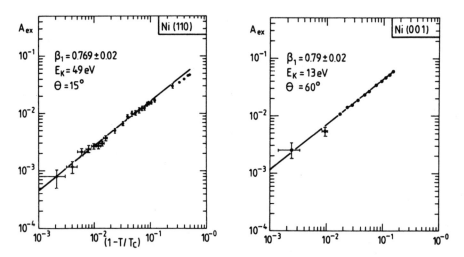

Fig. 8: Log-log plot of A_{ex} vs reduced temperature for Ni(001) and Ni(110) (from Alvarado et al 1982a).

Experimental values of β_1 for both Ni(001) and Ni(110) at different electron kinetic energies and angles of incidence are shown in table 1. A significant difference between the two surfaces cannot be seen. In the limits of accuracy, the experimental values for both surfaces fit to all three models. A slight preference for XY behavior, which was originally stated by Alvarado et al (1982b), or even Ising behavior, may be seen in the data. However, the relevance of surface anisotropies for surface critical behavior has been contradicted in a theoretical note of Diehl and Eisenriegler (1982). Some connexion with magnetic surface anisotropies, caused by lattice contractions between the first and second atomic layer, has been discussed by Alvarado et al (1982a,b), who conclude that this effect is irrelevant for the interpretation of the measured β_1.

Table 1 (from Alvarado et al 1982a)

Surface critical exponent β_1 of Ni(001) and Ni(110) surfaces. Experimental values from Alvarado et al (1982a), for different angles of incidence, θ , and electron energies E.

Experimental values	θ/degrees	E/eV	β_1
Ni(001)	15	46	$0.825^{+0.025}_{-0.040}$
	15	37	$0.805^{+0.045}_{-0.085}$
	60	13	0.79 ± 0.02
Ni(001), mean value			0.81 ± 0.02
Ni(110)	15	67	0.807 ± 0.035
	15	49	0.769 ± 0.02
Ni(110), mean value			0.79 ± 0.02
Theoretical models:	Ising model		$0.776 \ldots 0.80$
	XY-model		$0.79 \ldots 0.835$
	Heisenberg model		$0.81 \ldots 0.88.$

7.5. Enhanced Surface Curie Temperature and Surface Magnetic Reconstruction of Gd(0001)

From theoretical considerations concerning the surface of a semi-infinite ferromagnet it is expected that for a sufficiently enhanced coupling between neighbouring surface atoms a so-called pure surface transition can occur. Here the surface magnetic moments would undergo a critical ordering transition at a temperature T_{cs} larger than the bulk Curie point T_{cb} (cf. K. Binder, Chapter 3 of this book).

A first indication of this effect was observed by C. Rau and E. Eichner (1980) on polycrystalline Gd thin films

using electron capture spectroscopy. However, those measurements were performed in relatively high magnetic fields
and an extrapolation to zero magnetic field was used (C.
Rau, 1983) which is not unproblematic. The reevaluation of
the data by means of the proper scaling ansatz for m_1 as a
function of the reduced temperature and magnetic field h:

$$m_1(t,h) = |t|^{\beta_1} \sigma(h|t|^{-\Delta}) \tag{4}$$

where σ is an unspecified universal scaling function, gives
a value of $\beta_1 = 0.2-0.3$ which is taken as an indication
that the system is near a multicritical point (S. Eichner
1984). However, on the basis of these data it cannot be de-
cided whether the surface enhancement is strong enough to
produce a pure surface transition (Eichner 1984, Diehl
1985).

A definitive confirmation of the existence of surface
magnetic ordering above the bulk Curie temperature (T_{cb})
was achieved by temperature-dependent SPLEED measurements
on Gd(0001) thin films grown epitaxially on W(110) substrates (D. Weller et al, 1985a). The measurements were
performed with remanently magnetized samples using the same
technique as for the Ni measurements described in 7.2. Fig.
9 shows a typical measurement of the temperature dependence
of the exchange asymmetry A_{ex} of electrons scattered by an
epitaxial film of thickness d = 500 Å. It is found that A_{ex}
is different from zero at temperatures well above the bulk
Curie point T_{cb} = 293 K. This is taken as evidence for a
pure surface transition on Gd(0001). For this specific sample the surface Curie temperature is T_{cs} = 315 $\pm$ 1 K which
is 22 K higher than the bulk Curie temperature. This is the
maximum enhancement observed and was measured on several
samples. Exposure of the surfaces to doses of about 1 Lang-

muir of oxygen or hydrogen or to the accumulated residual gases of the measuring apparatus causes a decrease of T_{cs} to the bulk value T_{cb}. This strong sensitivity to surface contamination demonstrates that for $T_{cb} < T < T_{cs}$ the surface enhanced magnetic order is basically confined to the topmost Gd atomic layer. It should be noted that bulk Gd is a system whose critical behaviour conforms to the Heisenberg model (n = 3). As pointed out by Diehl and Eisenriegler (1984), if the surface region were also isotropic, it would not exhibit a surface transition. They show, however, that this transition can occur in the presence of an easy magnetization axis at the surface.

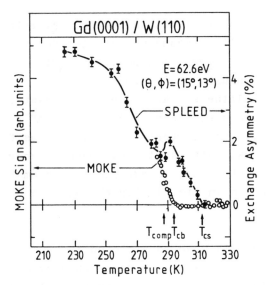

Fig. 9: Temperature dependence of A_{ex}, measured in a 500 Å thick Gd film. The bulk Curie temperature T_{cb} of typical film was determined in situ by a magnetooptical Kerr effect (MOKE) method. The Curie temperature of the surface is T_{cs} = 315 K, 22 K higher than T_{cb} = 293 K (from D. Weller et al, 1985a).

A characteristic feature of the A_{ex} vs T data is the

minimum of $|A_{ex}|$ at $T_{comp} \cong 289$ K which can be explained
from surface magnetic reconstruction where the surface magnetic moments are coupled antiparallel to the bulk magnetization. Actually a spin-polarized energy-resolved photoemission study of Gd(0001) (D. Weller et al, 1985a,b) shows
that the 4f levels of the surface layer atoms are antiferromagnetically coupled to the underlying atomic layers of
the bulk (cf. W. Gudat and E. Kisker, Chapter 12 of this
book).

7.6. Ferromagnetic Glasses

Spin-dependent elastic electron scattering from the
surface of metallic glasses was studied for the amorphous
ferromagnets $Ni_{40}Fe_{40}B_{20}$ and $Fe_{81.5}B_{14.5}Si_4$ by Pierce et al
(1982) and by Unguris et al (1984), respectively. In con-
trast to crystalline surfaces, where diffraction results in
well-defined beams with strong dependence of their intensities on energy, the scattering from amorphous ferromagnets
varies only slowly with angle and energy. Accordingly with
this diffuse scattering, an experimental geometry was chos-
en which is shown in fig. 10 for the example of
$Fe_{81.5}B_{14.5}Si_4$.

The sample is used as a closed loop of metallic glass
ribbon, which could easily be magnetized by a coil. For the
case of $Ni_{40}Fe_{40}B_{20}$, the sample formed a closed loop in
combination with a small electromagnet. For both cases, the
disturbance by magnetic stray fields could be neglected, as
shown by a detailed discussion. A transversally polarized
electron beam from a GaAs-source is incident with an angle
of incidence α. Scattered electrons are detected in the
plane of incidence only, using a Faraday cup forming a

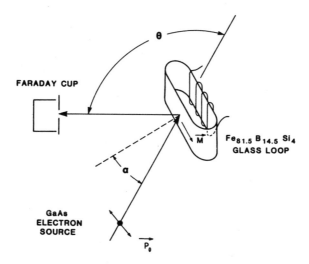

Fig. 10: Scattering geometry of Unguris et al (1984). The transversally polarized incident electron beam, the normal of incidence and the Faraday cup are lying in one scattering plane. Polarization P_o of the incident beam and magnetization $\vec{M}$ of the sample are lying in the scattering plane, too (from Unguris et al (1984)).

scattering angle θ with the incident beam. θ is varied independently of α. An enlarged aperture of $\Delta\theta = 9°$ is used for the cup. Spin-orbit effects are excluded by the orientation of spin polarization, which is in the scattering plane. Scattering asymmetries are caused only by the relative orientation of incident beam polarization $\vec{P}_o$ and magnetization $\vec{M}$, which is also in the scattering plane. The normalized asymmetry is defined as

$$A \equiv \frac{1}{P_o \cos\alpha} \frac{I_\uparrow - I_\downarrow}{I_\uparrow + I_\downarrow} \tag{5}$$

where $I_\uparrow$ and $I_\downarrow$ are intensities with the spins of incident electrons and of majority electrons in the target parallel and antiparallel, respectively.

The surfaces were cleaned by sputtering with 500 eV
Ar^+ ions, followed by annealing at 120°C for $Fe_{81}B_{14.5}Si_4$.
The $Ni_{40}Fe_{40}B_{20}$ was not annealed. In both cases, the chemical composition of the surface, as measured by Auger spectroscopy, presented approximately the bulk concentration of
the elements, however, with a significant additional amount
of C up to 11 at %. The variations in surface composition
resulted only in changed scaling of A without changing its
dependence on angles and energy.

Measuring A as a function of magnetic field H, good
square magnetization loops were found for the cleaned surfaces. They agreed with the bulk magnetization loop. Low
coercive field in the order of 50 mOe were found, characteristic for ferromagnetic glasses. Therefore, the main
measurements could be performed in the saturated remanent
state.

The elastically scattered intensity I is shown in
fig. 11 for the case of $Ni_{40}Fe_{40}B_{20}$ for a primary energy of
100 eV as a function of scattering angle θ, for different
angles of incidence α. After correction for electron attenuation, I does not depend on α. The same applies for the
scattering asymmetry $S \equiv A$, which is shown in fig. 12 as a
function of θ for $\alpha = 30°$, and as a function of α for $\theta =$
166°, correction for attenuation is not needed for A, as $I_\uparrow$
and $I_\downarrow$ are attenuated in the same manner, compare eq. 5.

The independence of both I and A on α confirms the
isotropic structure of the amorphous target. A further
far-reaching interpretation was proposed following previous
work of Schilling and Webb (1970), who made a comparative
study of low-energy electron scattering from liquid and
gaseous Hg; they found for the case of backscattering that

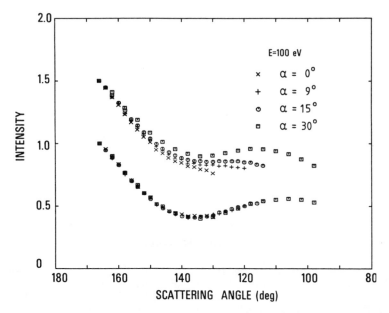

Fig. 11: Elastically scattered intensity from $Ni_{40}Fe_{40}B_{20}$ as a function of scattering angle θ for E = 100 eV and angles of incidence α of 0°, 9°, 15° and 30°. Curves are normalized at θ = 166° and displaced upwards from the lower curve, where corrections for electron attenuation have been applied (from Pierce et al 1982).

the angular and energy dependence of the scattering from the liquid is very similar to that of the free atom. It was proposed that the same applies for the ferromagnetic glasses, and that the data of figs. 11 and 12 can be interpreted as describing essentially atomic scattering. Some additional contribution of multiple scattering was estimated to be of the order of 30%. It was claimed that this information on atomic scattering might be useful in the analysis of multiple scattering processes from crystalline targets.

Because of the very low probing depth, which was estimated to 2.5 Å for $Ni_{40}Fe_{40}B_{20}$, at 90 eV, spin dependent elastic

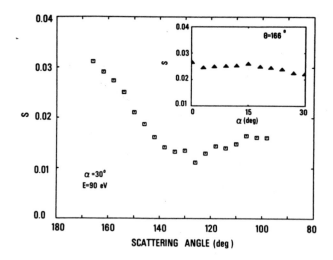

Fig. 12: Scattering asymmetry S ≡ A for elastic scattering of 90 eV electrons form $Ni_{40}Fe_{40}B_{20}$ with an angle of incidence $\alpha = 30°$, as a function of scattering angle θ. Inset shows S as a function of α for $\theta = 166°$ (from Pierce et al 1982).

electron scattering can be used as a probe of surface magnetism, as shown for the case of crystalline Ni(100) in section 7.4. Proportionality between A and surface magnetization M_s forms the basis of a straight-forward interpretation. For crystalline surfaces, this can be assumed only near T_c, as discussed by Feder and Pleyer (1982). In spin glasses, however, where single scattering dominates, $A \sim M_s$ can be assumed for all temperatures, M_s being a mean magnetization of the surface layer probed by the electrons, that means the topmost 1 or 2 atomic layers (whatever that means for an amorphous surface).

Fig. 13 shows M(T)/M(0) for surface of $Ni_{40}Fe_{40}B_{20}$ (from polarized electron scattering) and for the bulk, for comparison, in the low temperature region up to 300 K = 0.4 T_c. Apparently, the deviation from absolute saturation, 1 − M(T)/M(0), is proportional in both cases, being 3 times

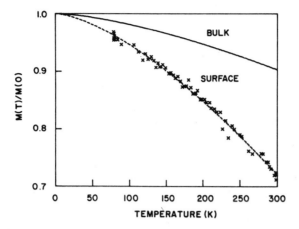

<u>Fig. 13</u>: Variation of surface magnetization in $Ni_{40}Fe_{40}B_{20}$, taken from stray asymmetry A of polarized electrons, in comparison with the bulk magnetization, measured by conventional methods (from Pierce et al 1982).

larger for the surface than for the bulk. The proportionality of both is shown clearly in fig. 14: $1 - M(T)/M(0)$ varies in the surface with T according the same power law as in the bulk, where the BLOCH law $1 - M(T)/M(0) \sim T^{3/2}$ is well known.

A behavior of this type was predicted by Mills and Maradudin (1967), who postulated

$$[1-M(T)/M(0)]_{surface} = \alpha\ [1-M(T)/M(0)]_{bulk} \qquad (6)$$

with $\alpha = 2$ for the low temperature region. This result, coming there from an extended theoretical analysis, can be interpreted easily as a result from the fact that spin waves show antinodes in the surface, if pinning is absent, and that the intensity in the antinodes is twice the mean intensity for a standing wave. However, Mills and Maradudin estimate that the surface disturbance, at 100 K, decays in about 2 atomic layers. From that, one should expect that

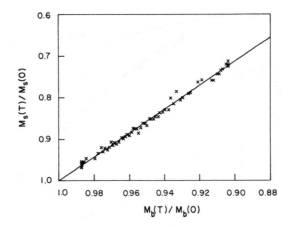

Fig. 14: Normalized magnetization $M_s(T)/M_s(0)$ of the surface of $Ni_{40}Fe_{40}B_{20}$ versus $M_b(T)/M_b(0)$ for the bulk, showing $(1-M_s(T)/M_s(0)) = \alpha(1-M(T)/M(0))$ with $\alpha = 3$. Temperature range is from 70 K to 300 K, which is about 0.09 to 0.40 T_c (from Pierce et al 1982).

the spin polarized electron data (probing depth ~ 1.5 atomic layers) should contain some bulk admixture, i.e. $\alpha < 2$ in eq. (2). So, whereas the experimental data are in agreement with the theoretical proportionality in eq. 2, the experimental value $\alpha = 3$ is in disagreement with Mills and Mardudin's theory. Note that Monte Carlo simulations of the Heisenberg model by Binder and Hohenberg (1974) resulted in a thermal decrease of surface magnetization even smaller than given by eq. (2) with $\alpha = 2$. Enhanced thermal decrease of the surface magnetization, as indicated by the present experiments, can be explained in principle from weakened exchange interaction in the surface, as has been discussed by Binder and Hohenberg (1973). A weakened exchange interaction seems not unreasonable in the surface of an amorphous ferromagnet.

Alvarado S F, Campagna M, Ciccacci F and Hopster H 1982a
    J. Appl. Phy. 53 7920
Alvarado S F, Campagna M and Hopster H 1982b
    Phys. Rev. Lett. 48 51
Alvarado S F, Feder R, Hopster H, Ciccacci F and Pleyer H
    1982c Z. Phys. B49 129
Alvarado S F, Hopster H and Campagna M 1982d Surf. Sci.
    117 294
Bayreuther G 1983a J. Magn. Mat. 38 273
    1983b J. Vac. Sci. Technol. A1 19
Bergter E, Gradmann U and Bergholz R 1985
    Solid State Commun. 53 565
Binder K and Hohenberg P C 1973 Phys. Rev. B9 2194
Binder K and Hohenberg P C 1974 Phys. Rev. B6 3461
Celotta R J, Pierce D T, Wang G C, Bader S D and Felcher
    G P 1979 Phys. Rev. Lett. 43 728
Davisson C and Germer L H 1929 Phys. Rev. 33 760
Diehl H W and Eisenriegler E 1982 Phys. Rev. Lett. 48 1767
Diehl H W and Eisenriegler E 1984 Phys. Rev. B30 300
Diehl H W 1985 private communication
Eichner S 1984 Ph. Dissertation, University of Munich
Feder R 1973 phys. stat. solidi (b) 58 K137
    1977 Phys. Rev. B15 1751
Feder R and Pleyer H 1982 Surf. Sci. 117 285
Feder R, Alvarado S F, Tamura E and Kisker E 1983
    Surf. Sci. 127 83
Freeman A J 1983 J. Magn. Magn. Mat. 35 31
Göpel W 1979 Surf. Sci. 85 400
Gradmann U 1974 Appl. Phys. 3 161
    1977 J. Magn. Mat. 6 173
    1984 Trans. Magn. 20 1840
Gradmann U, Waller G, Feder R and Tamura E 1983
    J. Magn. Magn. Mat. 31-34 883
Gradmann U and Bergholz R 1984 Phys. Rev. Lett. 52 771
Korecki J and Gradmann U 1985 to be published
Jepsen O, Madsen J and Andersen O K 1980
    J. Magn. Magn. Mat. 15-18 867
Kirschner J 1984a Surf. Sci. 138 191
    1984b Phys. Rev. B 30 415
Kuyatt C E 1975 Phys. Rev. B12 4581
Landolt M and Campagna M 1977 Phys. Rev. Lett. 39 568
Mills D L and Maradudin A A 1967 J. Phys. Chem. Solids
    28 1855
Noffke J and Fritsche L 1981 J. Phys. C14 89
Ohnishi S, Weinert M and Freeman A J 1984
    Phys. Rev. B30 36
Owens A H, Chien C L and Walker J C 1979 J. de Phys.
    40 C2-74
Palmberg P W, de Wames R E and Vredevoe L A 1968
    Phys. Rev. Lett. 21 682
Pierce D T, Celotta R J, Unguris J and Siegmann H C 1982
    Phys. Rev. B26 2566

Rau C and Eichner S 1980 Nuclear Methods in Materials
    Research edited by Bedge K, Baumann H, Jex H and
    Rauch F (Vieweg, Braunschweig) p. 354
Rau C 1982 J. Magn. Mat. 30 141
    1983 J. Magn. Mat. 31-34 874
Schilling J S and Webb M B 1970 Phys. Rev. B2 1665
Schmitt W, Hopster H and Güntherrodt G 1985 Phys. Rev. B,
    in print
Selwood P W, 1975 Chemisorption and Magnetization
Shih H D, Jona F, Jepsen D W and Marcus P M 1981
    Phys. Rev. Lett. 46 731
Tamura E and Feder R 1982 Solid State Commun. 44 1101
Tamura E and Feder R 1984 Surf. Sci. 139 L 191
Tamura E, Ackermann B and Feder R 1984 J. Phys. C17 5455
Unguris J, Pierce D T and Celotta R J 1984 Phys. Rev.
    B29 1381
Victora R H and Falicov L M 1984 Phys. Rev. B30 3896
Vredevoe L A and de Wames R E 1968 Phys. Rev. 176 684
Waller G and Gradmann U 1982 Phys. Rev. B26 6330
Wang C S and Freeman A J 1980 J. Magn. Magn. Mat. 15-18 869
Wang C S and Freeman A J 1981 Phys. Rev. B24 4364
Wang G C, Dunlap B I, Celotta R J and Pierce D T 1979 Phys.
    Rev. Lett. 42 1349
Weller D, Alvarado S F, Gudat W, Schröder K and Campagna M
    1985a Phys. Rev. Lett. 54 1555
Weller D, Gudat W, Schröder K and Campagna M
    1985b Z. Phys. B, in print

Chapter 8

Inelastic Electron Scattering by Ferromagnets

J. Kirschner

Institut für Grenzflächenforschung und Vakuumphysik,
KFA Jülich, Postfach 1913, D-5170 Jülich, FRG

8.1   Introduction

In this chapter we shall discuss the spin-dependence of inelastic scattering events. Spin polarization effects may appear in an energy loss spectrum by <u>elastic</u> spin-dependent interactions (via spin-orbit coupling or exchange interaction) of electrons that lost energy in spin-<u>in</u>dependent loss processes. These spin polarization structures shall not be discussed here, rather we will focus on explicitly spin-dependent energy loss processes. We will also not discuss the spin-dependence of core level excitations since these are treated in connection with Auger electron emission at another place in this book (see the article by M. Landolt). Our topic will be confined to 'shallow' excitations in the valence band region and the emphasis will be on excitations of electron-hole pairs with opposite spin (Stoner excitations). The Stoner continuum has largely remained 'terra incognita' and it is only very recently that it has become accessible to investigation by inelastic scattering of spin-polarized electrons. First we will give a brief discussion of the spin-dependence of the electron mean free path. Then two-electron processes with exchange are considered. The main part of the paper is devoted to single particle excitations in Fe and Ni.

8.2   Spin-dependence of·the Electron Mean Free Path

Electron spectroscopies owe their surface sensitivity to the strong inelastic interaction with the electronic system. Because of the efficient excitation of plasmons or inter- and intra-band transitions a free electron can travel only short distances in the solid without substantial energy loss, of the order of 1 eV or more. Quite naturally the question arises, whether in magnetic materials the attenuation length might depend also on the spin of the kinetic electron. Possible candidates for spin-dependent excitations are spin waves, plasmons, and electron-hole pairs. Spin waves have low energies, surface spin waves even less (Mills and Maradudin 1967). Estimates for the excitation cross section for electrons (Saldana and Helman 1977, Kleinman 1978, Feder 1981) show that their contribution to the loss-probability for electrons above ~ 10 eV should be an order of magnitude smaller than that of plasmons and electron-hole pairs. They have not yet been observed with electrons and there are even predictions that short wavelength spin waves may not exist at the surface (Egri 1985). Therefore we confine the discussion to the latter two elementary excitations.

When we allow for a spin-dependence of the mean free path we have to distinguish between $\lambda_\uparrow$, the mean free path for electrons aligned along the spin of the majority electrons, and $\lambda_\downarrow$, the mean free path for minority electrons. We may define an asymmetry

$$A_\lambda = \frac{\lambda_\uparrow^{-1} - \lambda_\downarrow^{-1}}{\lambda_\uparrow^{-1} + \lambda_\downarrow^{-1}} \tag{1}$$

for the inverse mean free paths which serves as a measure of the spin-dependent loss probabilities.

Plasmons are collective excitations of the electron gas, involving simultaneously majority and minority electrons. Generalizing the many-body treatment of Pines and Bohm 1952 to the case of a spin-polarized electron gas Helman and Baltensperger 1980 included the exchange interaction with the spin density modulation by the plasmon. From their results it follows that the asymmetry is rather small, of the order of a few percent, and of negative sign. Its energy dependence was neglected here. The negative sign infers $\lambda_\uparrow^{-1} < \lambda_\downarrow^{-1}$, which means that the mean free path for majority-type electrons should be larger than for minority-type electrons.

The same qualitative result was obtained from an atomistic argument considering electron-electron scattering (Bringer et al. 1979, Feder 1979, Feder 1981). The scattering between a 'hot' electron and a conduction-band electron is thought to be approximately isotropic (s-wave scattering) for energies comparable to the Fermi energy. Therefore the exchange amplitude is equal to the direct scattering amplitude and the triplet scattering amplitude vanishes ("singlet-only-scattering"). A hot majority electron should then only be scattered by minority-band electrons and vice-versa. For a quantitative estimate one may, to zeroth order, assume that the spin-dependent scattering cross sections are proportional to the number of available scattering partners, i.e. to the number $n_\uparrow(n_\downarrow)$ of the majority (minority) valence electrons per atom. The asymmetry of the mean free path then is

$$A_\lambda^0 = \frac{n_\downarrow - n_\uparrow}{n_\downarrow + n_\uparrow} \qquad (2)$$

which is negative, since $n_\uparrow > n_\downarrow$ by definition. This rather crude estimate leads to asymmetries of the order of -10 %, but any energy dependence is missing. With increasing energy of the hot electron the scattering involves more and more higher order phase shifts and the cancellation between direct and exchange amplitude in triplet-scattering should disappear gradually. Matthew 1982 considered a more detailed atomistic model, including energy dependence. He calculated the spin-dependence of inelastic scattering of polarized electrons ($E \gtrsim 100$ eV) from oriented atoms in the Born-Ochkur approximation. In the vicinity of inelastic thresholds the same expression (2) was derived, while for higher energies the result was

$$A_\lambda' = \frac{n_\downarrow - n_\uparrow}{n_\downarrow + n_\uparrow} \left(\frac{<\Delta E>}{2E}\right)^2 \qquad (3)$$

where $<\Delta E>$ is the mean excitation energy, weighing all possible transition energies by the corresponding cross section. It was estimated to be around 20-25 eV for transition metals. Thus, at energies $E \gtrsim 100$ eV the asymmetry A_λ' is around -0.3 % for iron, which is less by an order of magnitude than the low energy estimate A_λ^0, but the sign again is negative. A different point of view was chosen by Rendell and Penn 1980 who treated the conduction band electrons, including the d electrons as an electron gas in a kind of local density approximation. Extending the work of Ritchie and Ashley 1965 to the case of ferromagnetic materials and using the Born approximation they found similar results as Matthew 1982 for energies around 100 eV though the decrease of the asymmetry was not as fast with energy. At energies below 100 eV, however, they found a reversal of the sign of the asymmetry. The origin of this discrepancy is not clear at present (Matthew 1982). The magnitude of the asymmetry is material-dependent but mostly around a few percent. It was pointed out by Yin and Tosatti 1981 that in all the above treatments a particular exchange process leading to an apparent spin-flip was not considered. This process is discussed in detail below. As far as the spin-dependence of the mean free path is concerned they showed that, in particular at low energies, the result of Rendell and Penn 1980 might be modified substantially.

It appears that on the theoretical side further clarification and the use of more realistic models is necessary. From the theoretical results available so far one may conclude, however, that the asymmetry probably is larger at low energies than at high energies, ranging from perhaps up to 10 % at a few eV above E_F to a few percent or less at several hundred eV. The sign is most likely negative, which means that minority-type electrons have a larger chance to loose energy than majority-type electrons.

On the experimental side the situation is also far from being clear-cut. A systematic study of the spin-dependence of the mean free path has not yet been undertaken. Some evidence for a spin-dependence may be obtained from various experimental data on inelastic electron scattering (Siegmann et al. 1981, Siegmann 1984, Unguris et al. 1984), on photoemission (Feder et al. 1983), and inverse photoemission (Unguris et al. 1982, Scheidt et al. 1984). Though it is not possible to extract meaningful asymmetries $A_\lambda(E)$ from these data, there are indications that such an asymmetry exists. It must be small, however, of the order of 10^{-2} in the typical energy range of electron spectroscopy.

The electron mean free path is a rather global measure of energy loss processes since it integrates over all loss energies and all loss processes. Though the spin-dependence may be weak, this does not mean that there are not sizeable polarization effects for particular loss energies and/or primary energies. One example is the excitation of core levels when the primary energy is at or near the core level binding energy (Kirschner 1984, Mauri et al. 1984). Another possibility, discussed below in detail, is the excitation of electron-hole pairs in the valence band, where rather large effects have been found. This is no contradiction to the generally weak asymmetry of the mean free path since one relatively strong effect at a particular energy intervall may have little weight relative to all possible energy loss channels, or even the sign in one process might be opposite to the one in another process.

8.3   Two-electron Scattering Processes

The energy transfer of kinetic electrons to solids has traditionally been treated in the framework of the Dielectric theory (Ibach and Mills 1982). In this approximation the Fermion character is ignored and the electron is treated as a moving point charge. The response of the solid to its electric field bears close resemblance to the response to an electromagnetic wave. Indeed, it was shown that in transmission of high energy electrons through thin films there is a one-to-one correspondence between the energy loss function and the optical constants of the solid (Raether 1980). This works so well because an exchange between a 10 keV electron and an electron in the valence band is a very rare event and the electrons can essentially be considered distinguishable particles. At lower energies, say from ten times the Fermi energy downwards, the Fermion character can no more be ignored and exchange processes become increasingly important. The exchange amplitude in the matrix element for electron-electron scattering gains weight relative to the direct amplitude and this alters the energy loss function derived from the dielectric theory. As pointed out by Yin and Tosatti 1981 the exchange amplitude plays an important role in scattering from paramagentic systems as well. However, the effects of exchange become more directly visible in scattering from ferromagnets and they have mostly been studied in this context (Siegmann et al. 1984, Siegmann 1984). The spin of the electron is associated with its magnetic moment, but the interaction with the magnetic moments of a ferromagnet is a rather weak effect. The strongest spin-dependent interaction (left aside the spin-orbit coupling here) is essentially of a Coulombic nature, resulting from the requirement that the wavefunction of an n-electron system has to be antisymmetric with respect to permutations of the n electrons. When in an energy loss spectrum there is a region where exchange-related excitation processes occur, there are observable consequences even without explicit spin analysis of the electrons. It was observed recently by Modesti et al. 1985 that a particular structure in the loss spectrum from an iron-based metallic glass had a pronounced energy dependence. In agreement with the expected behaviour for exchange processes the feature vanished at high primary energy and was strong at low energy. Though there are some problems concerning the energy dependence of the background, this feature may be associated with the Stoner excitations to be discussed in more detail below.

In general, however, a detailed study of exchange processes requires a ferromagnetic sample and a source of polarized electrons or a detector for polarized electrons (see the chapter by Kirschner in this book). Let us assume that we have both and let us look more closely at the possible inelastic two-electron processes with explicit consideration of the electron spin.

Fig. 1 lists the possible configurations after we have sent in a primary electron of energy E_p and detected a scattered electron of energy $E_p - \varepsilon$. The energy ε is consumed in electron-hole pair production across the Fermi energy, and the possible momentum of the pair is reflected in the momentum change of the scattered electron. Since we exclude explicitly spin-dependent forces (such as dipole-dipole interaction) between the electrons, the spins of the individual electrons are conserved. For an incident majority-type electron (up-spin) we may detect either a down-spin electron (process (a)) with the rate $F^\uparrow$ or an up-spin electron (process (c)) with the rate $N^\uparrow$. Conversely, for primary down-spin electrons we may find up-spin electrons with the rate

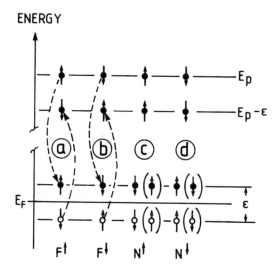

ENERGY

Fig. 1: Schematic representation of two-electron scattering processes
leading to an energy loss ε with explicit consideration of the
electron spin.

$F^{\downarrow}$ (process (b)) or down-spin electrons with rate $N^{\downarrow}$ (process (d)).
The transitions pertinent to (a) and (b) are called 'flip-rates'
$F^{\uparrow(\downarrow)}$, following Yin and Tosatti 1981, since the detected electron has
its spin opposite to the incident one. Correspondingly, processes (c)
and (d) are described by the 'non-flip rates' $N^{\uparrow(\downarrow)}$. For each of the
non-flip rates there are two possible electron-hole pair configura-
tions near the Fermi level. For the ones set in parantheses the primary electron and the one to be ejected from below the Fermi level
form a triplet state. For low primary energy we expect the transition
rate for triplet scattering to be small, since the direct and exchange
amplitude in the matrix element cancel for pure s-wave scattering
(Messiah 1961). We emphasize that in the 'flip' processes there is no
real spin reversal. Rather, e.g. in process (a), the primary up-spin
electron transfers all its energy (via Coulomb interaction) to a mi-

nority electron in an occupied band and finds itself a place in an empty majority-type band above the Fermi level. The energetic difference between the two band states corresponds to the measured energy loss, while their difference in $\underline{k}$-vectors is reflected in the momentum change $\underline{q}$ of the detected electron. The momentum transfer is measured from the specular beam or from one of the other diffracted beams since it is determined only modulo a reciprocal lattice vector. (The momentum needed to turn around the electron is provided by the crystal as a whole.) We note that when we neglect exchange processes all transitions occur only within the same spin system. Within this approximation electron energy loss spectroscopy and optical absorption are equivalent (for $\underline{q} \approx 0$) since the electromagnetic fields involved are far too weak to reverse the spin of an electron.

The full information on each of the flip and non-flip rates can, of course, only be obtained with polarized primary electrons and subsequent spin analysis. However, some valuable insight can already be gained from half the 'complete' experiment: either with a polarized source and subsequent intensity measurement by measuring an inelastic intensity asymmetry $A_{in}(E_0,\varepsilon,\underline{q})$, or with an unpolarized source and a polarization detector which gives the polarization $P_{in}(E_0,\varepsilon,\underline{q})$. Expressed in the rates we obtain following Yin and Tosatti 1981:

$$A_{in}(E_0,\varepsilon,\underline{q}) = \frac{(N^{\uparrow}+ F^{\uparrow}) - (N^{\downarrow}+ F^{\downarrow})}{(N^{\uparrow}+ F^{\uparrow}) + (N^{\downarrow}+ F^{\downarrow})} \qquad (4)$$

$$P_{in}(E_0,\varepsilon,\underline{q}) = \frac{(N^{\uparrow}+ F^{\downarrow}) - (N^{\downarrow}+ F^{\uparrow})}{(N^{\uparrow}+ F^{\downarrow}) + (N^{\downarrow}+ F^{\uparrow})} \qquad (5)$$

The dependence on E_0,ε and $\underline{q}$ is implicit in the partial rates. Though both quantities are obtained from the same rates they are in general not equal. Only in the case of exactly equal or vanishing spin-flip contributions we have $A_{in} = P_{in}$, the well-known result of elastic exchange scattering. This means in particular that both quantities have to agree for the zero loss peak in an energy loss spectrum. If we had both, A_{in} and P_{in} we still would not be able to extract all four rates separately, but we get information on the difference of the flip and non-flip rates respectively:

$$F^{\uparrow} - F^{\downarrow} = (A-P) \cdot \Sigma/2 \qquad (6)$$

$$N^{\uparrow} - N^{\downarrow} = (A+P) \cdot \Sigma/2 \qquad (7)$$

where $\Sigma = N^{\uparrow} + N^{\downarrow} + F^{\uparrow} + F^{\downarrow}$ is the total intensity.

Experiments have so far been carried out in both modes and the 'complete' experiment has also been carried out very recently. The flip processes allow to study Stoner excitations but they are not the only source of structure in asymmetry or polarization in an energy loss spectrum. An instructive example is shown in Fig. 2 (from Weller and Alvarado 1985). Fig. 2 a) shows an energy loss spectrum from Gd(0001) at 200 K, observed in the specular beam at a primary energy of 63.5 eV. The intensity spectrum shows a prominent loss peak at $\varepsilon = 4.5$ eV, which is associated with a positive asymmetry peak, riding on a slowly varying background. From the positive sign the authors concluded that the asymmetry is due to a non-flip process in which minority electrons from 5d levels close to the Fermi level are excited into a rather narrow 4f-type minority band, the counterpart of which lies deeply below the Fermi level. This example shows that non-flip processes also may generate structure in the inelastic asymmetry spectrum. Also of interest is panel c) which shows the elastic asymmetry for electrons of the same kinetic energy. The gross features are seen to be similar to the background curve in panel b). This means that the elastic asymmetry may show up to some extent in the inelastic spectrum at the same kinetic energy. An inelastic diffraction event may conceptually be decomposed into two separate processes (interferences neglected): First energy loss and then diffraction with the reduced energy, or first diffraction with the full energy and then energy loss. Since the majority of energy loss processes involve zero momentum transfer the two channels are undistinguishable in the specular beam. The measured asymmetry results from both processes which are a priori

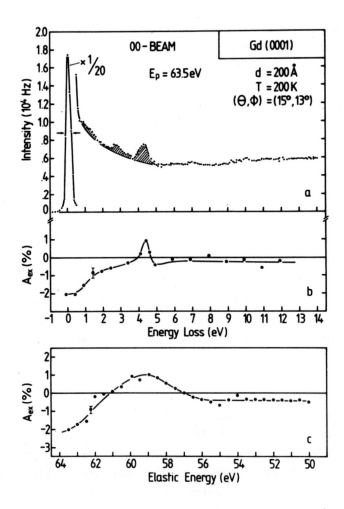

<u>Fig. 2:</u> Intensity (a) and asymmetry (b) versus the energy loss ε of
inelastically scattered electrons of primary energy E_p, observed in the (0 0) beam from Gd{0001}. Panel c) shows the
asymmetry of elastically scattered electrons with energy E_p-ε.

of similar probability. Thus one may expect to find a weakened and modified image of the elastic asymmetry underlying the proper inelastic asymmetry. For beams other than specular the loss-before-diffraction processes are removed since the diffraction occurs with a larger wavelenght into a larger deflection angle. For measurements outside diffraction spots this problem of higher order effects is much alleviated since the elastic peak in the spectrum is much reduced. (Remember that for a perfect surface and a perfect source there is no elastic scattering other than into the diffraction spots.) Momentum change and energy loss then occur in one step to good approximation.

8.4   Stoner Excitations

Stoner excitations are electron-hole pairs with opposite spin character and definite quasi-momentum q. A schematic picture of the space of magnetic excitations is shown in Fig. 3 (after Cooke et al.

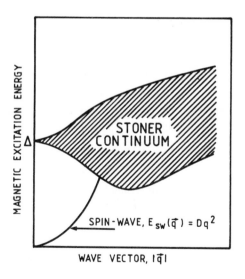

Fig. 3: Schematic representation of the space of elementary magnetic excitations, showing the spin wave region and the Stoner continuum.

1980). In the lower left corner of the ε-q diagram is the existence region of spin waves with quadratic dispersion and the spin wave stiffness D. At larger wave vectors the spin wave dispersion curves meet the region of the single particle excitations, the Stoner continuum, and both types of elementary excitations are indistinguishable. One may also say that the spin wave decays into a Stoner excitation by analogy to plasmons which decay into electron-hole pairs. On the left hand edge at q = 0 the Stoner continuum has a singularity at the energy Δ, which is the mean exchange splitting. At this point electron and hole of opposite spin have the same wavevector $\underline{k}$ in the Brillouin zone and the energy of this pair is given by the (mean) exchange splitting Δ of the bands. At larger wavevectors $\underline{q}$ hole and electron have different $\underline{k}$ and since the bands have dispersion a continuum of energy differences exists within certain limits. The figure was drawn with Ni in mind where the lower energy edge of the Stoner continuum does not touch the horizontal axis, since the upper edge of the majority density of states in Ni has a distance δ (the 'Stoner gap') from the Fermi energy. The figure is schematic by necessity since there is very little experimental information about the Stoner continuum available. Stoner excitations play an important role in the theory of the 'itinerant ferromagnets', since in these materials (e.g. Fe, Co, Ni) the magnetic moment is carried by the 'itinerant' electrons forming the conduction band. In spite of its importance the Stoner continuum has remained virtually unexplored, with one special exception: MnSi (Ishikawa et al. 1977, Ishikawa et al. 1985). In this material the exchange splitting is very small and the Stoner continuum comes into the range of common spin wave energies ($\lesssim$ 100 meV). Spin waves have been much studied by inelastic neutron scattering but for experimental reasons the upper limit for the excitation energy is around 300 meV and at large wave vectors. Thus, the Stoner continuum of Ni can just be touched and that of most other itinerant ferromagnets is out of reach.

It was only fairly recently that an alternative experimental approach to the study of Stoner excitations was successful (Rebenstorff 1983): the study of exchange processes in inelastic scattering of spin-polarized electrons. The basis for this approach is (see Fig. 1) that the final state in an exchange process (a) and (b) in Fig. 1 is characterized by a hole below E_F and an electron above E_F with opposite spin character. This is precisely a Stoner excitation! Since the momentum transfer can be controlled in the experiment, the full Stoner continuum can be studied by electron loss spectroscopy. It is necessary, however, that flip processes can be discriminated against non-flip processes. The experimental results available so far are discussed in the following.

8.4.1 Amorphous Glasses

In their measurements with the ferromagnetic glass $Fe_{82}B_{12}Si_6$ Hopster et al. 1984 used the approach of equ. (5). They bombarded the magnetized sample with unpolarized electrons from a thermal cathode and measured the spin polarization of the scattered electrons. Though the experiment was basically momentum-resolving, the amorphous structure of the sample made the measurements essentially integrate over all momentum transfers. An example of their results is reproduced in Fig. 4. The spin polarization shows a maximum between 2.0 and 2.5 eV energy loss, though the intensity distribution curve shows no significant structure in this region. Later intensity measurements by Modesti et al. 1985 with an almost identical sample ($Fe_{80}B_{15}Si_5$) but a somewhat different geometry show a slight hump at 200 eV primary energy. On the low-loss side the polarization decays rapidly to the small value for elastic scattering which may be positive or negative depending on the primary energy. On the high-loss side the distribution decays approximately to the polarization of elastically scattered electrons of the same kinetic energy (c.f. Fig. 2). The shape of the polarization curve is very similar for both primary energies but its height is significantly less at high primary energy. From the polarization curve alone one cannot decide whether the non-flip processes or the flip processes are responsible for the existence of the polarization, but the dependence on the primary energy provides strong evidence for the exchange-type flip processes to play a dominant role. In

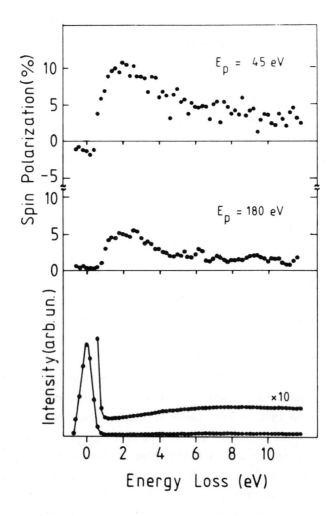

Fig. 4: Electron energy loss spectroscopy results for the iron-based amorphous metal $Fe_{82}B_{12}Si_6$. Upper panel: spin polarization versus energy loss for two primary energies E_p. Lower panel: energy distribution curve for $E_p = 180$ eV.

fact, the authors interpreted their data in terms of Stoner excitations between the occupied majority density-of-states and the empty minority DOS above E_F. If we assume the non-flip rates $N^{\uparrow}$ and $N^{\downarrow}$ to contribute about equally, the sign of the spin polarization depends on the relative magnitude of the two flip rates $F^{\uparrow}$ and $F^{\downarrow}$. If the above exchange process is the dominant one we read $F^{\downarrow} > F^{\uparrow}$ from Fig. 1 and the spin polarization should be positive according to equ. (5). This is in agreement with the experimental result. The assumption $N^{\uparrow} \sim N^{\downarrow}$ receives a posteriori support from more detailed results for crystalline Fe to be presented below.

As far as theory is concerned, there is at present no sufficiently realistic theory of electron energy loss spectroscopy including exchange to compare with the experiments quantitatively. Based on earlier work by Yin and Tosatti 1981, Glazer and Tosatti 1984 carried out theoretical calculations with a drastically simplified model of the bandstructure of iron. Assuming slightly split parabolic bands, the d-bands were mimicked by 'windows' in the free electron bands with the effect that all processes, where either the initial or the final state falls inside the window, are enhanced by an adjustable parameter. For the calculation of the matrix elements plane waves were assumed for initial and final states and a statically screened Coulomb potential. With reasonable assumptions for a number of bandstructure related parameters a reasonable fit to the experimental data could be obtained, see Fig. 5. The peak at ~ 2 eV was found to result mainly from exchange processes simulating Stoner-type excitations. The magnitude of this peak was found to decrease rapidly for large primary energies E_p in agreement with the experiment of Hopster et al. 1984 and with expectation. It would be interesting to see whether similar agreement might also be obtained with suitably convoluted spin-resolved densities of states for Fe, which would amount to non-conservation of $\underline{k}$ in the Stoner excitation.

Experiments with amorphous samples allow some important simplifications for theory, but one cannot expect to obtain very detailed information on the $\underline{q}$-dependence of the Stoner continuum.

370

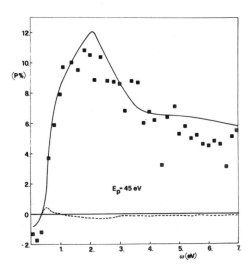

Fig. 5: Fit of spin polarization (solid line) calculated by Glazer and Tosatti 1984 to experimental data (dots) of Hopster et al. 1984 (see Fig. 4, E_p = 45 eV). The dashed line is the theoretical result without the 'windows' described in the text.

8.4.2 Single Crystals

In this section we will discuss experiments with single crystals which provide a much more detailed insight into the Stoner continuum. As seen in Fig. 1 the Stoner continuum has a singularity at $q = 0$ at the energy of the mean exchange splitting. Zero momentum q of the electron-hole pair means that in a Stoner excitation an electron makes a vertical transition in the optical sense at some point in the Brillouin zone into a band of opposite spin character. In a single particle picture the energy of the electron-hole pair equals the exchange splitting of the spin-split bands at this particular point in the Brillouin zone. The exchange splitting of occupied bands can also be measured in photoemission from the energetic separation of the corresponding two peaks in the photoelectron spectrum. Experimental data for Ni, taken at various points in the Brillouin zone, cluster around $\Delta = 0.3$ eV, while most theoretical predictions based on the local spin density approximation agree on a value around 0.6 eV. There is a continuing discussion in the literature whether this discrepancy is due to the influence of many-body effects on the energy of the hole created in photoemission, or due to a subtle failure of the local spin density approximation (see Kirschner et al. 1984 and references therein). It is therefore of interest to measure the singularity of the Stoner continuum in Ni at $q = 0$ and to compare with photoemission results.

In the experiments by Kirschner et al. 1984 the approach of equ. (4) was used, i.e. the measurement of intensity asymmetries for polarized primary electrons. To obtain an adequate energy resolution the electrons from a GaAsP cathode were monochromatized by an electrostatic analyzer and an overall resolution of 35 meV was achieved (see Fig. 6). The observations were made in the specular beam from Ni(110), which corresponds to a momentum transfer $q = 0$ since the momentum needed to reflect electrons is provided by the crystal. The electron spin polarization was normal to the scattering plane and parallel (up) or antiparallel (down) to the spin orientation of the majority electrons (see the inset in Fig. 6). A typical result for the energy loss spectra with opposite primary beam spin orientation is shown in Fig. 6 a). Evidently, over some portion of the loss spectrum minority-type primary electrons have a substantially higher probability to loose energy than majority type electrons. Correspondingly the asymmetry is negative over this range. In Fig. 6 b) the asymmetry is plotted for the two magnetizations and the mirror symmetry of the asymmetry curves proves the magnetic origin of the effect. A close inspection shows that the crossing of the two curves is not exactly on the zero line. This is due to the magnetization-independent contribution from spin-orbit coupling. By measurements with a 'demagnetized' sample, i.e. by averaging over many microdomains, this contribution was found to be sufficiently small to be neglected.

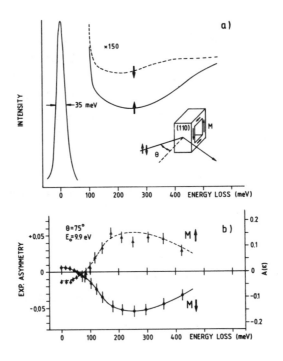

Fig. 6: a) Typical electron-energy-loss spectra for spin-up and spin-down primary electrons, corrected for the finite polarization of the source. The scattering geometry is shown in the inset. The measurements are made in the specular beam at room temperature.

b) Intensity asymmetry as a function of energy loss for the two orientations of the magnetization, showing the predominantly magnetic origin of the asymmetry. The asymmetry $A(\varepsilon)$ is normalized to complete polarization of the source.

The evaluation of the data in terms of flip and non-flip rates is substantially facilitated in Ni since it is a saturated ferromagnet. There are only very few majority-type empty bands, which allows to neglect the flip-rate $F^\uparrow$. Furthermore, it was concluded from the experimental results and from other information on optical data that the non-flip rates are approximately equal and weakly energy-dependent over the range of interest. Therefore from equ. (4) one obtains that the negative of the measured asymmetry is proportional to the flip rate $F^\downarrow$. The magnitude of the flip rate was estimated to be about 20 % of the sum of the two non-flip rates, neglecting $F^\uparrow$. Within the statistical scatter the asymmetry curve was found to be independent of the angle of incidence, and for a limited variation in primary energy (5-18 V) a pronounced energy dependence was not found. Therefore a number of data points was lumped together and the result is shown in Fig. 7 b). The curve peaks around 300 meV energy loss and has a width of about the same value. The significance of this curve is discussed with reference to Fig. 7 a) showing a semi-empirical band structure of Ni due to Weling and Callaway 1982. In the above approximation the asymmetry curve describes the energy dependence of the flip-down rate $F^\downarrow$ at q ~ 0. Since this rate is pertinent to the Stoner excitations with a majority hole and a minority electron above E_F it reflects the spectrum of vertical Stoner excitations across the Fermi-surface. In the experiment the momentum transfer is fixed, but the region in k-space where the vertical transition occurs is not limited but extends over the whole Brillouin zone. Looking at the bandstructure we see that the energetic separation of the spin-split bands near E_F is not constant. The summation over all vertical Stoner excitations thus reflects the abundance distribution of the exchange splittings, integrated over the Fermi surface, if we neglect a possible wavevector- and energy-dependence of the transition matrix element. This experimental result is important in two aspects: first, the width of the distribution proves that the Stoner continuum does not have a sharp singularity at q = 0. This would occur only if the bands were rigidly split in energy by a constant amount, which is not the case in reality. Rather, the distribution is fairly broad (~ 0.3 eV FWHM), with its variance approximately equal to the mean value. Secondly, the most probable value of the exchange splitting agrees with that found by

374

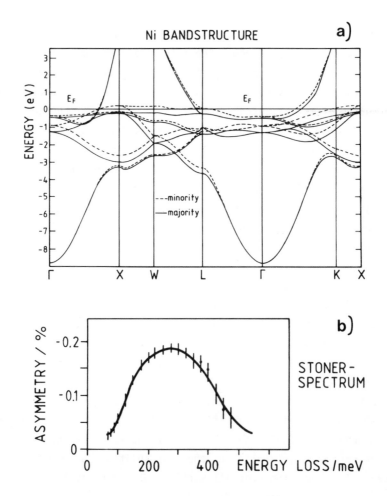

Fig. 7: a) Semi-empirical bandstructure of Ni, after Weling and Callaway 1982. Full lines represent majority states, broken
lines minority states.

b) Intensity asymmetry curve reflecting the spectrum of vertical Stoner excitations (q ~ 0) across the Fermi level, integrated over the Brillouin zone.

photoemission. This is quite remarkable since two quite different experiments give essentially the same answer. In electron energy loss spectroscopy a rather 'soft' excitation is made, where the electronic system absorbs a minute amount of energy only, one to two orders of magnitude less than in photoemission. In contrast to photoemission charge neutrality is preserved and the hole and electron states are rather long-lived since they are close to the Fermi level. Thus one might suspect that many-body effects in photoemission, though undoubtedly present, might be overestimated quantitatively. Recent results for high-energy spin waves in Ni and their broadening when crossing the Stoner continuum boundary (Mook and McK. Paul 1985) also infer that the lower boundary is significantly lower in energy than one would expect for an exchange splitting of 0.6 eV.

If the matrix elements in the exchange process were constant, the experimental result would be directly comparable to the 'Stoner density of states' at q = 0. The Stoner density of states is a kind of joint density of states with the additional conditions of a fixed momentum transfer q and of opposite spin character of the bands. Such Stoner DOS's have been calculated for Ni by Cooke et al. 1980 on the basis of a semi-empirical bandstructure which is somewhat different from that of Weling and Callaway 1982. Some of their results are shown in Fig. 8. The dominant spike at ~ 0.45 eV for q = 0 is due to the flat bands of predominantly d-character. In the experiment such a spike would show up as a feature of ~ 40 meV width with low peak height since the area under the spike is rather small. The data points at present are not sufficiently dense to discern such a feature on the experimental curve in Fig. 7 b). One sees, however, that already at q = 0 the Stoner DOS is more than 0.2 eV wide, due to the wave vector dependent exchange splitting near the Fermi surface. The right hand panel in Fig. 8 shows a calculated Stoner DOS at finite momentum transfer ($|q|$ = 0.173 · $2\pi/a_o$ along <111>). The structure is seen to broaden and to change shape. When comparing the theoretical results to experiment one has to keep several points in mind. Besides the question of the constancy of the matrix elements (on which we have abso-

376

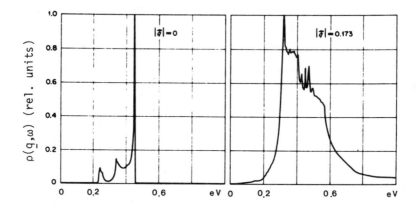

<u>Fig. 8:</u> Calculated Stoner densities of states $\rho(\underline{q},\omega)$ in Ni for two
different momentum transfers. Left hand panel: $|\underline{q}|$ = 0; right
hand panel: $|\underline{q}|$ = 0.173 · $2\pi/a_0$ along <111> (after Cooke et
al. 1980).

lutely no information at present), the experimental parameters of en-
ergy- and momentum-resolution are of importance. The energy resolution
of 35 meV would act in the sense discussed above. The angular resolution of the experiment is such that q = 0 is uncertain to within
plus/minus a few percent of the Brillouin zone. As seen in Fig. 8 this
leads qualitatively to a broadening of the Stoner DOS, but probably
not to a large one. It would be interesting to compare to Stoner DOS's
on the basis of other bandstructure calculation and to see how they
are affected by details of the wavevector-dependent exchange splitting.

The study of Stoner exciations in Ni at q = 0 requires good en-
ergy- and angle-resolution but is also alleviated by two facts: (i)
The loss intensity in the specular beam is high compared to that at
finite momentum transfer, and (ii) the small number of empty majority
states simplifies the data analysis and allows to use the intensity

asymmetry technique. In the general case, at finite q and with other materials, e.g. iron, the assumptions on the non-flip rates are probably not justified and the flip-up rate $F^{\uparrow}$ is probably not negligible. In order to make inelastic electron scattering a generally applicable tool for the study of Stoner excitations it is therefore mandatory to measure all four rates simultaneously. This can only be done in an experiment with a polarized source and a spin polarization detector to analyze the inelastically scattered electrons. However, the intensity loss in the polarization analysis and the low loss intensity at finite q pose serious intensity problems with a monochromatized source like used in the experiments on Ni. This difficulty has recently been overcome (Kirschner 1985) by making use of a previous observation by Kirschner et al. 1983, that the energy width of the photoemitted electrons from GaAsP can be made rather small by adjusting the work function. This does not affect the spin polarization and less than 50 meV FWHM have been reported. The intensity gain by omitting the monochromator, together with an efficient detector made possible the experiments on Fe at finite q on which is reported below.

The apparatus incorporated a source and detector system similar to the ones discussed in the chapter on polarized sources and detectors and shall not be described here. The angular resolution was about $\pm\ 2^{\circ}$, the energy resolution ~ 0.4 eV FWHM. The polarization and magnetization vectors both lie in the scattering plane which coincides with a mirror plane of the crystal. In this way spin-orbit polarization effects are effectively suppressed. The momentum transfer q is measured from the specular beam (see inset in Fig. 9). The measurements were made on a clean Fe(110) crystal which was kept in a single magnetic domain state by an external soft iron yoke. For each loss energy ε and each orientation of the primary spin polarization relative to the magnetization of the sample (upper index) the spin detector yields two intensities $j_{\uparrow}^{\uparrow(\downarrow)}(\varepsilon)$ and $j_{\downarrow}^{\uparrow(\downarrow)}(\varepsilon)$ of the two spin-sensitive diffracted beams (lower index). If source and detector were ideal, these four intensities would directly yield the desired four spin-dependent scattering rates. The non-ideal behaviour of source and detec-

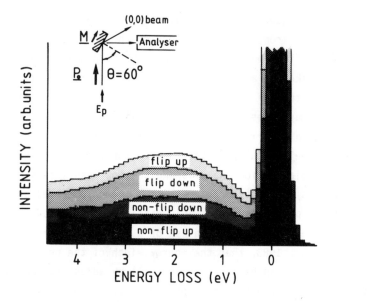

Fig. 9: Flip and non-flip rates in Fe as a function of energy loss at momentum transfer $|g| \sim 1.2$ $Å^{-1}$. The inset shows the scattering geometry. The sample magnetization $\underline{M}$ is along <100> and the primary energy E_p = 21 eV.

tor can be described by (2x2) matrices $\underline{Q}$ and $\underline{D}$, respectively. With these and the unknown transition rate matrix $\underline{\underline{R}}(\varepsilon)$ the resultant intensity matrix $\underline{\underline{J}}(\varepsilon)$ is given by

$$\underline{\underline{J}}(\varepsilon) = \underline{\underline{D}} \cdot \underline{\underline{R}}(\varepsilon) \cdot \underline{\underline{Q}} \tag{8}$$

The matrices are

$$
\underline{\underline{J}}(\varepsilon) = \begin{pmatrix} j_\uparrow^\uparrow(\varepsilon) & j_\uparrow^\downarrow(\varepsilon) \\[2mm] j_\downarrow^\uparrow(\varepsilon) & j_\downarrow^\downarrow(\varepsilon) \end{pmatrix} \quad ; \quad \underline{\underline{R}}(\varepsilon) = \begin{pmatrix} N^\uparrow(\varepsilon) & F^\downarrow(\varepsilon) \\[2mm] F^\uparrow(\varepsilon) & N^\downarrow(\varepsilon) \end{pmatrix}
$$

$$
\underline{\underline{D}} = \frac{1}{2} \begin{pmatrix} 1 + A & 1 - A \\[2mm] 1 - A & 1 + A \end{pmatrix} \quad ; \quad \underline{\underline{Q}} = \frac{1}{2} \begin{pmatrix} 1 + P & 1 - P \\[2mm] 1 - P & 1 + P \end{pmatrix} \tag{9}
$$

A is the polarization sensitivity of the detector (A = -0.27 ± 0.02) and P the effective polarization of the source (P = $P_0 \sin\Theta$ with P_0 = 0.35 ± 0.03). This system of linear equation is solved by

$$
\underline{\underline{R}}(\varepsilon) = \underline{\underline{D}}^{-1} \cdot \underline{\underline{J}}(\varepsilon) \cdot \underline{\underline{Q}}^{-1} \tag{10}
$$

which yields the desired flip and non-flip rates as a function of the energy loss ε. An experimental result, after having applied the above corrections is shown in Fig. 9 for $\Theta = 60^0$, corresponding to a momentum transfer of $|q| \sim 1.2$ $Å^{-1}$. Each of the shaded areas represents the contribution from the process indicated and the sum of all rates gives the total scattered intensity.

First, we note that in the loss region the sum of the non-flip rates is nearly equal to the sum of the flip rates. This shows that the Stoner excitations, i.e. those involving a 'spin-flip', play a substantial role in the energy loss process, also at finite momentum transfer. Secondly, we see that the two non-flip rates $N^\uparrow$ and $N^\downarrow$ are similar. Since electron and hole have the same spin this means that the electron-hole pair production within each of the two spin systems is, to first order, indpendent of the spin of the incoming electron. This was inferred for the case of Ni from other experimental and theoretical results and is shown here directly. A closer inspection, however, reveals that the non-flip up rate $N^\uparrow$ is somewhat larger than the non-flip down rate $N^\downarrow$. This means that an incoming electron of majority type has, to second order, a larger probability to loose energy in a non-flip process than a minority electron. The inequality of the non-flip rates may also cause structure in an intensity measurement as discussed above with reference to Fig. 2. Finally, turning to the flip rates $F^\uparrow$ and $F^\downarrow$ we see, first of all, that the flip-up rate $F^\uparrow$ is by no means negligible with respect to $F^\downarrow$. This we did expect since in contrast to Ni there are quite a number of majority bands above the Fermi level in iron. The ratio of the two flip rates depends also on the momentum transfer as further experiments told.

Since the energy loss of the electron is small relative to the primary energy, Fig. 9 essentially represents a 'constant-q scan' or vertical cut through the Stoner continuum at about half-way to the zone boundary. The broad structure is consistent with the schematic picture of Fig. 3. There is one calculated Stoner DOS for Fe (Cooke et al. 1980), which is not directly comparable to the present results since the momentum transfer is different. The width of the Stoner DOS is, however, consistent with the experimental result.

One final remark shall be made on the surface sensitivity of the present experiments. The mean free path of electrons with typically 10 eV kinetic energy is of the order of 10 monolayers (Seah and Dench 1979). By working at non-normal incidence the effective depth of information is reduced to perhaps half this value. On the other hand, the surface-induced changes of the electronic structure and magnetization are limited to the first two monolayers at zero temperature. At finite temperature the magnetization at the surface is lower than in the bulk, but at room temperature only the first one or two layers are significantly lowered, at least for the single crystals. Thus the results are expected to be relevant to the bulk properties of the samples. It will be interesting to see, how e.g. in Ni the Stoner spectrum looks like for the top layer.

8.5 Outlook

The study of high energy Stoner excitations is just beginning to be established. The basic experiments discussed here have shown that Stoner excitations can in general be probed via inelastic scattering of spin-polarized electrons and that a great deal of detailed information on two-electron scattering processes can be obtained. At present there is a lack of theoretical calculations with sufficiently realistic models for the transition rates in two-electron scattering including exchange. There is also a lack of calculated Stoner densities of states on the basis of self-consistently calculated bandstructures with many-body effects included. Perhaps one will be able to assess the range of validity of the single particle picture adopted here and the importance of many-body effects. On the experimental side the greatest challenge will be to study the high-temperature properties of the Stoner continuum. For example, it will be interesting to see whether the Stoner spectrum for Ni shows any changes near the Curie temperature...

Perhaps spin-polariezd electrons may become for Stoner excitations what neutrons have been for spin waves.

382

References

Bringer A, Campagna M, Feder R, Gudat W, Kisker, E and Kuhlmann E 1979 Phys. Rev. Lett. 42 1705

Cooke J F, Lynn J W and Davis H L 1980 Phys. Rev. B 21 4118

Egri I 1985 (to be published)

Feder R 1979 Solid State Commun. 31 821

Feder R 1981 J. Phys. C 14 2049

Feder R, Gudat W, Kisker E, Rodriguez A and Schröder K 1983 Solid State Commun. 46 619

Glazer J and Tosatti E 1984 Solid State Commun. 52 905

Helman J S and Baltensperger W 1980 Phys. Rev. B 22 1300

Hopster H, Raue R and Clauberg R 1984 Phys. Rev. Lett 53 695

Ibach H and Mills D L 1982 Electron Energy Loss Spectroscopy and Surface Vibrations (New York: Academic Press)

Ishikawa Y, Noda Y, Uemura Y J, Majkrzak C F and Shirane G 1985 Phys. Rev. B 31 5884

Ishikawa Y, Shirane G, Tarvin J A and Kohgi M 1977 Phys. Rev. B 16 4956

Kirschner J 1984 Solid State Commun. 49 39

Kirschner J 1985 (submitted to Phys. Rev. Lett.)

Kirschner J, Oepen H P and Ibach H 1983 Appl Phys. A 30 177

Kirschner J, Rebenstorff D and Ibach H 1984 Phys. Rev. Lett. 53 698

Kleinman L 1978 Phys. Rev. B 17 3666

Matthew J A D 1982 Phys. Rev. B 25 3326

Mauri D, Allenspach R and Landolt M 1984 Phys. Rev. Lett. 52 152

Messiah A 1969 Quantum Mechanics, Chap. 14 (Amsterdam: North Holland)

Mills D L and Maradudin A A 1967 J. Phys. Chem. Solids 28 1855

Modesti S, Della Valle F, Rosei R, Tosatti E and Glazer J 1985 Phys. Rev. B 31 5471

Mook H A and McK. Paul D 1985 Phys. Rev. Lett. 54 227

Pines D and Bohm D 1952 Phys. Rev. 85 338

Raether H 1980 Excitation of Plasmons and Interband Transitions by Electrons, Springer Tracts in Modern Physics Vol. 88 (New York: Springer)

Rebenstorff D 1983 PhD thesis, Rheinisch-Westfälische Technische Hochschule Aachen, Germany (unpublished)

Rendell R W and Penn D R 1980 Phys. Rev. Lett. 45 2057

Ritchie R H and Ashley J C 1965 J. Phys. Chem. Solids 26 1689

Saldana X I and Helman J S 1977 Phys. Rev. B 16 4978

Seah M P and Dench W A 1979 Surf. Interface Anal. 1 2

Siegmann H C 1984 Physica 127B 131

Siegmann H C, Meier F, Erbudak M and Landolt M 1984 Advances in Electronics and Electron Physics, Vol. 62 (New York: Academic Press)

Siegmann H C, Pierce D T and Celotta R J 1981 Phys. Rev. Lett. 46 452

Unguris J, Pierce D T and Celotta R J 1984 Phys. Rev. B 29 1381

Unguris J, Seiler A, Celotta R J, Pierce D T, Johnson P D and Smith N V 1982 Phys. Rev. Lett. 49 1047

Weling F and Callaway J 1982 Phys. Rev. B 26 710

Weller D and Alvarado S F 1985 Z. Physik (in press)

Yin S and Tosatti E 1981 Report IC/81/129 International Centre for Theoretical Physics, Miramare, Trieste, Italy

Chapter 9

Spin Polarized
Secondary Electron Emission
from Ferromagnets

M. Landolt

Laboratorium für Festkörperphysik, Eidgenössische Technische
Hochschule, CH-8093 Zürich, Switzerland

9.1. Introduction

Magnetic phenomena near or at surfaces of magnetically ordered solids are of still growing interest both from a fundamental point of view with respect to critical behaviour (see Binder, Chapter 2) or electronic structure (see Freeman et al., Chapter 1), as well as for many applications, the latter particularly with respect to forthcoming magnetic information technologies (for a recent state-of-the-art report see Kryder and Bortz 1984). The motivation to study the spin polarization of true secondary electrons arose from the demand for an efficient and simple surface magnetometer. True secondary electrons at very low kinetic energies are emitted with high intensity, typically two orders of magnitude higher than elastically scattered electrons or valence band photoelectrons, and they were expected to carry a spin polarization proportional to the average magnetization near the surface of the sample.

For an overview, let me briefly describe the spectrum of second-ary electrons ejected from a solid which is irradiated with a beam of monochromatic primary electrons of moderate kinetic energy (up to a few keV). One can characterize, somewhat arbitrarily, different class-es of electrons, each of which constitutes a separate regime of inves-tigation and interest and to some of which I devote a section of the present Chapter. The classes are as follows: (i) True secondary elect-rons at very low kinetic energies, which are mainly conduction elect-rons excited in the cascade processes. Their spectral spin polarizat-ion as well as their application for magnetometry with either lateral or depth resolution are the subjects of section 9.2. (ii) Auger elect-rons: the spin polarization of Auger electrons provides a signal of the local, element specific magnetization in composite systems. Detailed analysis of specific Auger lines furthermore yields informat-ion on electron correlation effects, as well as on inter-shell exchange interactions of an atom in a magnetic solid. Spin polarized

Auger spectroscopy is described in section 9.3. (iii) Characteristic electron energy losses through inner-shell excitations: They directly reveal the spin dependence of electron-electron scattering cross sections in solids, and are the subject of section 9.4. (iv) Shallow electron energy losses due to conduction-electron excitations: plasmons and single electron excitations. Plasmon excitation and decay was found to be essentially unpolarized. Single electron intra-3d excitations in transition metal ferromagnets, on the other hand, provide us with basic insights into magnetic properties. The Stoner-excitations, which are exchange scattering processes promoting an electron from a filled majority into an empty minority state, give rise to spin polarized electron energy losses. The observation of these polarized losses will eventually allow us to determine the complete Stoner spectrum (see Kirschner, Chapter 8 and references therein). (v) Elastically backscattered electrons peaked at the high energy end of the secondary electron spectrum: Elastic scattering of electrons from a solid is weakly spin polarized because of two different effects, namely spin-orbit coupling in the scattering process, which is particularly large for large Z materials, and magnetic exchange asymmetry in the case of magnetically ordered solids, arising from the exchange interaction between the scattered particles and the conduction electrons in the solid. Spin polarized elastic scattering or spin polarized low energy electron diffraction (SPLEED) has been well developed both experimentally and theoretically during the last decade, and important results concerning crystallographic and magnetic structures at surfaces have been obtained (see Feder, Chapter 4; Dunning and Walters, Chapter 6; Gradmann and Alvarado, Chapter 7; and references therein).

The spin polarization in the elastic scattering processes, arising from spin-orbit coupling, also governs the spin polarization in the reemission of true secondary electrons from nonmagnetic materials (for a review see Siegmann, Meier, Erbudak and Landolt 1984; Müller and Erbudak 1984).

Throughout this Chapter the spin polarization is denoted as P and defined, according to the convention, as $P = (n\uparrow - n\downarrow)/(n\uparrow + n\downarrow)$, where $n\uparrow(\downarrow)$ is the number of electrons with magnetic moment parallel (antiparallel) to the sample magnetization. Experimentally, in most cases the polarization value consists of an average of two measurements with opposite magnetization directions. In this way the spin polarization arising from spin-orbit coupling in the scattering processes is eliminated and magnetic information only is retained.

9.2. Spin Polarization of True Secondary Electrons

9.2.1. Energy Distribution of the Secondary Electron Spin Polarization

Two pioneering studies of the spin polarization of true secondary electrons emitted from ferromagnets showed that a fairly large spin polarization can be observed (Chrobok and Hofmann 1976) and that it exhibits a rather unexpected energy distribution at very low kinetic energies (Unguris, Pierce, Galejs and Celotta 1982).

Aside from particular excitations the secondary electron intensity is generally characterized at low energies by the large peak of true secondary electrons, and at the high energy and by the elastically reflected and inelastically rediffused primary electrons. In the intermediate region a mixture of cascade electrons and rediffused primaries is emitted. The polarization spectrum reflects this situation with a monotonical decrease for increasing secondary electron kinetic energy, see Fig. 1 (Mauri, Allenspach and Landolt 1985). The data in Fig. 1 are presented on a reduced energy scale, and details below 20 eV are left out for clarity. The general behaviour of P, which reflects the relative weight of polarized true secondary electrons and nearly unpolarized rediffused primaries, bears upon the understanding of the cascade processes which govern the hot electron production. Exchange scattering is believed to be important only in the close vicinity of E_F (Rendell and Penn 1980) and thus does not determine the general spin polarization behaviour.

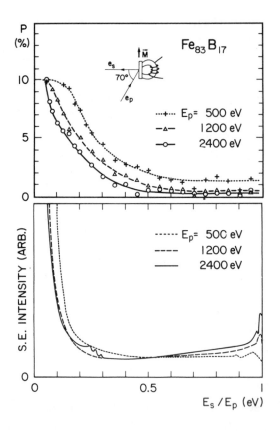

<u>Fig. 1</u> Secondary electron spin polarization P (upper panel) and
intensity I (lower panel) vs reduced secondary electron
kinetic energy E_s/E_p for various primary energies
E_p from $Fe_{83}B_{17}$. The intensities are normalized to
coincide at energy 0.5.

The principal question now is: How large is the spinpolarization of the true secondary electrons ? A naive approach is to assume
that the cascade process excites electrons with equal probability from
each conduction band ground state. We then expect at low energies a
constant spin polarization identical to the average band polarization
or magnetization of the sample averaged in space over the range of

true secondary electron emission. In contrast, however, the experiment shows a pronounced spectral distribution which currently is believed to manifest the importance of Stoner excitations at low kinetic energies. Experimental spectra are depicted on Figs. 2 and 3. The steep decrease of P at energies between 0 and 10 eV, followed by a plateau of approximately the band polarization, has been observed on every 3d-ferromagnet investigated up to now: amorphous Fe-B alloys (Unguris et al. 1982, Landolt and Mauri 1982, Mauri et al. 1985, Glazer et al. 1985), and single crystal Fe, Co and Ni faces (Kisker, Gudat and Schröder 1982, Hopster et al. 1983, Landolt, Mauri and Allenspach 1985, Allenspach and Landolt 1985).

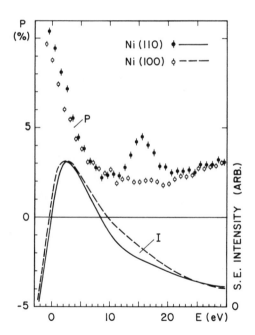

Fig. 2: Secondary electron spin polarization P and intensity I vs kinetic energy from Ni(100) and Ni(110) faces, recorded with primary electrons of energy E_p = 600 eV.

We note that the structures at ≈16 eV in Ni and ≈12 eV in Fe, respectively, are not understood at present, and we focus on the general trend of P. One possible explanation (Glazer and Tosatti 1984) of the unexpected increase of P to values much larger than the band

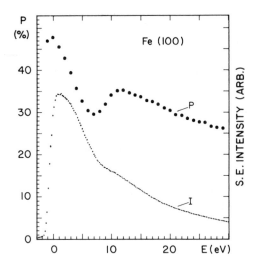

Fig. 3: Secondary electron spin polarization P and intensitiy I vs kinetic energy from Fe(100), recorded with primary electrons of energy E_p = 2500 eV.

polarization at energies below 10 eV rests upon the strong spin polarization of the unfilled d-states in these materials. The 3d-holes are predominantly of minority spin, and minority spin electrons can be scattered into these states. At sufficiently low kinetic energies the hot electrons undergo inelastic exchange scattering with the 3d-electrons. The most flashy process is the one in which a minority spin hot electron of energy E is scattered into a d-state near E_F while a majority spin d-electron is excited to the energy E-ω. It corresponds to a "spin flip" scattering or Stoner excitation of energy ω. This exchange scattering process produces the polarization enhancement, and the strong dependence of the exchange scattering cross section on the hot electron kinetic energy E is responsible for the rapid falloff of polarization. By this mechanism the low-energy secondary-electron spin polarization spectrum yields information on the Stoner spectrum in the ferromagnetic material as well as on the energy dependence of exchange scattering amplitudes. We note that this information, in a more direct and unique way, is also obtained from spin polarized electron energy loss spectroscopy (ELS). Glazer and Tosatti (1984, Glazer 1984) have presented a quantitative analysis for a simple free-electron band model with exact scattering matrix elements for a screened Coulomb interaction. They were able to reproduce the Stoner excitations observed in spin polarized ELS (see Kirschner, Chapter 8 and references therein) as well as the true-secondary-electron spin polarization spectra (Hopster et al. 1985). A similar calculation using more realistic densities of states but parameterized scattering matrix elements was presented by Penn, Apell and Girwin (1985).

Another consequence of the exchange scattering is that the inelastic mean free path is largely spin dependent. One finds that the inelastic mean free path asymmetry $A = (\lambda\uparrow - \lambda\downarrow)/(\lambda\uparrow + \lambda\downarrow)$ is always positive, it increases with decreasing kinetic energies and reaches, in the case of Fe, as much as 30% at the vacuum level (Glazer 1984, Penn, Apell and Girwin 1985). This calculated results strongly contradict the latest of the earlier calculations (Rendell and Penn 1980), which indicates that the question of spin dependence of the inelastic mean free path of electrons in transition metal ferromagnets is by no means settled. The strong asymmetry should also be considered with the interpretation of other low-energy polarized-electron spectra, such as threshold photoemission (for a review see Siegmann et al. 1984) or inverse photoemission (see Dose and Glöbl, Chapter 13 and references therein). We note that in elastic scattering from amorphous ferromagnetic Fe-Ni-B alloys (Pierce et al. 1982) at 2 eV an asymmetry of only 1% was observed, which is in apparent contrast to the calculated inelastic mean free path asymmetry.

An alternative explanation (Mauri 1984) describes the polarization enhancement as a band structure effect without involving exchange scattering. The key is a strong assumption about the cascade process: small-energy excitations dominate in the cascade, which leads to preferential emission of electrons from states near the Fermi level. An enhancement of the polarization of true secondary electrons at low kinetic energies is then to be expected since in 3d-ferromagnets the polarization near E_F exceeds the mean band polarization.

Finally, a very important question remains: what is the relation between the polarization of the true secondary electrons and the sample magnetization ? The exchange scattering amplitude and consequently the inelastic mean free path asymmetry rapidly go to zero for increasing kinetic energy. In that model, the polarization at $E \simeq 10$ eV approaches the value of the band polarization. Furthermore, a computer simulation of the cascade processes (Glazer 1984) shows that the entire polarization spectrum approximately scales with the band polarization. The integrated polarization of the true secondary electrons thus can be identified as approximately proportional to the sample magnetization, averaged in space over the range of secondary electron emission, for all 3d-ferromagnets. This result is particularly important for the use of secondary electrons for magnetometry.

9.2.2.Magnetic Domain Microscopy

The structure and distribution of magnetic domains in a solid or thin film is of interest for a number of applications: the performance and characteristics of hard magnetic materials in permanent magnets, dynamics of domains in soft magnetic materials to minimize magnetic losses in transformers and generators, and most recently in magnetic information technology for the development of thin film data storage media. Most of these problems require a technique which operates in a reflection mode. Spatial resolution is important and should be, for the study of domain walls, in the 100 Å range. Since optical techniques have a wavelength limited resolution of $\simeq 1 \mu m$, electron spectroscopy is necessary.

The spin polarization of secondary electrons, produced by the electron beam of a scanning electron microscope, now can provide a strong magnetic contrast. This new imaging technique is superior to Lorentz microscopy, which is based on the deflection of secondary electrons by the Lorentz force of fringing magnetic fields, in two major aspects: First, the spin polarization directly reveals the distribution of the magnetization vector in sign and magnitude, which is not the case in Lorentz microscopy, and second, the magnetic contrast is independent of topological contrast. The variation of secondary electron intensity caused by surface roughness can obscure the magnetic contrast in Lorentz microscopy, whereas the spin polarization $P = (n\uparrow - n\downarrow)/(n\uparrow + n\downarrow)$ is normalized in intensity and thus provides magnetic contrast alone. The topological contrast is simultaneously available by measuring the secondary electron intensity or the absorbed current at the sample.

The use of spin polarized secondary electrons in scanning electron microscopy for magnetic domain viewing was suggested by DiStefano (1978) and Unguris et al. (1982), and a first pioneering instrument was built by Koike and Hayakawa (1984a, 1985). They used a scanning 10 μm / 10 keV electron beam in conjunction with a Mott detector, where one component of the magnetic vector in the surface plane was measured, and the polarization signal, after signal processing, was displayed on a CRT. Koike and Hayakawa (1984a, 1985) produced the first domain patterns of Fe(100) as well as of Co(1210), and demonstrated the capability of the microscope to determine magnetization directions as well as to observe magnetic and topological contrasts separately.

A next generation instrument has now been completed by Unguris et al. (1985). They use an ultrahigh vacuum scanning electron microscope with a field emission source producing beam diameters of 10 nm or less, with 3×10^{-12}A. The initial samples were (100) faces of Fe-3% Si. Fig. 4 shows polarization images at two magnifications of the striped magnetic domains in these crystals. These 64 x 64 pixel pictures were obtained in 5 minutes. The polarization difference between light and dark domains is approximately 60%. The best resolution achieved was 50 nm, which was limited by stage vibrations and can be improved to the beam diameter limit. The data acquisition rate is limited by the quantum noise and can, owing to the novel spin analyzer after some improvements of the electronics, be driven as high as 1.5×10^3 pixels/sec or more (Unguris et al. 1985). The spin analyzer of new design utilizes the asymmetry of low energy elastic scattering

from polycrystalline gold, integrated over large emission angles. Two
components, normal to the beam momentum, of the polarization vector
are measured simultaneously. Because of its small size, orthogonal polarization analyzers can be used so that all three components of the
magnetization vector can be resolved (Unguris et al. 1985).

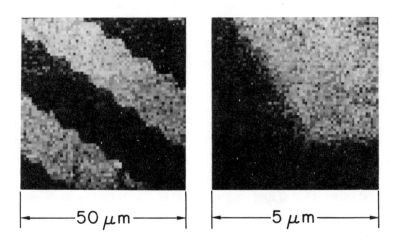

Fig. 4: Spin polarization images of magnetic domains in Fe-3%Si.
(from Unguris et al. 1985).

Sub-domain-size lateral resolution of a magnetometer is not only
of importance for the study of domain structures and walls, but also
for magnetic studies on materials which can not easily be magnetized
to a single domain state, which is frequently the case in magnetic-ma-
terials science. Another interesting prospect is the fact that a secondary electron microscope readily can be modified to a scanning
Auger microscope. The combination of magnetic and chemical distributions, together with topological contrast, along a sample surface will
evidently be a most valuable information for specific studies of magnetic materials.

In scanning electron microscopes the focussing of the primary beam requires kinetic energies of ≃ 5 keV. The secondary electrons, on the other hand, which are collected for spin analysis, are of kinetic energies below 5 eV. Under these circumstances the depth below the surface of secondary electron production and emission is of the order of ≃50 Å, and the magnetic contrast obtained in the microscope can be considered a bulk-like information. In the following section we will discuss how the probing depth of secondary electrons can be reduced to perform magnetic depth profiling.

9.2.3. Magnetic Depth Profiling

The depth dependence of magnetization at solid-vacuum interfaces, including surface chemistry, is an active area of research (see Freeman et al., Chapter 1; Gradmann and Alvarado, Chapter 7; and references therein). Of particular interest are finite temperature effects where thermodynamics plays the key role. Since the magnetic excitations near the surface can strongly differ from the ones in the bulk one expects large effects, which, particularly near T_C, can extend deeply into the solid. The thermodynamics of magnetism of semi-infinite systems has been treated theoretically to a great extent (see Binder, Chapter 3 and references therein), but experimentally, however, only a very limited number of techniques are available. Here, the possibility of a simple magnetometer utilizing secondary electrons is described, which has the unique feature of a tunable probing depth up to ≃50 Å. The possibility of recording magnetic depth profiles or comparing "surface-like" with "bulk-like" magnetic properties in the same experiment is a valuable complementation of the surface techniques described by Gradmann and Alvarado (Chapter 7).

The lateral resolution of ≃100 Å of the secondary electron magnetometer (see sect. 9.2.2.) can be turned into depth resolution when the primary beam energy is reduced to below a few hundred eV. We note that at these low primary electron energies the lateral resolution is lost, since even field-emitter sources do not provide us with sufficient brightness.

Fig. 5 shows the primary energy dependence of two quantities: the true secondary electron yield δ and the spin polarization of true secondaries P, both from amorphous ferromagnetic $Fe_{83}B_{17}$ (Mauri, Allenspach and Landolt 1985, Mauri 1984). A similar observation on polycrystalline permalloy was reported by Koike and Hayakawa (1984b). The variations of δ and P both arise from varying surface sensitivity of the secondary electron emission, and the $P(E_p)$ curve represents a magnetic depth profile.

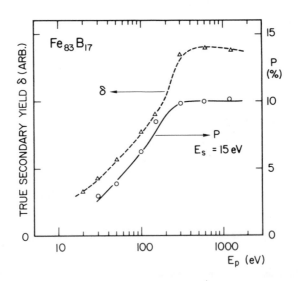

Fig. 5: Secondary electron spin polarization P at E_s = 15 eV and total true secondary electron yield δ vs primary electron energy E_p from $Fe_{83}B_{17}$.

Universal yield curves are well understood in the frame of elementary theories (Dekker 1960). In particular, the increase of δ at low energies, below its maximum at E_p^{max}, is due to the fact that the penetration depth of the primary electrons is shorter than the escape depth of the secondary electrons. The penetration depth increases with increasing primary energy. The yield reaches its maximum at E_p^{max} where the penetration depth becomes comparable to the escape depth, and at $E_p > E_p^{max}$ the yield is limited by the escape depth of the secondary electrons. Thus the probing depth of the experiment is determined by the penetration depth of the primary electrons at $E_p < E_p^{max}$, and it is given by the escape depth and independent of E_p at $E_p > E_p^{max}$. E_p^{max} for amorphous $Fe_{83}B_{17}$ was found to be about 300 eV. Hence, the increase of $P(E_p)$ at $E_p <$ 300 eV must be ascribed to an increase of sample magnetization when going from the surface towards the bulk. The fact that P still increases when approaching E_p^{max} indicates that in the present example the sheet of reduced magnetization is thicker than the escape depth of secondary electrons at 15 eV which is of the order of 10-20 Å (Seah and Dench 1979). Consistently the P value of 10% at $E_p > E_p^{max}$ is smaller than the bulk band polarization of 21% of $Fe_{83}B_{17}$ at room temperature. (P_{bulk} = 21% is obtained from μ = 2.1 μ_B/Fe and T_C = 650 K (Luborsky 1980)).

In order to obtain a magnetic depth profile M(z) from the $P(E_p)$ curve on Fig. 5 it is necessary to convert E_p into depth z. Within a Monte Carlo simulation Ganachaud and Cailler (1979a,b) studied the depth dependence of the energy transfer from an incoming electron to the solid. They calculated for Al the total energy dissipated at the depth z. Under the simplifying assumption that the number of hot electrons at a given E_s is proportional to the total energy transfer, we can identify the depth distribution of hot electrons in the solid with the energy dissipation profiles. These profiles exhibit a strong z dependence. They are strongly peaked in z with a steep increase at low z and a tail towards higher z. This means that secondary electron emission at $E_p < E_p^{max}$ even provides us with a certain depth resolution. Of particular interest is the E_p dependence of the depth z^{max} of maximum energy dissipation. It occurs at 7.5 Å for E_p = 200 eV, 10 Å for 300 eV and 30 Å for 600 eV, respectively (Ganachaud and Cailler 1979b). This strong energy dependence can be approximated by $z^{max} = 10^{-2} \times E_p^{1.4}$ (z in Å, E_p in eV), and indicates that the $P(E_p)$ curve on Fig. 5 represents a depth profile of magnetization in the range from 2 to 20 Å below the surface.

We note that the observed magnetization reduction near the surface of the $Fe_{83}B_{17}$ sample is not of purely thermodynamic origin, since it was found to be essentially temperature independent (Mauri, Allenspach and Landolt 1985). It is ascribed to structural and chemical effects such as segregation or decomposition of the glass near its surface. Metalloid segregation has been already observed and very likely is the cause of the low surface magnetization revealed by the $P(E_p)$ curve of Fig. 5.

The depth resolution of a secondary electron surface magnetometer can further be enhanced by varying the energy E_S of the secondary electrons which are spin analyzed. Spectral features as depicted in Figs. 2 and 3 of course will obscure the magnetization calibration. Magnetic field dependences, on the other hand, can be recorded at variable probing depths, which reveals fundamental magnetic properties. Particularly the questions of magnetic domain nucleation at surfaces or the change of coercivity near surfaces can experimentally be addressed. Fig. 6 shows a first example of the depth dependence of a magnetic hysteresis loop (Allenspach, Taborelli, Landolt and Siegmann 1985).

The model system is the (100) Fe surface magnetized along the easy direction [001] in the surface plane. The upper panel of Fig. 6 shows a hysteresis $P(H)$ recorded at $E_S = 1$ eV, which corresponds to a probing depth of approximately 40 ± 20 Å below the surface. The rectangular hysteresis loop is characteristic of the irreversible movement of a 180^0 domain wall. The same hysteresis is observed using Kerr rotation with a probing depth of $\simeq 200$ Å, which can be considered as bulk information. When reducing the probing depth of the secondary electrons to the outermost 1-3 atomic layers by recording at $E_p = 100$ eV and $E_S = 50$ eV, one finds that the coercivity H_c is unchanged but the hysteresis loop is rounded, see lower panel of Fig.6. The magnetization continuously goes to zero for H between 0 and $-H_c$. This steady change corresponds to the nucleation of a 180^0 domain at a (100) face of Fe magnetized along [001] parallel to the surface. The new result is that the depth of nucleation is less than 20 Å which is smaller by one or two orders of magnitude than the thickness of a domain wall in bulk Fe.

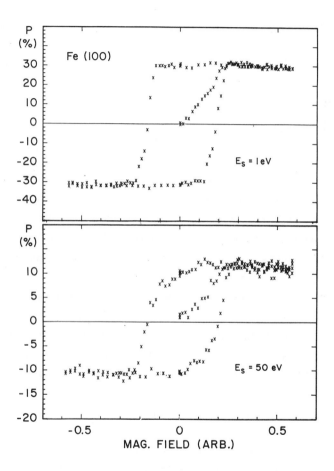

Fig. 6: Magnetic hysteresis loops of Fe(100), magnetized along [001], of varying probing depth, recorded with secondary electron spin polarization at different secondary electron energies of 1 eV (upper panel) and 50 eV (lower panel) with a primary electron energy of 100 eV.

The experimental equipment required for secondary electron depth profiling is fairly simple: it includes a commercial electron gun suitable for Auger spectroscopy, a spin analyzer, and a retarding grid for secondary electron energy discrimination. A retarding stage is sufficient since the low energy true secondary electrons strongly dominate in intensity which makes the use of a low pass filter unnecessary. The data of Figs. 5 and 6, however, have been recorded on an instrument designed for spin polarized secondary electron spectroscopy (Landolt, Mauri and Allenspach 1985).

9.3. Spin Polarized Auger Spectroscopy

Auger spectroscopy today is a very popular standard technique for chemical analysis in surface physics, mainly because it is highly element specific, and absolutely non- destructive. Beyond that, high resolution Auger spectroscopy (for a review see Fuggle 1981) is recognized as a powerful complement to photoemission and electron or photon absorption techniques for the study of electron correlation and screening effects in highly excited electronic systems. Magnetic or nearly magnetic solids, evidently, are favorable candidates for such investigations since magnetism certainly is the most important appearance of electron exchange and correlation. Spin polarized electrons, on the other hand, are well established probes for magnetic solids, and, consequently, spin polarized Auger spectroscopy appears to be a very promising development.

The net spin polarization of Auger electrons from 3d ferromagnets was found to be large, on the order of 20%, and it carries a wide information depending on the nature of the particular two-hole final state of the Auger transition (Landolt and Mauri 1982).

If valence states are involved in the Auger decay, information on the local magnetization at the site of a particular element in an alloy or compound can be extracted. If two core holes are left behind after the Auger decay, the polarization measures the coupling of partly filled inner shells with the net spin of the magnetic 3d-electrons.

In some cases, the polarization is able to discern the deexcitation of two different excitations of a core-hole: the so-called self-screened or resonantly excited state where the core-hole is strongly spin polarized, and the normal "ionized" state of a screened core-hole, where the core-hole is only weakly polarized via exchange interaction. If the core-hole is regarded as a Z + 1 impurity, the measurement addresses the problem of screening of an impurity in a magnetic host and particularly the existence of a magnetic moment at the impurity site.

9.3.1. The MMM Auger Process

The Super-Coster-Kronig $M_{23}M_{45}M_{45}$ Auger emission in a 3d-ferromagnet is a very intense decay channel of an excited 3p-hole. The spectrum of the emitted 3d-electron in a naive picture reflects the self-convolution of the occupied part of the 3d-bands. Thus, in a magnetic material, one expects to observe a spin polarization which relates to the band polarization belonging to the particular atom under consideration.

Two somewhat finer points have to be considered which modify the observed spectra and, if properly analyzed, will contribute to the investigation of screening and correlation. First, the presence of a 3p-hole leads to a rearrangement of 3d and 4s,p electrons at the site in order to screen the positive charge. If the screening is carried out by a 4s,p - electron the modification of the 3d states is only minor and the magnetic moment remains unchanged. Otherwise an extra 3d-electron can be bound at the site, and the magnetic moment will then be reduced. Second, the correlation among the two extra d-holes left behind after a $M_{23}M_{45}M_{45}$ transition can be strong, and the two-hole excitation spectrum dominates over the one of the band like extended hole states (Antonides, Janse and Sawatzky 1977, Tréglia et al. 1981). The correlation shifts the Auger lines towards lower energies. In the following, two examples are presented, namely single crystalline Fe and Ni, which exhibit fairly different behaviour.

Fig. 7 presents spin polarization and intensity spectra of secondary electrons emitted from Fe(100) (Allenspach and Landolt 1985). The features between 35 eV and 95 eV are attributed to Auger decays. The upper panel represents raw data, from which the effective Auger polarizations and intensities on the lower panel are generated by background subtraction. The choice of the background introduces some ambiguity of the absolute values of P_{eff} within a factor of $\simeq 1.3$. The errorbars include statistical errors only.

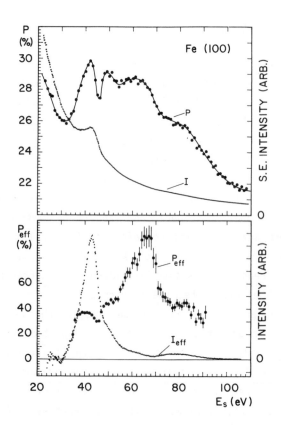

Fig. 7: Spin polarization P and intensity I vs kinetic energy of secondary electrons from Fe(100), excited with primary electrons of energy E_p = 2500 eV. Upper panel: raw data, lower panel: effective Auger electron signals after background subtraction.

The leading peak in the intensity at 42 eV is the Auger main line $3p^63d^n \rightarrow (3p^53d^n)_{screened} + \epsilon l \rightarrow 3p^63d^{n-2})_{screen} + \epsilon l + \epsilon_{Auger}$. The corresponding P_{eff} spectrum exhibits a minimum at the high energy end of the line. It is satisfactorily reproduced by a self-convolution of spin-split bands confirming that the valence band Auger lines of Fe can be described with a small effective corre-lation energy (Antonides, Janse and Sawatzky 1977). The spin polari-zation of this very intense MMM main line scales with the element spe-cific local band polarization and is large enough, + 40%, to be used as a monitor for sign and magnitude of the element specific magneti-zation in composite magnetic systems.

Just above the M_{23} threshold, which is at 47 eV, a first gain satellite occurs at 51 eV. This satellite was observed in intensity spectra and is interpreted as autoionization emission of a resonant $3p \rightarrow 3d$ excitation $3p^63d^n \rightarrow 3p^53d^{n+1} \rightarrow 3p^63d^{n-1} + \epsilon_{autoion}$ (Bader et al. 1983, Zajak et al. 1983, Zajak et al. 1984). This emission includes all final state configurations $3p^63d^{n+1-2} + \epsilon_{resonant}$ (with $\epsilon_{resonant} \equiv \epsilon_{autoion}$) of the super-Coster-Kronig decay of the resonantly excited 3p-hole configuration $3p^53d^{n+1}$. This emission, we call it resonant Auger emission, is identical to the emission in the d-band satellite observed in resonant photoemission at the 3p → 3d resonance, e.g. in Ni (Guillot et al. 1977). The new in-formation in this spectrum is that the resonant Auger emission in Fe is strongly spin polarized, P_{res} = 55%. A similar observation, less quantitative however, was reported for ferromagnetic amorphous $Fe_{83}B_{17}$ (Landolt and Mauri 1982). The strong polarization clearly indicates that the weak-correlation description for Fe breaks down for resonant excitation processes. Regarding main line and resonant Auger emission, the spin polarization reveals the deexcitation of two diffe-rent electron-excited 3p-hole states.

The polarization maximum of $P_{eff} \simeq$ 80% at 64 eV in Fig. 7, with a weak bump in intensity, possibly is an $M_{23}M_{45}N$ decay of a resonant 3p → 3d excitation accompanied by a 3d → 4d(s,p) shake-up process. Its large spin polarization requires a resonant 3p-hole excitation, which leads to a strongly polarized hole, and a decay which predominantly goes to a singlet state.

An alternative explanation for the observed spin polarization above the M_{23} threshold is presented by Nesbet (1985): The secondary electrons of the background undergo inelastic scattering, in particular $3p \rightarrow 3d$ losses. These were found to be strongly spin-dependent near threshold (Mauri, Allenspach and Landolt 1984) leading to positive spin polarization because of strong exchange-scattering (see sect. 9.4). Nesbet suggests that the spin polarization enhancement above the M_{23} threshold is accounted for by $3p \rightarrow 3d$ energy losses of the background secondary electrons. The intensity, however, of these losses right at threshold is extremely weak and therefore is not likely to be sufficient to cause the strong polarization enhancement.

The broad peak at the high energy end of Fig. 7 is the $M_1 M_{45} M_{45}$ Auger line of Fe centered around 82 eV. The observed polarization reasonably well corresponds to the one of the $M_{23} M_{45} M_{45}$ main line which has the same final state.

In Ni, on the other hand, the situation is considerably different. There the resonantly and nonresonantly excited 3p-hole states are degenerate: In resonant photoemission one finds that the MMM Auger emission and the d-band satellite occur at the same kinetic energy if the photon energy corresponds to the $3p \rightarrow 3d$ excitation energy (Guillot et al. 1977). Therefore it is assumed that the screening of a 3p-hole induces a local $3d^{10}$ configuration, i.e. $(3p^5 3d^9)_{screened} = 3p^5 3d^{10}$ in Ni. Starting from a $3d^{10}$ configuration, the normal Auger emission away from resonance is expected to be unpolarized, which evidently is not observed experimentally, see Fig. 8. An effective polarization of $\simeq 10\%$ in the Auger main line is found. The polarization arises from the resonant process, from the fraction of $3p \rightarrow 3d$ resonantly excited p-holes. These holes are fully down spin polarized, since only minority d-holes exist in ferromagnetic Ni and the exchange scattering amplitude is weak at the high primary electron energy used in the experiment (Mauri, Allenspach and Landolt 1984). The correlation among the d-holes now is

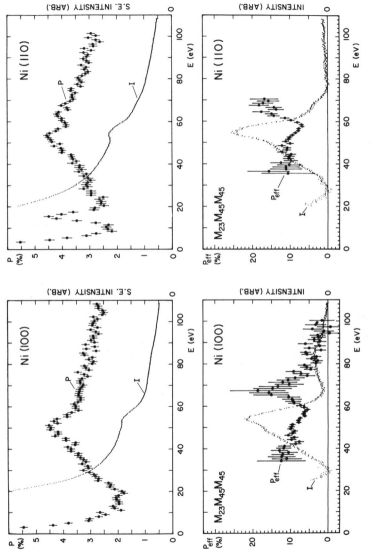

Fig. 8: Spin polarization P and intensity I vs kinetic energy of secondary electrons from Ni(100) and Ni(110), excited with primary electrons of 600 eV. Upper panels: raw data; lower panels: effective Auger electron signals after backgr. subtraction.

important. The effective Coulomb interaction U_{eff} is of the same magnitude as twice the bandwidth (Antonides, Janse and Sawatzky 1977), and atomic features dominate the d-band Auger spectra as in Cu or Zn. With fully polarized 3p-holes and $3d^{10}$ as initial states, one finds spin polarizations of + 100% and -33% for the singlet and triplet final states, respectively, of the $3p^6 3d^8$ final state configuration (Feldkamp and Davis 1979). The spectral distribution of the effective polarization P_{eff} to the left of the polarization minimum is well reproduced when combining the calculated polarizations with a calculated Auger intensity spectrum (Mc Guire 1977). From the absolute value of P_{eff} one obtaines the ratio of resonant 3p → 3d to nonresonant 3p-hole excitations to be 1/6, for primary electron kinetic energies between 300 eV and 2500 eV.

The pronounced peak in P_{eff} at 66 eV lies above the M_{23} threshold energy of 62 eV. At present this structure is thought to be identical to the strongly spin polarized upper gain satellite in Fe, namely an $M_{23}M_{45}N$ decay of a resonant 3p → 3d excitation accompanied by a 3d → 4d(s,p) shake-up process. Also for Ni it was proposed by Nesbet (1985) that the spin polarization enhancement above the M_{23} threshold is caused by inelastic exchange scattering of the background secondary electrons involving 3p → 3d excitations.

The intensity structures at 68 eV in Ni(110) and at 84 eV in Ni(100), respectively, are tentatively ascribed to the final-state band-structure of Ni.

The spin polarization of the MMM main lines in both cases, Fe and Ni, reflect the local band polarization, in completely different ways, however, as described above. Spin polarized Auger emission from valence-band lines therefore will be a useful technique for the study of magnetic behaviour near surfaces, particularly of alloys, or with respect to surface chemistry on magnetic thin films. As an illustration we present the polarization of the valence-electron KLL Auger line of B in ferromagnetic $Fe_{83}B_{17}$ in Fig. 9 (Landolt and Mauri 1982).

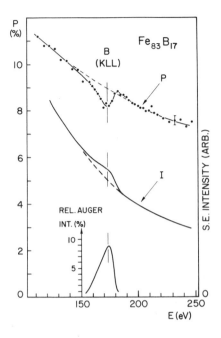

Fig. 9: Spin polarization P and intensity I vs kinetic energy of secondary electrons from $Fe_{83}B_{17}$, excited with electrons of E_p = 2900 eV.

The spin polarization spectrum exhibits a minimum in a strongly spin polarized secondary electron background. A simple background subtraction yields that the KLL Auger electrons are unpolarized: $P_{eff} = P_0 + \Delta P \cdot I_{tot}/I_{Auger} = 0$, with P_0 = background polarization and $\Delta P = P - P_0$. This concurs with the fact that the magnetization is zero at the B site and suggests how spin polarized Auger spectroscopy reveals local magnetizations.

9.3.2. The LMM Auger Process

Spin polarization and intensity spectra of the LMM Auger transitions of Fe and Ni are presented in Figs. 10 and 11. The major difference between the spectra of the two elements is that the Ni spin polarization is much smaller than the one of Fe. We note that the structures on Fig. 11 are weak, of the order of 0.5%.

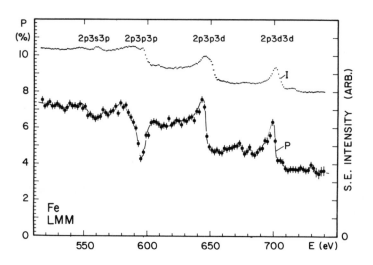

Fig. 10: Spin polarization P and intensity I vs kinetic energy of secondary electrons from Fe(100), excited with electrons of E_p = 2500 eV.

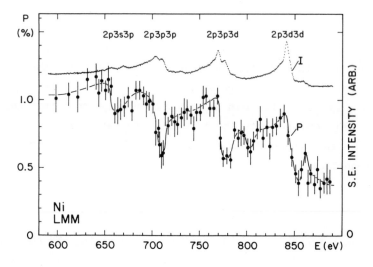

Fig. 11: Spin polarization P and intensity I vs kinetic energy of secondary electrons from Ni(100), excited with electrons of E_p = 2100 eV.

The weak polarization of Ni LMM lines corroborates the assumption that the 2p-hole again is screened by a local $3d^{10}$ configuration. As in the MMM main line, the polarization arises from $2p \rightarrow 3d$ resonant excitations, which produce down-spin 2p-holes. Comparison of the absolute P_{eff} values with a calculated polarization spectrum (Allenspach 1985), using atomic Auger transition probabilities starting from $2p^5 3d^{10}$ (Antonides, Janse and Sawatzky 1977) and fully polarized 2p-holes, shows good overall agreement and yields a ratio between resonant $2p \rightarrow 3d$ and non resonant 2p-hole excitations of 1/30. The relative strengths of $2p \rightarrow 3d$ and $3p \rightarrow 3d$ excitations of 1/30 and 1/6, respectively, are found to be consistent with directly observed electron energy loss intensities of $2p \rightarrow 3d$ and $3p \rightarrow 3d$ excitations. This in fact strongly supports the interpretation in terms of resonant excitation processes.

In Fe LMM emission, on the other hand, the observed spin polarization is large: P_{eff} is of the order of 30%, which is about 10 times larger than in Ni LMM. The major fraction of this large polarization must arise directly from the net 3d spin, and the resonant $2p \rightarrow 3d$ excitation processes can, because of their weakness, be neglected for a qualitative discussion.

The LVV line, $L_3 M_{45} M_{45}$, leads to the same final state as the MMM main line, and the spin polarization also reflects the local band polarization in the form of a self-convolution of occupied bands. The autoionization gain-satellite does not occur since the $2p \rightarrow 3d$ resonance is much weaker than the $3p \rightarrow 3d$ resonance, as is directly observed in ELS. The weak lines above the L_3MM emissions are known to be L_2MM transitions. They are weak since the L_2 hole has the strong Coster-Kronig $L_2 L_3 M_{45}$ decay channel in competition.

Of particular interest is the spin polarization structure observed in the core-hole-only transition $L_3 M_{23} M_{23}$, where two holes in the 3p shell are left behind. This decay provides the opportunity to directly determine exchange interactions between partly filled shells of an atom in a magnetic system. The intensity spectrum has two peaks, corresponding to singlet and triplet states of the 2p-hole pair, and the polarization is positive for the singlet and negative for the triplet final states, respectively. Bennemann (1983) gave a first description using multiplicities. Kotani and Mizuta (1984, Kotani 1985) presented a model calculation for Fe where the d-band ferromagnetism is described by a molecular field acting on localized spins $S^{3d} = 1$, and where they introduce the exchange interactions among the two 3p-holes, between 3d and 3p spins, and between 3d and 2p spins, respectively. The molecular field and the intra-3p exchange integral are obtained from T_c and the singlet-triplet splitting observed in intensity, respectively. The remaining two coupling constants are determined by a best fit to spin polarization spectra. Interestingly, they find that the 3d-2p coupling J_2, which yields negatively polarized 2p holes, is essential to reproduce the data: the singlet-emission is then positively polarized. The 3d-3p coupling J_3, on the other hand, lifts the degeneracy of the triplet final state, which leads to strongly negative polarization at the high energy end of the triplet emission. The actual values are $J_2 = 0.05$ eV and $J_3 = 0.5$ eV. (Kotani and Mizuta 1984). This result illustrates how spin polarized Auger spectroscopy reveals magnetic interactions between partly filled shells of an atom in a ferromagnetic solid. Detailed analysis will provide further information on the localized spin of the 3d-shell.

9.4. Electron Spin Polarization in Inner Shell Excitations in a Solid

This section addresses the virtue of spin polarization in electron energy loss spectroscopy (ELS), and therefore strongly relates to Chapter 8 by Kirschner on inelastic electron scattering. Chapter 8 mainly deals with beautiful high resolution ELS experiments which, because of the spin analysis, eventually will reveal the complete spectrum of Stoner excitations. The Stoner continuum, which is the ensemble of electron-electron exchange scattering processes in the conduction band, is a fundamental quality of itinerant magnetism. In the present section, the ferromagnetism is not primarily the subject to be studied, it rather is part of a more general experiment in the area of electron-electron interactions in solids.

The asymmetry of the wave function required by the Pauli Principle is known to induce spin dependent forces between electrons. As a consequence the scattering of electrons from atoms, molecules or solids depends on the relative orientation of the spins of the projectile and the bound electrons in the target. A direct way to reveal the spin dependence in the scattering is to conduct an experiment where a spin polarized electron beam interacts with a spin polarized target.

A ferromagnetic transition-metal sample can be used as an array of spin polarized target atoms, and by measuring the spin polarization of the inelastically scattered electrons the spin-dependence of a $3p \rightarrow 3d$ excitation can be observed. The goal is to quantify the importance of exchange scattering processes in ELS at primary energies approaching the excitation threshold. This indicates the breakdown of optical transitions in ELS and detailed analysis will provide a handle on the nature of the screened scattering potential.

The well studied $3p^63d^n \rightarrow 3p^53d^{n+1}$ excitation furthermore exhibits interferences, known as Fano resonances, with $3p^63d^n \rightarrow 3p^63d^{n-1} + \epsilon l$ one-step excitations through autoionization of the excited state $3p^53d^{n+1} \rightarrow 3p^63d^{n-1} + \epsilon l$ via Auger decay (Fano 1961; Dietz et al. 1974, 1980; Davis and Feldkamp 1976, 1977; Jach and Girwin 1983). The spin polarized approach described in this section also provides information towards a more detailed understanding of the Fano processes in transition metals.

Unpolarized primary electrons of variable kinetic energy are scattered from a single domain $Fe_{83}B_{17}$ amorphous ferromagnet (Mauri, Allenspach and Landolt 1984). Fig. 12 shows the intensity I and spin polarization P versus secondary electron kinetic energy E_s in the vicinity of the $3p \rightarrow 3d$ loss line of Fe, measured with a primary electron energy Ep = 550 eV. The spin polarization exhibits an asymmetric $3p \rightarrow 3d$ energy loss structure, which is shifted relative to the intensity line to approximately 5 eV higher kinetic energy. Similar behaviour was observed for different primary electron energies ranging from 150 to 2500 eV (Mauri, Allenspach and Landolt 1984).

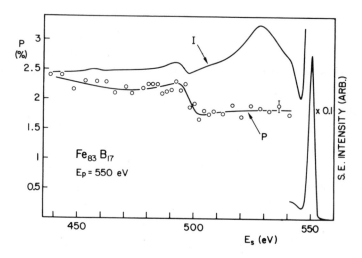

<u>Fig. 12:</u> Spin polarization P and intensity I **vs** kinetic energy of secondary electrons from $Fe_{83}B_{17}$, excited with electrons of $E_p = 550$ eV.

The fact that the $3p \rightarrow 3d$ loss line gives a polarization signal implies that the cross section for $3p \rightarrow 3d$ excitation is spin dependent: $\sigma\uparrow\uparrow \neq \sigma\uparrow\downarrow$. In order to simplify the discussion let us assume that only down-spin 3d-holes exist, denoted as $d*\downarrow$. In the case of an atom the different multiplets of the final state configuration have to be considered and two scattering amplitudes for each magnetic quantum number are needed to describe the process. For the present qualitative discussion we restrict ourselves to transitions between one-electron levels, and two complex quantities, namely the direct and exchange scattering amplitudes f and g, respectively, completely describe the problem. The following three scattering processes occur for unpolarized incoming electrons e_p and fully polarized $d*\downarrow$ holes (Kessler 1976):

$$e_p\uparrow + 3_p\downarrow \rightarrow e_s\uparrow + d*\downarrow : \text{amplitude } f$$
$$e_p\downarrow + 3_p\uparrow \rightarrow e_s\uparrow + d*\downarrow : \text{amplitude } -g \qquad (1)$$
$$e_p\downarrow + 3_p\downarrow \rightarrow e_s\downarrow + d*\downarrow : \text{amplitude } (f-g)$$

The spin polarization of the scattered electrons P_{loss} then amounts to

$$P_{loss} = (|f|^2 + |g|^2 - |f-g|^2)/(|f|^2 + |g|^2 + |f-g|^2) \qquad (2)$$

which alternatively can be written as

$$P_{loss} = (\sigma\uparrow\downarrow - \sigma\uparrow\uparrow)/(\sigma\uparrow\downarrow + \sigma\uparrow\uparrow) \qquad (3)$$

to illustrate the meaning of the experiment. From equation (2) one expects P_{loss} to vanish when the exchange amplitude g goes to zero. This is known to occur at primary energies large compared to the excitation energy. Spin polarized ELS now allows to quantify this property by varying the primary electron kinetic energy. The corresponding results are compiled in Fig. 13.

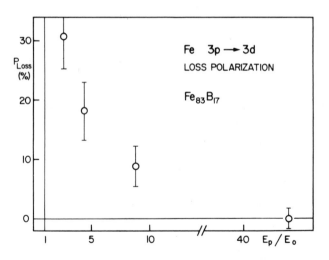

Fig. 13 Spin polarization P_{Loss} of the inelastically scattered electrons after an Fe 3p → 3d excitation in $Fe_{83}B_{17}$ vs primary electron energy E_p reduced to the 3p → 3d excitation threshold E_0.

To extract the net spin polarization P_{loss} of the 3p → 3d core excitation from the measured spectra requires a theory of the polarization line shape. The formalism of Fano (1961) must be extended to the case of many discrete states mixing with many continua as was carried out by Davis and Feldkamp (1977). Still in the simplest case there are five different pairs of interfering channels. The polarization line shape is then described by three direct and three exchange scattering matrix elements corresponding to the discrete excitation, the autoionization via Auger decay, and the continuum excitation. In this way the full spin information in the ELS line-shape will ultimately reveal the relative strength of exchange and direct Auger decays and continuum excitations, a task still open for theoretical efforts.

As a starting point, however, the Fano interferences have been ignored and P_{loss} is approximated by peak-to-peak heights of polarization ΔP and intensity ΔI. $P_{loss} \simeq P_0 + I_{tot} \cdot \Delta P / \Delta I$, with P_0 = background polarization. A preliminary analysis of the line shape (Allenspach 1985) suggests that this data analysis correctly reveals the trends of P_{loss}. Fig. 13 shows that P_{loss} monotonically decreases from high values near threshold to zero at high primary electron kinetic energies. It provides direct evidence that the exchange scattering is still considerably large at primary electron energies ten times larger than the threshold.

A number of studies have been published in recent years which emphasize the importance of exchange scattering in ELS spectra from solids (Powell and Erickson 1983; Bertel, Stockbauer and Madey 1983; Netzer, Strasser and Matthew 1983; Netzer et al. 1984, Strasser et al. 1984, Sakisaka, Hiyano and Onchi 1985). One particular study (Powell and Erickson 1983) concerns 3p → 3d excitations in vanadium, where additional loss features are observed at primary energies below approximately ten times the threshold value. They have been interpreted as spin non-conserving or optically forbidden transitions arising from exchange scattering. The spin polarized experiment now clearly corroborates the identification of exchange scattering in ELS.

Using a source of spin polarized primary electrons and measuring the asymmetry $S = (i\uparrow - i\downarrow)/(i\uparrow + i\downarrow)$ of the electrons inelastically scattered from a ferromagnet ($i\uparrow$ and $i\downarrow$ denote scattered intensities for up and down spin incoming electrons, respectively) is an alternative way to carry out spin polarized ELS measurements. Asymmetry data in the $3p \rightarrow 3d$ energy loss range of Fe in a ferromagnetic amorphous Fe-B alloy (Siegmann Pierce and Celotta 1981) have already been published, however, without emphasizing the $3p \rightarrow 3d$ loss asymmetry. Combining a polarized source with a spin and energy analyzer makes the concept applicable also to non-magnetic samples, which then allows systematic studies of electron-electron scattering potentials in solids.

9.5. Summary

"Secondary electrons" is a very general term including all electrons ejected from a solid upon irradiation with primary electrons.

The vast majority of primary electrons dissipate their energy in the solid via multiple electron-electron scattering, be it single-electron or collective excitations, which leads to the emission of true secondary electrons near zero kinetic energy. The application of true secondary electron spin polarization is twofold: First, in connection with a scanning electron microscope it provides very good magnetic contrast in reflection mode, strong enough to allow imaging of magnetic domain structures at a rate of more than 10^3 pixels/sec and with a lateral resolution of 10 nm. Second, the true secondary electron spin polarization can be utilized for non-destructive magnetic depth-profiling. Magnetic hysteresis loops can be recorded at various depths below the sample surface. The probing depth is tuned by varying the kinetic energies of incoming primary and analyzed secondary electrons. It ranges from approximately 2 to 50 Å, which allows to study thermodynamics of surface magnetism or magnetic properties of surface chemistry, segregation or adsorbate induced decomposition.

Another energy loss mechanism for the primary electrons is core-hole excitation. This is manifest in the secondary electron spectrum either by the appearance of Auger electrons or as characteristic energy-loss lines of inelastically scattered primary electrons. In both cases the spin polarization adds new perspectives to secondary electron emission.

The Auger electrons carry a wide variety of information depending on the nature of the two-hole final state. The most important aspect is that from valence-band Auger lines the local, element specific magnetization can be extracted. This is very attractive for materials science investigations. From core-hole only transitions the coupling between partly filled shells on an atom can be obtained. Furthermore different strongly polarized gain satellites on MMM transitions are revealed which are interpreted as either autoionization emission or shake-up excitation of resonantly excited 3p-holes.

The characteristic energy-loss lines, in the case of $3p \rightarrow 3d$ excitations, were shown to be polarized at low primary electron energies. This observation implies that the 3p-hole excitation cross section is spin dependent: $\sigma\uparrow\uparrow \neq \sigma\uparrow\downarrow$, and the significance of exchange scattering processes in electron energy loss spectroscopy can be quantified.

A small fraction of the primary electrons are elastically reflected from the solid. The spin polarization of elastic scattering today is a well established technique for magnetic or non-magnetic surface structure analysis. It is reviewed in Chapters 6 and 7 by Dunning and Walters, and by Gradmann and Alvarado, respectively. Some of the primary electrons are scattered quasielastically generating collective or single-electron excitations in the solid. Among the single-electron excitations the intra 3d-band energy losses include the Stoner excitations, which are important in itinerant ferromagnetism. High resolution electron energy loss spectroscopy and Stoner excitations are discussed in Chapter 8 by Kirschner.

Most of the examples presented in this Chapter are pilot studies of experiments in their state of development. Spin polarized secondary electron emission from ferromagnets is a new experimental area and this Chapter is ment to show its great potential for both fundamental research as well as materials-science applications.

Acknowledgements

It's a pleasure to acknowledge the many stimulating conversations in a very pleasant and fruitful collaboration with H.C. Siegmann, D. Mauri, R. Allenspach, M. Taborelli and K. Brunner. Financial support by the Schweizerischer Nationalfonds is acknowledged.

References

Allenspach R 1985 Ph.D. Thesis ETH, Zürich

Allenspach R and Landolt M 1985 Surf. Sci. in press

Allenspach R, Taborelli M, Landolt M and Siegmann H C 1985 Proc. ICM'85 San Francisco in press

Antonides E, Janse E C and Sawatzky G A 1977 Phys. Rev. B15 1669

Bader S D, Zajak G and Zak J 1983 Phys. Rev. Lett. 50 1713

Bennemann K H 1983 Phys. Rev. B28 5304

Bertel E, Stockbauer R and Madey T E 1983 Phys. Rev. B27 1939

Chrobok G and Hofmann M 1967 Phys. Lett. 57A 257

Davis L C and Feldkamp L A 1976 Solid State Commun. 19 413

Davis L C and Feldkamp L A 1977 Phys. Rev. B15 2961

Dekker A J 1960 Solid State Physics (Englewood Cliff N.J: Prentice Hall)

Dietz R E, McRae E G, Yafet Y and Caldwell C W 1974 Phys. Rev. Lett. 33 1372

Dietz R E, McRae E G and Wheaver J H 1980 Phys. Rev. B21 2229

DiStefano T H 1978 IBM Techn. Discl. Bull. 20 4212

Erbudak M and Müller N 1984 J. Phys. E: Sci. Instrum. 17 951

Fano U 1961 Phys. Rev. 124 1866

420

Feldkamp L A and Davis L C 1979 Phys. Rev. Lett. $\underline{43}$ 151

Fuggle J C 1981 Electron Spectroscopy Vol 4 p. 85, ed. C R Brundle and A D Baker (London: Academic Press)

Ganachaud J P and Cailler M 1979a Surf. Sci. $\underline{83}$ 498

Ganachaud J P and Cailler M 1979b Surf. Sci. $\underline{83}$ 519

Glazer J 1984 Ph.D. Thesis Intl. School for Advanced Studies, Trieste

Glazer J and Tosatti E 1984 Solid State Commun. $\underline{52}$ 905

Glazer J, Tosatti E, Hopster H, Kurzawa R, Schmitt W, Walker K H and Güntherodt G 1985 Phys. Rev. B in press.

Guillot C, Ballu Y, Paigné J, Lecante J, Jain K P, Thiry P, Pinchaux R, Pétroff Y and Falicov L M 1977 Phys. Rev. Lett. $\underline{39}$ 1632

Hopster H, Raue R, Kisker E, Güntherodt G and Campagna M 1983 Phys. Rev. Lett. $\underline{50}$ 71

Jach T and Girwin S M 1983 Phys. Rev. $\underline{B27}$ 1489

Kessler J 1976 Polarized Electrons (Berlin: Springer Verlag)

Kisker E, Gudat W and Schröder K 1982 Solid State Commun. $\underline{44}$ 623

Koike K and Hayakawa K 1984a Appl. Phys. Lett. $\underline{45}$ 585

Koike K and Hayakawa K 1984b Japan. J. Appl. Phys. $\underline{23}$ L85

Koike K and Hayakawa K 1985 J. Appl. Phys. $\underline{157}$ 4244

Kotani A 1985 J. Appl. Phys. $\underline{57}$ 3632

Kotani A and Mizuta H 1984 Solid State Commun. $\underline{51}$ 727

Kryder H and Bortz A B 1984 Physics Today $\underline{37/12}$ 20

Landolt M and Mauri D 1982 Phys. Rev. Lett. $\underline{49}$ 1783

Landolt M, Allenspach R and Mauri D 1985 J. Appl. Phys. $\underline{57}$ 3626

Luborsky F E 1980, Ferromagnetic Materials, Vol 1 p. 451, ed. E P Wohlfarth (Amsterdam: North Holland Publ. Comp.)

Mauri D 1984 Ph.D. Thesis ETH, Zürich

Mauri D, Allenspach R and Landolt M 1984 Phys. Rev. Lett. $\underline{52}$ 152

Mauri D, Allenspach R and Landolt M 1985 J. Appl. Phys. $\underline{58}$ 906

McGuire E J 1977 Phys. Rev. $\underline{A16}$ 2365

Nesbet R K 1985 Phys. Rev. $\underline{B32}$ 390

Netzer F P, Strasser G and Matthew J A D 1983 Phys. Rev. Lett. $\underline{51}$ 211

Netzer F P, Strasser G, Rosina G and Matthew J A D 1985 J. Phys. F: Met. Phys. $\underline{15}$ 753

Penn D R, Apell S P and Girwin S M 1985 Phys. Rev. Lett. 55 518

Pierce D T, Celotta R J, Unguris J and Siegmann H C 1982 Phys. Rev. B26 2566

Powell C J and Erickson N E 1983 Phys. Rev. Lett. 51 61

Rendell R W and Penn D R 1980 Phys. Rev. Lett. 45 2057

Sakisaka Y, Miyano T and Onchi M 1985 Phys. Rev. Lett. 54 714

Seah M P and Dench W A 1979 Surface and Interface Analysis 1 1

Siegmann H C, Pierce D T and Celotta R J 1981 Phys. Rev. Lett. 46 452

Siegmann H C, Meier F, Erbudak M and Landolt M 1984 Adv. Electron. Electron. Phys. 62 1

Strasser G, Rosina G, Matthew J A D and Netzer F P 1985 J. Phys. F: Met. Phys. 15 739

Tréglia G, Desjonquères M C, Ducastelle F and Spanjaard D 1981 J. Phys. C: Solid State Phys. 14 4347

Unguris J, Pierce D T, Galejs A and Celotta R J 1982 Phys. Rev. Lett. 49 72

Unguris J, Hembree G G, Celotta R J and Pierce D T 1985 J. Microscopy in press.

Zajak G, Zak J and Bader S D 1983 Phys. Rev. Lett. 50 1713

Zajak G, Bader S D, Arko A J and Zak J 1984 Phys. Rev. B29 5491

Chapter 10

Spin Polarized Photoemission by Optical Spin Orientation in Semiconductors

F. Meier

Laboratorium für Festkörperphysik, ETH Hönggerberg,
CH-8093 Zürich, Switzerland

10.1 Introduction

Historically, the prime motivation for setting up a spin polarized photoemission (SPP) experiment was the study of magnetically ordered solids (Busch et al 1969). Using natural, unpolarized light the
spin polarization of the photoelectrons is related or, in the simplest
case, even equal to the polarization of the electrons before excitation. Then, SPP offers the possibility to measure the energy- and
wave-vector dependent polarization of the electrons in their ground
state, an experiment which has in fact been realized in the last few
years (see the chapter by Gudat and Kisker of this volume). In particular, the spin-dependent electronic structure of the ferromagnetic
transition metals has been clarified to a large extent, and much will
be learnt in the future from application of SPP to other magnetic materials.

However, magnetism is not the only leg on which SPP rests. Even
when starting from an unpolarized electronic ground state polarized
electrons can be emitted. The only requirement is that the photoemission process is not "even-handed" with respect to all spin directions
but preferentially enhances (or suppresses) one of them. Numerous possibilities exist to achieve this. In the present chapter the attention
is restricted to the case where the polarization of the photoelectrons
is created by the optical excitation process: This mechanism is called
optical spin orientation. Much of the chapter will be reserved for
applications of this technique. Hopefully, the examples chosen help to
support the claim that optical spin orientation is not an aim in itself
but a tool to reveal new and basic physics hardly accessible to other
methods.

Optical spin orientation has been introduced by Kastler and co-
workers (Kastler, 1972) in order to produce spin polarized atomic
beams. The principle of the experiment is simple: an unpolarized vapor
e.g. of alkali atoms with all atoms in the ground state $ns_{1/2}$ is ex-

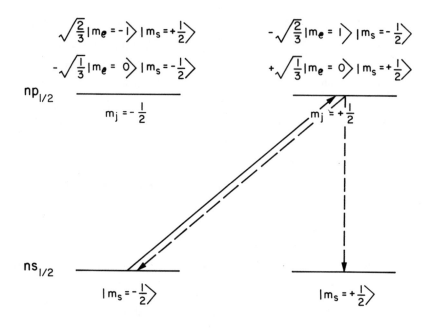

Fig. 1: Transition scheme for left-hand circularly polarized light from a $us_{1/2}$ to a $up_{1/2}$ level. The symmetry of the wave functions - expressed by their angular parts $|m_\ell\rangle$ and spin parts $|m_s\rangle$ - are indicated. The full line represents the excitation process and the broken lines show the two possible channels for deexcitation. In photoemission, the polarization of the excited state is measured.

posed to circularly polarized light, inducing transitions into an empty $np_{1/2}$-level. A $p_{1/2}$-state can be decomposed in the following way, see Fig. 1.:

$$|p_{1/2}; \; m_j = +\tfrac{1}{2}\rangle = -\sqrt{\tfrac{2}{3}}|m_\ell = 1\rangle \, |m_s = -\tfrac{1}{2}\rangle + \sqrt{\tfrac{1}{3}} \; |m_\ell = 0\rangle |m_s = +\tfrac{1}{2}\rangle \quad (1)$$

$|m_s = \pm \tfrac{1}{2}\rangle$ are the spin functions, $|m_\ell\rangle$ is the orbital angular momentum function with azimuthal quantum number m_ℓ; m_j denotes the z-component of total angular momentum. Let the circularly polarized light change Δm_j by +1, the z-direction being defined by the direction of propagation of the radiation. Then only one transition is allowed, namely from the $m_s = m_j = -\tfrac{1}{2}$ ground state to the $m_j = +\tfrac{1}{2}$ excited $p_{1/2}$-state. As is evident from eq. (1) the $m_j + \tfrac{1}{2}$ state is spin polarized. The z-component of the spin polarization is

$$P_z = \langle p_{1/2}; \; m_j = \tfrac{1}{2} \, | \, \sigma_z \, | \, p_{1/2}, \; m_j = \tfrac{1}{2}\rangle = 33\%. \quad (2)$$

σ_z is the Pauli matrix

$$\sigma_z = \begin{pmatrix} 1 & 0 \\ 0 & 1 \end{pmatrix}$$

acting only on the spin functions: $\sigma_z \, \big| \, m_s = +\tfrac{1}{2}\rangle = (+1)|m_s = +\tfrac{1}{2}\rangle$ and $\sigma_z |m_s = -\tfrac{1}{2}\rangle = (-1)|m_s = -\tfrac{1}{2}\rangle$.

Consequently, the atoms excited by circularly polarized light into a $p_{1/2}$-final state are partially polarized. Of course the life time of the excited state is finite. For deexcitation two channels are open. First, the electron may return to the level it came from, namely to the $m_s = -\tfrac{1}{2}$ ground state, thereby emitting a quantum of circularly polarized light. The second possibility is deexcitation into the $m_s = +\tfrac{1}{2}$ ground state, emitting linearly polarized light. Observing that spin must be conserved in these transitions, it follows from eq. (1) that deexcitation by channel 1 is two times more probable than by channel 2. However, the net effect of the whole operation is to transfer electrons from the $m_s = -\tfrac{1}{2}$ ground state into the $m_s = +\tfrac{1}{2}$ ground state. This process was fittingly called optical pumping; the excited state mediating the transfer from one spin state to the other is called the "pump level" or just the "pump".

This simple example contains also much of the physics of optical spin orientation in solids. It should be noted that the polarization of the wave function eq. (1) is entirely independent of its radial part. The experiment can be done as well with sodium as with cesium: the polarization will not be affected at all. Evidently, the polarization of eq. (2) is determined by the coefficients which appear on the right hand side of eq. (1) in front of the products of the angular- and spin-dependent wave functions. These are the well-known Clebsch-Gordan coefficients of atomic physics. As the example shows they also determine the relative intensities of the deexcitation transitions into the ground state levels $m_s = -\frac{1}{2}$ and $m_s = +\frac{1}{2}$. The Clebsch-Gordan coefficients depend only on the symmetry of the system. Atoms possess full rotational symmetry, whereas the space-group symmetry of a cristalline solid contains only a finite number of symmetry elements (n - fold rotations, mirror planes, inversion) apart from the translations. Therefore, in order to tackle the problem of optical spin orientation in solids, knowledge of the Clebsch-Gordan coefficients for cristalline symmetries is required.

The most important feature of the polarization obtained by optical spin orientation is its exclusive dependence on the symmetry of the electronic states, or, more precisely, on the Clebsch-Gordan coefficients. Therefore the polarization obtained by optical spin orientation is in fact an extremely simple quantity and it is exactly this feature which makes this experiment so worthwile. This will be illustrated by the examples presented in section 3.

Accidentally, not only the first experiments on optical pumping of atoms were performed in Paris, but also the first optical pumping experiments in solids (Lampel, 1968). In silicon optically aligned electron spins were detected by observing the alignment of the nuclear spins via the hyperfine interaction. Shortly later, optical spin orientation in GaSb (Parsons, 1969) and $Ga_{1-x}Al_xAs$ (Ekimov et al, 1970) has been achieved. Since that time the III-V-compound semiconductors have played a most prominent role in optical spin orientation, one reason being that they are used as very efficient sources of polarized electrons (Pierce et al, 1976, chapter by Kirschner and Gudat of this volume). However, it should be noted that optical spin orientation can in principle be applied to any system. If used in connection with photoemission the spin polarization thus obtained serves to get new and unique insights into the electronic structural properties of solids.

10.2 Principles of Optical Spin Orientation in Solids

This section intends to illustrate the mechanism of optical spin orientation, neglecting the point of view of its "usefulness" for spin polarized photoemission from solids. The latter is emphasized in the following section.

For simplicity, the 3-step model of photoemission is adopted (Cardona et al, 1978). The polarization of the excited electrons is induced in the first step, namely the optical excitation by circularly polarized light. The aim is to find the final state polarization of the electrons along the direction of propagation of the radiation.

In the bulk of the crystal, the electron is assumed to make a transition between Bloch states. They are of the form

$$\psi_{\vec{k},\sigma}(\vec{r}) = u_{\vec{k},\sigma}(\vec{r})e^{i\vec{k}\vec{r}}$$

The wave function $\psi_{\vec{k},\sigma}(\vec{r})$ possesses certain symmetry properties imposed by the symmetry of the crystal lattice. The symmetry operations of the crystal can in the simplest case (Tinkham, 1964) be divided into translations by multiples of primitve lattice vectors and the operations forming the point group of the crystal. The latter may contain various rotations, mirror planes, and possibly also the inversion. Any of these operations maps the crystal into itself. In mathematical terminology, the symmetry operations form a group.

Obviously, the Hamiltonian H of the electrons is invariant under the symmetry operations of the crystal. If ψ is an eigenfuction of the equation

$$H \cdot \psi = E \cdot \psi$$

and O a symmetry operation of the crystal,

then $H \cdot O \psi = O \cdot H \psi = E \cdot (O \psi)$ (3)

Evidently, $O\psi$ is also an eigenfunction of the Hamiltonian, belonging to the same energy E. Applying all symmetry operations O of the crystal to ψ, all degenerate eigenfunctions $\{\psi_i\}$ belonging to the energy value E are found, except for accidental degeneracy. Then each transformed function $O\psi_i$ can be expressed as a linear combination of all the $\{\psi_j\}$.

$$O\psi_i = \sum_j a_{ij} \psi_j$$ (4)

Because there is no smaller set of wave functions among $\{\psi_i\}$ which transforms into itself under all symmetry operations, the $\{\psi_i\}$ are said to form the basis of an irreducible representation Γ^n of the symmetry group of the crystal, the number of functions $\{\psi_i^n\}$ being the dimension of the representation. For the full rotation group - the symmetry of the atom - the basis functions of the odd-dimensional irreducible representations Γ^ℓ are the spherical harmonics $\{Y_\ell^m\}$, $m = -\ell \ldots\ldots +\ell$.

How many irreducible representations of the crystal symmetry groups do exist ? First consider the translations $T(\vec{a})$ by a primitive lattice vector $\vec{a}$. According to Bloch's theorem:

$$T(\vec{a}) \, \psi_{\vec{k}}(\vec{r}) = \psi_{\vec{k}} (\vec{r} + \vec{a}) = \psi_{\vec{k}}(\vec{r})e^{i\vec{k}\vec{a}}$$ (5)

Therefore, by a translation each wave function is transformed into itself modulo a phase factor e^{ika}. All the representations of the group of translations are therefore 1-dimensional, i.e. no extra degeneracy is introduced by them. The degeneracy of the states $\psi_{\vec{k}}$ is entirely determined by the point group symmetry of the crystal. In some cases extra degeneracy may be introduced by time-reversal symmetry (Bassani et al, 1975). The degenerate wave functions $\left\{\psi_{\vec{k}_i}^E\right\}$ belonging to the eigenvalue E contain generally many different wave vectors $\vec{k}_i$, namely all those which are transformed into each other by the point group symmetry operations. Because of k-conservation in optical transitions we are interested in the degeneracy of those wave functions $\left\{\psi_{\vec{k}}^E\right\}$ which belong to the same wave vector $\vec{k}$. The degeneracy of these wave functions is determined by those symmetry operations of the point group which leave $\vec{k}$ invariant. This subgroup of the point group is called the group of the wave vector $\vec{k}$. In case of minimum symmetry it contains only one element, namely the identity. However, along symmetry lines or at symmetry points of the reciprocal lattice, the group of the wave vector contains nontrivial symmetry elements producing degeneracy of the state $E(\vec{k})$.

So far, the spin was not considered explicitly. Since it is the crucial quantity in spin polarized photoemission, it must be taken into account in the symmetry arguments, as is evident from the atom-example of the introduction. The essential symmetry property of the spin $\frac{1}{2}$ is that it transforms into itself only by a rotation of 4π. Thus a new symmetry element is introduced by the spin, doubling the number of symmetry operations of the crystal point group or the group of the wave vector. Accordingly, the symmetry group including spin is named "double group" in contrast to the single group without spin.

Assume we know the basis functions of wave-vector $\vec{k}$ of an irreducible representation Γ of the spin-independent group of the wave vector: they are denoted by u^i, in accordance with the notation of

Ref. (Koster et al, 1963). The spin functions are made up of the two orthonormal spinors $v_{+1/2}$ and $v_{-1/2}$. The irreducible representations of the symmetry group with these spinors as basis functions are at most 2-dimensional. If there is a linear combination of the $v_{+1/2}$ and $v_{-1/2}$ transforming into itself under all symmetry operations the corresponding representation is one-dimensional.

Having found an irreducible representation Γ^{ℓ} of the spatial wave functions and Γ_s of the spin functions new irreducible representations with the spin-dependent wave functions as basis functions may be constructed. For this purpose the product group $\Gamma^k \otimes \Gamma_s$ must be reduced into its irreducible parts:

$$\Gamma^{\ell} \times \Gamma_s = \sum_j \Gamma_j^{\ell} \tag{6}$$

The superscript indicates the representation of the spatial wave function without spin (single group representation), j indicates the new irreducible representation including spin (double group representation).

An important result of group theory is, that once the basis functions of Γ^{ℓ} and Γ_s are known, new basis functions of Γ_j^{ℓ} are constructed from them using symmetry arguments alone. The basis functions of Γ_j^{ℓ} are linear combinations of product functions $u^k \cdot v_{\pm 1/2}$. The coefficients appearing in these linear combinations are the Clebsch-Gordan coefficients. For all relevant symmetry groups (i.e. point groups) they are tabulated in Ref. (Koster et al, 1963).

Once the symmetry of the initial and final states is known, the matrix element of the wave functions with the operator of circularly polarized light is of interest. The operator for circularly polarized lights is

$$0^{\pm}_{\text{light}} = p_x \pm i p_y$$

where $\oplus$ refers to left-hand circularly polarized light ($\Delta m_j = +1$ in atomic transitions) and Θ to right-hand circularly polarized light ($\Delta m_j = -1$). Assume that $0^{\pm}_{\text{light}}$ transforms according to the irreducible representation Γ^{op}, i.e. p_x and p_y are basis functions of the same irreducible representation, then transitions between the levels considered are forbidden by symmetry (selection rule!) if

$$\Gamma^{op} \otimes \Gamma(\text{initial state})$$

does not contain the final state representation $\Gamma(\text{final state})$. Otherwise, transitions may be possible. For p_x and p_y to transform according to the same irreducible representation a minimum amount of symmetry of the group of the wave vector is required, as pointed out by Borstel and Wöhlecke, namely the presence of at least one 3- or higher fold rotation axis (Wöhlecke et al, 1984).

In many cases there will be just one allowed transition between the set of (possibly) degenerate initial and final states. However, as e.g. for excitations across the energy gap of GaAs at Γ, more than one transition may occur. Then, the final state polarization is the weighted average

$$P_f = \frac{\sum P_{f,n} I_n}{\sum I_n} \qquad (8)$$

of the final state polarization $P_{f,n}$ of the individual transitions which occur with intensity I_n. A corner-stone of group theory, the Wigner-Eckart theorem, determines the relative intensities I_n. In fact, it gives the matrix elements M_{nm} of the transitions

$$\left\langle \psi_f^n \left| 0_{\text{light}} \right| \psi_i^m \right\rangle = M_{nm} \cdot c(\Gamma_i, \Gamma_j) \qquad (9)$$

up to a common factor c (Γ_f, Γ_i), which depends - for a given operator - only on the irreducible representations Γ_f and Γ_i, of which $\{\psi_f^n\}$ and $\{\psi_i^m\}$ form a basis. Obviously this common factor cancels in the expression of the polarization.

The matrix elements M_{nm} are obtained using the Clebsch-Gordan coefficients alone. This will be illustrated by an example below. Therefore, the Clebsch-Gordan coefficients make possible (1) to set up properly symmetrized basis functions of the double group representation and (2) to determine the relative matrix elements of simultaneous transitions between degenerate initial and final state levels.

Example: Consider the transitions induced by circularly polarized light between the top of the valence band and the bottom of the conduction band of GaAs. The symmetries of the respective states are Γ_8^{15} and Γ_6^1 (Γ_8^5 in the notation of Ref. (Koster et al, 1963)). Γ_8^{15} is 4-fold, Γ_6^1 twofold degenerate, see Fig. (2).

The basis functions of Γ_6^1 are $\left\{ \psi_{-1/2}^6, \psi_{+1/2}^6 \right\}$ and of Γ_8^{15} they are $\left\{ v_{-3/2}^8, v_{-1/2}^8, v_{+1/2}^8, v_{+3/2}^8 \right\}$. The operator for circularly polarized light transforms according to Γ^{15}. Right-hand circularly polarized light is represented as

$$x - iy = u_{yz}^5 - i u_{xz}^5 \tag{10}$$

A typical matrix element is of the form

$$M_{fi} = \left\langle \psi_f^6 \left| u_{yz}^5 - i u_{xz}^5 \right| v_i^8 \right\rangle \tag{11}$$

To be definite consider $\psi_f^6 = \psi_{-1/2}^6$ and $v_i^8 = v_{-3/2}^8$.

Table 83 of Ref. (Koster et al, 1963) lists the Clebsch-Gordan coefficients for developing ψ^6 into product functions $u^5 v^8$. For instance

$$\psi_{-1/2}^6 = \frac{i}{\sqrt{12}} u_{yz}^5 v_{-3/2}^8 + \frac{i}{2} u_{yz}^s v_{1/2}^8 + \frac{1}{\sqrt{12}} u_{xz}^5 v_{-3/2}^8$$

$$- \frac{1}{2} u_{xz}^5 v_{+1/2}^8 - \frac{i}{\sqrt{3}} u_{xy}^5 v_{3/2}^8 . \tag{12}$$

Introducing this expression for $\psi_{-1/2}^6$ into M_{fi} (eq. 11), all those product functions which do not contain $v_{-3/2}^8$ will give zero contribu-

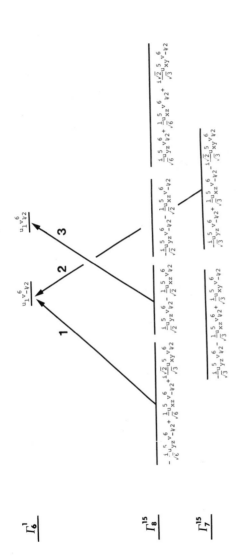

Fig. 2: Transition scheme for GaAs at Γ. Γ_6^1 are the lowest conduction band states, Γ_8^{15} is the top of the valence band and Γ_7^{15} is the spin-split valence band. Indicated are the transitions for right-hand circularly polarized light. The relative intensities are given by the numbers. The two simultaneous transitions $\Gamma_8^{15} \rightarrow \Gamma_6^1$ result in $P = 0.5$.

tion. Therefore:

$$M_{fi}(-1/2,-3/2) = \left\langle \frac{i}{\sqrt{12}} u^5_{yz} + \frac{1}{\sqrt{12}} u^5_{xz} + \ldots \,\middle|\, u^5_{yz} - iu^5_{xz} \right\rangle = -\frac{i}{\sqrt{3}} \quad (13)$$

As basis of Γ^1_6, the spin polarization of the final state $\psi^6_{-1/2}$ is $P_f = -1$.

The only other transition giving a nonzero matrix element is

$$M_{fi}(+1/2, -1/2) = \left\langle \psi^6_{+1/2} \,\middle|\, u^5_{yz} - iu^5_{xz} \,\middle|\, v^8_{-1/2} \right\rangle$$

$$= \left\langle (i/2) u^5_{yz} v^8_{-1/2} - (1/2) u^5_{xz} v^8_{-1/2} + \ldots \,\middle|\, u^5_{yz} - iu^5_{xz} \,\middle|\, v^8_{-1/2} \right\rangle$$

$$= \frac{i}{2} + \frac{i}{2} = i. \quad (14)$$

The polarization of the final state $\psi^6_{+1/2}$ is $P_f = +1$. Consequently, the total spin polarization is with $I_{fi} = |M_{fi}|^2$:

$$P_{tot} = \frac{1 - 1/3}{1 + 1/3} = 0.5, \quad (15)$$

i.e. the two simultaneous transitions result in a total polarization of 50%.

Consider now the spin-split valence band Γ^{15}_7. There the only non-vanishing matrix element is

$$M_{fi}(-1/2,1/2) = \left\langle \psi^6_{-1/2} \,\middle|\, u^5_{yz} - iu^5_{xz} \,\middle|\, v^7_{+1/2} \right\rangle = \frac{2i}{\sqrt{3}}. \quad (16)$$

The value of this matrix element is not related to those of the transitions $\Gamma^{15}_8 \to \Gamma^1_6$, because the Wigner-Eckart theorem, see Ref. (Tinkham, 1964) will contain now the constant $c(\Gamma^1_6, \Gamma^{15}_7)$ instead of $c(\Gamma^1_6, \Gamma^{15}_8)$. Physically, this is easily recognized by the fact, that the combined transitions $\Gamma^{15}_8 \to \Gamma^1_6$ and $\Gamma^{15}_7 \to \Gamma^1_6$ should result in zero polarization because for vanishing spin-orbit splitting the polarization must vanish, too (Meier et al, 1984). However, using the matrix element (16) for $M_{fi}(-1/2,1/2)$ together with the others for $\Gamma^{15}_8 \to \Gamma^1_6$ given by (13) and (14), P will not be zero. This contradiction is only apparent, because it simply reflects that the Wigner-Eckart constant for $\Gamma_7^{15} \rightarrow \Gamma_6^1$ is different from the one $\Gamma_8^{15} \rightarrow \Gamma_6^1$.

This difficulty may be circumvented in the following way: one goes back from the double group representations to the single group representations. I.e. we expand v^8 and v^7 into product functions $u^{15}v_{\pm 1/2}$, because both are derived from the same single group Γ^{15}.

E.g.: $v_{-3/2}^8 = -\dfrac{i}{\sqrt{6}} u_{yz}^5 v_{-1/2}^6 + \dfrac{1}{\sqrt{6}} u_{xz}^5 v_{-1/2}^6 + i \sqrt{2/3}\, u_{xy}^5 v_{1/2}^6$, (17)

as indicated in Fig. 2. Then the matrix element

$M_{fi}(-1/2,-3/2) = \left\langle \Psi_{-1/2}^6 \left| w_{yz}^5 - iw_{xz}^5 \right| v_{-3/2}^8 \right\rangle$ can be expressed as

$\left\langle u^1 v_{-1/2}^6 \left| w_{yz}^5 - iw_{xz}^5 \right| - \dfrac{1}{\sqrt{6}} u_{yz}^5 v_{-1/2}^6 + \dfrac{1}{\sqrt{6}} u_{xz}^5 v_{-1/2}^6 + i \sqrt{2/3}u_{xy}^5 v_{1/2}^6 \right\rangle$

Because of the orthonormality of the spin functions this is equal to

$\left\langle u^1 \left| w_{yz}^5 - iw_{xz}^5 \right| - \dfrac{i}{\sqrt{6}} u_{yz}^5 + \dfrac{1}{\sqrt{6}} u_{xz}^5 \right\rangle$

Now u^1 is expanded in terms of $w^5 \cdot u^5$

$u^1 = \dfrac{1}{\sqrt{3}} w_{yz}^5 u_{yz}^5 + \dfrac{1}{\sqrt{3}} w_{xz}^5 u_{xz}^5 + \dfrac{1}{\sqrt{3}} w_{xy}^5 u_{xy}^5$

Then

$M_{fi}(-1/2,3/2) = -\dfrac{i}{\sqrt{18}} - \dfrac{i}{\sqrt{18}} = -\dfrac{2i}{\sqrt{18}} ; \left| M_{fi} \right|^2 = 2/9$ (18)

It should be noted that the unknown factor of this matrix element is the one between the single group representations Γ^1 and Γ^{15}. Therefore the same factor applies to the top of the valence band (Γ_8^{15}) and to the spin-split valence level Γ_7^{15}, because both are derived from the same single group representation. Then, the correct relative intensities for the three transitions $\Gamma_7^{15} \rightarrow \Gamma_6^1$ and $\Gamma_8^{15} \rightarrow \Gamma_6^1$ can be calculated. The derivation of these matrix elements proceeds exactly as shown above for $M_{fi}(-1/2,-3/2)$, see Ref.(Meier et al, 1984). It turns out that for zero spin-orbit splitting, P = 0 as it should be. This example shows that working back from the double group representations to the single groups, the Wigner-Eckart theorem becomes applicable to a larger set of wave functions - all those, which are derived from the same orbital representation by the spin-orbit interaction.

10.3 Applications of Optical Spin Orientation

10.3.1 Hybridization of Energy Bands

Applying the formalism described in section 2. by using the tables of Ref. (Koster et al., 1963), a strange phenomenon is noticed: practically all transitions induced by circularly polarized light should result in final state polarizations equal to -1, 0, or +1. Exceptions are points of very high symmetry, like the Γ-point of GaAs, where P = 0.5. When Reyes and Helmann predicted in their pioneering work the polarization of the transitions along W(100) (Reyes et al, 1977) they got a spectrum fluctuating wildly between +1 and -1. In this paper they showed for the first time how the group theoretical formalism is applied to optical spin orientation in its full generality. The experiment later showed, that the transitions along W(100) were polarized indeed; however, the maximum polarizations measured were drastically lower than the ones anticipated, not exceeding 10% (Zürcher et al, 1979). An analysis of this experiment showed that an intrinsic mechanism was responsible for the result and not an experimental artifact. Subsequently, the same phenomenon occured in many other situations. It finds a natural explanation in the concept of band hybridzation. A very transparent example of band hybridization has been studied recently for transitions near Γ in Ge (Allenspach et al, 1984). In this case, no complications arise from the fact that Ge does not belong to a symmorphic space group (Allenspach et al, 1984).

Consider the band structure of Ge near the energy gap. Fig. 3 shows the symmetry of the single group energy bands. Band Δ^5 is doubly

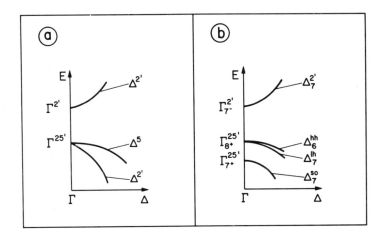

Fig. 3: Band structure of Ge along the Δ-direction near Γ.

a: without spin-orbit interaction (single group symmetry)

b: with spin-orbit interaction (double group symmetry)

degenerate. Turning on the spin-orbit interaction the 3-fold degeneracy of $\Gamma^{25'}$ is lifted: with spin, it splits into a 4-fold degenerate $\Gamma_{8+}^{25'}$ level and a $\Gamma_{7+}^{25'}$ level, see Fig. 3b. Similarly, the Δ^5 valence band splits into a heavy hole Δ_6^{hh} and a light hole $\Delta_7^{\ell h}$ band. The splitting, however, is very small near Γ and therefore it cannot be experimentally resolved. The spin orbit split valence band has symmetry Δ_7^{so}.

As for GaAs, the polarization of the transition $\Gamma_{8+}^{25'} \to \Gamma_{7-}^{2'}$ is $P = 0.5$; for $\Gamma_{7+}^{25'} \to \Gamma_{7-}^{2'}$ it is $P = -1$ with half the intensity of the previous one. The interesting point is what happens when the transitions occur not at Γ but slightly off Γ, along the Δ - direction.

The point group symmetry of the states with k along $\Delta = (100)$ is C_{4v}. The appropriate tables in Ref. (Koster et al, 1963) show that Δ^5 goes over into bands of symmetry Δ_6^5 and Δ_7^5, and $\Delta^{2'}$ becomes $\Delta_7^{2'}$.

Furthermore

$$P(\Delta_6^5 \rightarrow \Delta_7^{2\prime}) = +1 \qquad \text{intensity: } I_0$$

$$P(\Delta_7^5 \rightarrow \Delta_7^{2\prime}) = -1 \qquad \text{intensity: } I_0$$

$$\Delta_7^{2\prime} \rightarrow \Delta_7^{2\prime} \qquad \text{forbidden.}$$

Accordingly, only transitions from the heavy and light hole band are allowed and both have the same intensity but opposite polarization. Since the splitting of the Δ_6^{hh} and $\Delta_7^{\ell h}$ - band is experimentally not resolved the total polarization of the two transitions should be zero. This is a very surprising and physically improbable result: at Γ, the polarization originating from $\Gamma_{8+}^{25\prime} \rightarrow \Gamma_7^{2\prime}-$ is 50%, but as soon as the transition takes place at an infinitesimal distance off Γ, P should drop to zero discontinuously. The experiment shows clearly, that this is not the case at all.

Fig. 4 displays a polarization spectrum obtained from a Ge(001) crystal. At photothreshold, P = 50%. For higher photon energies the polarization drops rapidly before some additional structure appears, which will not be discussed here.

The important point is that the energy gap of Ge is only 0.9 eV at the measuring temperature of 40 K. Then, the excitation for the threshold photon energy hν = 1.6 eV takes place not at Γ but away from Γ. Therefore, we are exactly facing the situation where - according to the preceding argument - the polarization should vanish. The fact that it does not is due to hybridization.

When an orbital degeneracy is lifted by spin-orbit interaction, states of the same double group symmetries become mixed or hybridized (Heine, 1969). States of different double group symmetry cannot mix, their matrix element with the spin-orbit interaction is strictly zero for symmetry reasons (Tinkham, 1964). For the example studied hybridi-

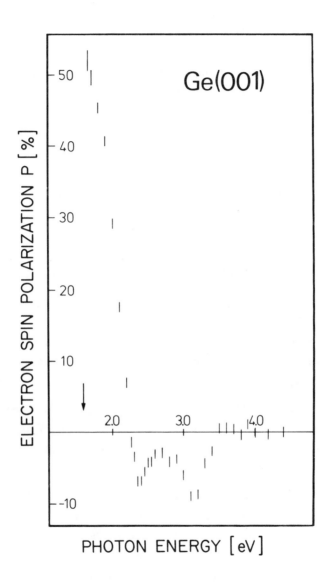

Fig. 4: Spectrum of the spin-polarization of Ge(001) at T = 40 K.
The photothreshold, indicated by the arrow, is 1.6 eV

zation is possible between the light hole band Δ_7^5 and the spin-orbit split valence band $\Delta_7^{2'}$, but the Δ_6^5-state cannot mix with neither of the two. Then the following interesting question arises: Is it possible to get a polarization of 50% by suitable hybridization of the Δ_7^5 and $\Delta_7^{2'}$-states even along k = (100)? Obviously it is possible: by adding $\Delta_7^{2'}$ to Δ_7^5 the intensity of the transition $\Delta_7^{\ell h} \rightarrow \Delta_7^{2'}$ is reduced (since $\Delta_7^{2'} \rightarrow \Delta_7^{2'}$ is forbidden), whereas the intensity of $\Delta_6^{hh} \rightarrow \Delta_7^{2'}$ remains unaffected. Consequently, there will be a net polarization. It is easily verified that the polarization becomes +50% if the mixing occurs in the following way

$$\Delta_7^{\ell h} = \left(\sqrt{1/3}\right)\Delta_7^5 + \left(\sqrt{2/3}\right)\Delta_7^{2'} \tag{19a}$$

Similarly the spin-orbit split valence band acquires Δ_7^5 - character:

$$\Delta_7^{so} = \left(\sqrt{2/3}\right)\Delta_7^5 + \left(\sqrt{1/3}\right)\Delta_7^{2'} \tag{19b}$$

Then, the polarization $\Delta_7^{so} \rightarrow \Delta_7^{2'}$ is -1 with intensity 2/3. Regarding the polarization, the same values are obtained along Δ as at Γ, a result which is now in accordance with the experiment.

This example shows that hybridization of energy bands can quantitatively be determined by optical spin orientation. This unique application will certainly be exploited much more intensively in future band structure investigations.

Our interest has been focused on the (100)-direction of Ge; due to the isotropy of the energy bands near Γ, any other direction would have served the purpose as well. Since in our experiment no angular resolution of the escaping photoelectrons is made, the dependence of P on the angle Ω between the propagation of the radiation and the $\vec{k}$-vector of the electron should be investigated. It turns out, that for transitions near Γ P does not depend on Ω as long as the hybridization is of the form (19), see Ref. (Meier et al, 1984). It is due to this particular effect that for Ge as well as for GaAs a polarization of 50% is observed. As a further consequence the spectrum of the spin polarization - Fig. 4 - is independent of crystal orientation since all structure is due to transitions near Γ.

It is clear that the band hybridization derived above does not extend throughout the whole Brillouin-zone. If k becomes too large, the anisotropy of the bands will show up and band separations become so wide that the matrix elements of the spin-orbit-operator become very small. In a series of experiments on GaAs the dehybridization of the energy bands with increasing k-vector could be experimentally observed. Near Γ the hybridization of the bands is identical for GaAs and for Ge. In order to study the dehybridization, GaAs samples were prepared by suitable cesiation with work function Φ ranging from 1.5 eV to 3.7 eV, see Fig. 5. For Φ = 1.5 eV a polarization of 33% at photothreshold is measured at T = 300 K. Since the energy gap of GaAs is 1.5 eV the transition takes place just at Γ. The fact that the polarization is less than 50% at room temperature is well known. Many experiments (Allenspach et al, 1984) showed that the threshold polarization of GaAs depends strongly on temperature and various explanations have been put forward to explain this effect. In the present context the precise value of P at hν = Φ = 1.5 eV is of no concern. What is important is the dependence of P on Φ. If Φ increases, the transitions take place at increasing distance from Γ. Nevertheless, Fig. 5 shows that by varying Φ from 1.5 to 2.5 eV the threshold polarization changes only slightly. For Φ > 2.5 eV the polarization drops rapidly and vanishes at Φ = 3.7 eV. Therefore, qualitatively, the hybridization found near Γ extends in fact up to that distance into the Brillouin zone where initial and final states are separated by 2.5 eV. Along two symmetry directions these locations are indicated in Fig. 6. In order to follow precisely the dehybridization of a given band as function of $\vec{k}$, energy- and angle resolved spectra are clearly highly desirable.

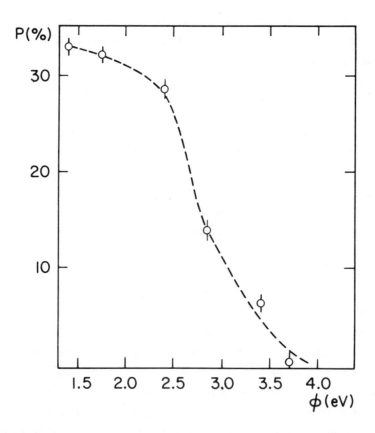

<u>Fig. 5:</u> The spin polarization of the electrons excited at photo-
threshold of GaAs at room temperature for various values of
the photothreshold Φ. For Φ = 1.5 eV, the transitions induced
by circularly polarized light occur at the Γ - point. The maximum polarization is only slightly above 30%, due to temperature
dependent depolarization. Up to Φ = 2.5 eV, P changes only
slightly due to hybridization as described in the text. Only
for Φ > 2.5 eV strong dehybridization occurs producing a
corresponding drop of P.

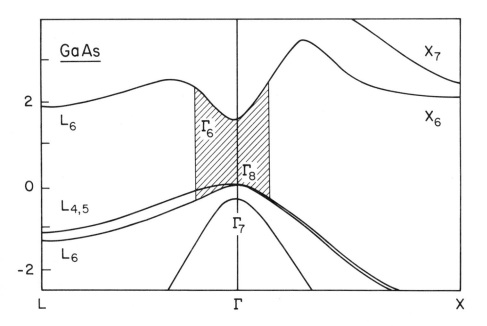

Fig. 6: Extension of band hybridization into the Brillouin zone. The shaded area covers the region where the final state polarization is nearly constant according to Fig. 5

10.3.2 Lifting of the Spin-Degeneracy in Non-Centrosymmetric Materials

For crystal structures without inversion symmetry the spin degeneracy of the energy bands is lifted throughout most of the Brillouin zone. Assuming the system to be invariant with respect to time reversal, the energy of the electronic state $E(k,\uparrow)$ is equal to $E(-k,\downarrow)$; applying the inversion, $-k$ is transformed into $+k$, whereas the spin - as an axial vector - is not affected. Thus $E(-k,\downarrow) = E(k,\downarrow)$ or $E(k,\uparrow) = E(k,\downarrow)$. However, this relation does not hold generally in the absence of inversion symmetry. As an example, the (110)-direction of GaAs may be considered. The symmetry of the electronic states along (110) is C_s.

For the corresponding centrosymmetric structure - i.e. the diamond lattice - it is C_{2v}. Table 9 of Ref. (Koster et al, 1963) shows that the double-group representations of GaAs(110) are 1-dimensional. Since states with $-\vec{k}$ and $\vec{k}$ are not connected to each other along (110) by a spatial symmetry operation, time reversal does not introduce an extra degeneracy at $\vec{k}$, i.e. the spin-degeneracy is lifted. Contrary, the only double group representation along (110) of the diamond lattice is 2-dimensional, as it should be. Along the special directions (100) and (111) the spin-degeneracy of GaAs persists.

Although the breaking of the spin-degeneracy in crystal structures lacking inversion symmetry has been recognized since a very long time, it was noticed only much later that it gives rise to an interesting effect which has been experimentally observed recently (Riechert et al, (1984), Riechert et al, 1985) in GaAs. For the lowest conduction band the single-electron Hamiltonian contains a term cubic in k and is of the form

$$H = \frac{\vec{p}}{2m} + \frac{h}{2} \vec{\sigma} \cdot \vec{\Omega} (\vec{k})$$

where $\Omega_x(\vec{k}) = \alpha h^2 (m_e \sqrt{2m_e E_G})^{-1} k_x (k_y^2 - k_z^2)$ with $\alpha = 6 \times 10^{-2}$ a dimensionless parameter. $\vec{\Omega}$ acts on the electron like an external magnetic field with direction perpendicular to $\vec{k}$' since $\vec{k} \cdot \vec{\Omega} = 0$. Consequently, the spin precesses around $\vec{\Omega} (\vec{k})$. It has been proposed that this wave vector dependent precession could lead to a depolarization of the optically oriented electrons in GaAs (D'yakonov et al, 1972, Pikus et al, 1984). The spin precession has been verified directly in the following way: For p-doped, negative-electron affinity GaAs (see chapter by Gudat and Kirschner of this volume) there is a band bending region at the surface where the electrons in the conduction band gain several tenths of an eV in kinetic energy with respect to the bottom of the band. For sufficient p-doping, the band bending region is thin enough that the electrons pass through it without scattering, i.e.

ballistically.

Therefore, if they have the same energy and crystal momentum along (110) they all precess by the same amount in the same direction perpendicular to $\vec{k}$ and $\vec{\Omega}$. Then, for electrons polarized initially along (110) by optical spin orientation, the direction of polarization will turn by a certain amount out of the $(\vec{k},\vec{\Omega})$-plane. With a Mott-detector measuring the spin-component along (110) the spin precession results in a decrease of the polarization. Directing the electrons through a solenoid, the precession in the band bending region can be compensated by the opposite precession in the external magnetic field. This effect was observed for Ga(110), see Fig. 7.

The quasi-magnetic field $\vec{\Omega}$ defines the natural quantization direction for the electron spin in the band bending region: An electron with spin parallel to $\vec{\Omega}$ is energetically separated by an electron with spin antiparallel to $\vec{\Omega}$ by a quasi-Zeeman splitting. Therefore, at constant electron energy, the density of states for 'up' and 'down' electrons will not be exactly equal. This results in a polarization of the electrons along $\vec{\Omega}$ even if they are not plarized initially, i.e. before entering the band bending region. Also this effect was verified experimentally, see Fig. 8. Clearly, the spin precession and spin orientation in the band bending region presents a very elegant means for investigating the consequences of the removal of the spin-degeneracy in non-centrosymmetric crystals.

10.3.3 Structural Phase Transitions

Photoemission has been used successfully not only for obtaining information on the electronic structure of solids but also for clarifying the geometry of adsorbate positions (Liebsch, 1976). For this purpose, the intensity distribution of the photoelectrons is measured and subsequently a LEED-type ("single-step photoemission") formalism is applied to fit the data. Clearly, this procedure involves heavy use of

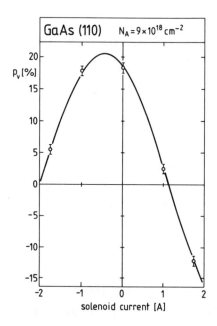

Fig. 7: Compensation of the spin precession in the band bending region by an external solenoid. Before entering the band bending region the electrons are polarized along the direction of the exciting circularly polarized light, i.e. normal to the surface. P_v is the projection of the polarization on this direction after the electrons have passed through the solenoid. $N_A = 9 \times 10^{18}$ cm^{-3} is the zinc doping level.

data processing and a more straightforward approach would be highly welcome.

Since the polarization created in optical spin orientation depends sensitively on the symmetry of the wave functions and therefore on the symmetry of the atomic arrangement of the sample, this technique should prove very useful for obtaining structural information. This has been tested by an admittedly very drastic example: the phase transition bet-

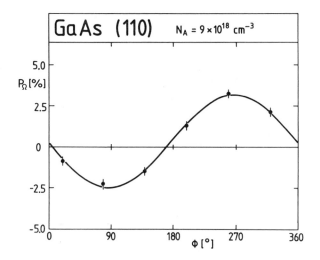

Fig. 8: Transverse polarization along 1$\bar{1}$0 acquired by initially unpolarized excited electrons with $\vec{k}$ 110 in the band bending
region of GaAs. The measured component of the polarization
depends on the azimuthal angle φ of the sample, see Riechert
et al., 1985

ween amorphous and crystalline germanium.

It is well known that for tetrahedrally bonded semiconductors
(Shevchik et al, 1972) even in the amorphous phase the nearest neighbor
configuration of the atoms is very similar to the one of the crystalline
state. Therefore, if the local geometry of the atoms were responsible
for the polarization, amorphous germanium could be imagined as consisting of an assembly of tetrahedral microcrystallites. Since the polarization is independent of crystal orientation for transitions at or near Γ,
it would then be anticipated that the P-spectrum of amorphous Ge should
be similar to the one of the cristalline state.

Amorphous germanium was prepared by evaporation onto a molybdenum
substrate held at room temperature. High resolution electron diffraction

showed that the sample was indeed amorphous.

Quite contrary to cristalline germanium, the polarization of the photoelectrons was constantly zero for the photon energy range investigated 1.6 < hν < 3.8 eV, see Fig. 9.

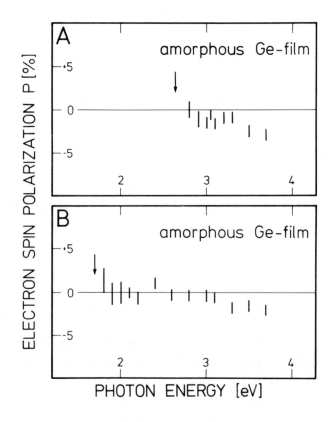

Fig. 9: Spin polarization of electrons emitted from amorphous germanium.
A: Photothreshold Φ = 2.6 eV. B: Φ = 1.6 eV.
T = room temperature.

On annealing the sample at 250°C for a few minutes polycrystals of 100 - 1000 Å lateral dimensions were formed. For the annealed samples, the P-spectrum is identical to the one of single crystalline Ge, see Fig. 10. Again, the maximum polarization at photothreshold Φ = 1.5 eV is 30% only instead of 50% because the sample was held at room temperature.

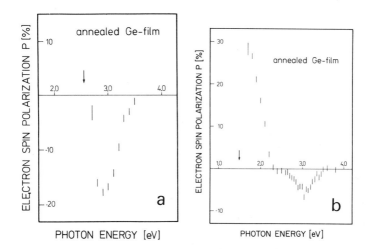

<u>Fig. 10:</u> Spin polarization of electrons emitted from initially amorphous Ge films which have been annealed at 250°C for a few minutes. a: Photothreshold Φ = 2.6 eV. b: Φ = 1.6 eV. T = room temperature.

The outcome of the experiment is interesting for two reasons: First, the polarization is affected very severely by the atomic structure of the solid. Generally, intensity spectra of photoelectrons (Greuter et al, 1979) or the optical reflectivity (Aspnes et al, 1982) do not change very much when a material is transformed from an ordered to a disordered state. Therefore, spin polarized photoemission by optical spin orientation is a valuable, sensitive probe of crystalline order. Secondly, more specific to Ge, the result shows that it is not the short range atomic order which gives rise to the polarization, but a property of the Bloch functions of the quasi-infinitely periodic lattice. It is of course very interesting to know at which crystallite size the typical properties of Bloch states appear. By careful preparation of the samples it should be possible to monitor the transition amorphous → crystalline by the spin polarization. This would be most remarkable contribution to a long-standing problem of small particle physics (Perenboom, 1981), namely at which critical size an agglomerate of atoms gets the properties of the infinite crystal.

10.3.4 Enhancement of Resolution in Photoelectron Spectroscopy by Measuring the Spin Polarization

Although photoemission has become a very highly developed technique and proved its versatility for the most different purposes, there are fundamental limitations to this experiment which severely restrict its range of applicability. Probably the most serious limitation is set by the resolving power. The resolution in photoemission from solids is generally not an instrumental problem, e.g. of the energy analyzer. The crucial problem is life-time broadening of the individual transitions. Whereas the life-time of the 2p state in atomic hydrogen is $\sim 10^{-8}$ sec, (Kuhn, 1969) the valence band hole life-time in a solid is usually $10^{-14} - 10^{-15}$ sec only giving rise to a level broadening of the order of 0.1 - 5 eV (Pendry et al, 1977). Clearly, many phenomena involving e.g. the valence band states of a solid concern only states within a much narrower energy interval, often around the Fermi level. Such phenomena cannot be observed directly by photoemission but become at best measurable using indirect evidence.

Since level-broadening is a fundamental quantum mechanical effect nothing can be done to reduce it externally. However, as will be shown subsequently, the resolution obtainable in a photoemission experiment is significantly improved by measuring the spin polarization of a transition - which is a differential quantity - instead of the intensity. As an example, the spin-orbit splitting of the top valence bands in silicon is considered at Γ. It amounts to 0.044 eV at room temperature.

The principle of the enhancement of resolution is illustrated in Fig. 11. Fig. 11a shows the superposition of two transitions I_1 and I_2 in an intensity measurement. Obviously, from the superposition of I_1 and I_2 it is by no means obvious that two transitions are hidden behind

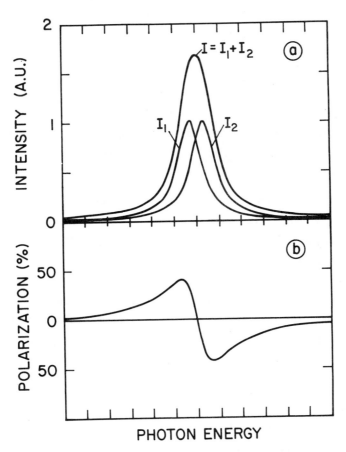

Fig. 11: a: Superposition of intensity spectra of two closely spaced transitions

b: Superposition of polarization spectra of two closely spaced transitions

454

the resulting curve. It could at best be revealed by line shape analysis, which however is often not very satisfactory since it requires a-priori knowledge of the individual line shapes I_1 and I_2 which is generally not available.

Suppose that the transitions I_1 and I_2 are polarized for excitation with circularly polarized light. For the example of Si, to be discussed in the following, $I_1 = 2 I_2$ and $P_1 = 50\%$, $P_2 = -100\%$. However, the general argument does not depend on these specific numbers. For simplicity, in Fig. 11b, the intensities I_1 and I_2 are taken to be equal, and the respective polarizations $P_1 = +1$ and $P_2 = -1$. The total polarization $P = (I_1 P_1 + I_2 P_2)/(I_1 + I_2)$ with its clear structure reveals at once the presence of two nearly degenerate transitions. Qualitatively, it is this simple property which enhances the resolution of the photoemission experiment.

In the experiments described in this paper the spin-polarization of the total photoyield is measured, without external angle- or energy discrimination of the electrons, as for all materials the photoyield of silicon is quite structureless, see Fig. 12. In contrast the spectrum of the polarization shows prominent structure. Fig. 13a gives the P-spectrum of the transitions from the top valence bands $\Gamma_{8+}^{25'}$ and $\Gamma_{7+}^{25'}$ (separated by the spin-orbit splitting Δ_0) to the conduction band $\Gamma_{7-}^{2'}$, whereas in Fig. 13b the polarizations of the transitions from the same initial states to the final states Γ_8^{15-} and Γ_6^{15-} is displayed. A schematic band structure of Si is shown in Fig. 14.

Although the spin-orbit splitting Δ_0 is 44 meV only it produces clearly observable polarization peaks. For the two transitions giving rise to the P-spectrum of Fig. 13a the polarizations and relative intensities are exactly the same as for the corresponding transitions of GaAs (Γ_8^{15}, $\Gamma_7^{15} \rightarrow \Gamma_6'$), see the preceeding section. However, due to the small Δ_0 of Si the two transitions overlap and the P-values are much smaller, but for the detection of the closely spaced transitions this

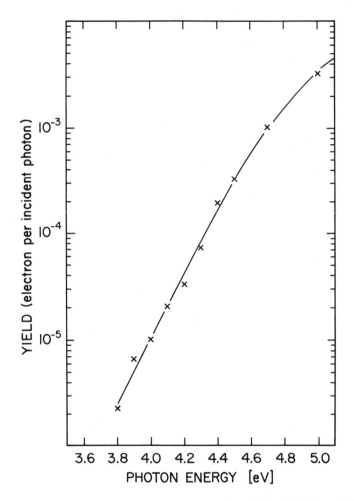

Fig. 12: Quantum yield of silicon sample with photothreshold
Φ = 3.7 eV, adjusted by cesiation. The corresponding
polarization spectrum is shown in Fig. 13

is not essential.

It is possible to fit the data points of Fig. 13a using the measured yield curve and the well-known material parameters of silicon

(Bona et al, 1985). Working in reverse order, such a procedure could be used for determining the spin-orbit splitting experimentally from photoelectron spectra.

To take full advantage of the differential nature of the polarization measurement, combination with energy- and angular resulution will be indispensable. Then, it seems possible to observe transitions separated by only a few milli-eV. Obviously, with a routinely achievable resolution of kT (= 0.025 eV at room temperature) or better, there will

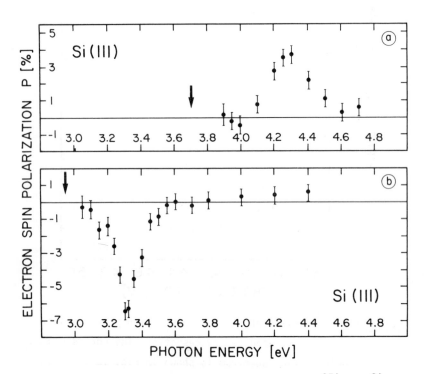

Fig. 13: a: Polarization spectrum of transitions $\Gamma^{25'} \to \Gamma^{2'}$
for silicon sample with photothreshold $\Phi = 3.7$ eV
b: Polarization spectrum of transitions $\Gamma^{25'} \to \Gamma^{15}$
for silicon sample with $\Phi = 2.9$ eV

Sequence of energy levels of Si at Γ (schematic)

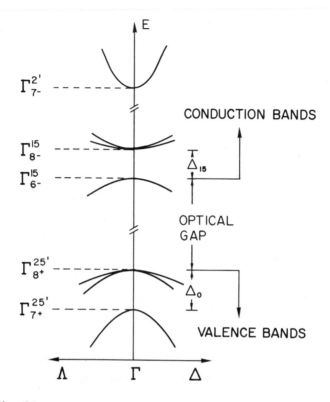

Fig. 14: Schematic band structure of silicon near Γ

be a whole new range of applications be opened for photoelectron spectroscopy.

10.3.5 Spin Exchange Scattering

Photoemission spectroscopy profits - or suffers - from the strong interaction of the electrons with each other. Among the welcome proper-

ties is the surface sensitivity thus achieved; generally less welcome
is the fact that only a fraction of the excited primary electrons
reaches the vacuum without change of energy and momentum by elastic or
inelastic scattering. In spin polarized photoemission, interest is
naturally focused on the problem as to which degree the spin polarization of the photoelectrons is conserved during the emission process.

Certainly, there are examples where the polarization is not affected to any noticeable degree during emission. For transitions across
the band gap of germanium and gallium arsenide, circularly polarized
light produces 50% polarized electrons and this is what is measured for
samples held at low temperature. Similarly, in Nickel electrons which
are -100% polarized in the ground state appear as -100% polarized at
the Mott detector (Kisker et al, 1980). Thus, under favorable conditions, the spin polarization is a well conserved quantity. However
there are examples where the situation is less clear, and, finally,
there are situations where it is absolutely certain that the spin polarization of the photoelectrons has undergone considerable alteration
during emission. It is this latter class of experiments which will be
dealt with in this subsection. Again, it turns out that an effect which
appears as a nuisance at first sight may be finally exploited for obtaining very direct evidence on the energy-dependent spin exchange interaction of excited electrons.

The first clear-cut evidence for non-conservation of the spin
polarization was found in the early experiments on EuO (Sattler et al,
1972). At low temperatures, the spins of the electrons in the half
filled 4f shells are all aligned. Therefore, well below the Curie
temperature ferromagnetic EuO should produce 100% polarized f-electrons - a feature never observed, see Fig. 15. Even more surprising
was the absence of magnetic saturation in an external applied field.
Only a change of slope indicates magnetic bulk saturation, but P continues to rise with increasing external field. It was concluded that

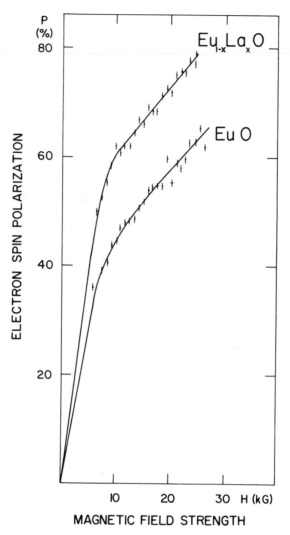

<u>Fig. 15:</u> 4f-Electron spin polarization of $Eu_{1-x}La_xO$ for x = 0 and
x = 0.02 as function of the external magnetic field. The bulk
of the sample saturates where the slope of P(H) changes,
around 10 kG. The increase of P with higher fields is attributed to the more complete alignment of surface magnetic
moments, making spin-flip scattering less effective. The
measurement was made at T = 10 K.

not ferromagnetically aligned 4f-moments at the surface depolarize by spin exchange scattering the outgoing stream of 4f-electrons excited in the bulk of the sample. The external magnetic field tends to align the disordered surface moments and therefore the spin-exchange scattering becomes less effective with increasing H.

Very direct evidence for this mechanism was found in an experiment, where the postulated properties of the EuO surface were actually built into the sample artificially. For this purpose, an overlayer of gadolinium was applied to a germanium substrate. By optical spin orientation, a transition of known polarization was induced in the Ge crystal. The $\Gamma_{25'} \rightarrow \Gamma_{15}$ (hν = 3.05 eV) transition was chosen, since the polarization P = 23 $\pm$ 1% turns out to be rather insensitive to the exact position of the vacuum level, which was adjusted to lie in the interval 2.4 - 2.9 eV by cesium deposition on top of the sample. The thickness of the Gd-overlayer was determined from Auger spectroscopy, using reference spectra of clean Ge and Gd measured for identical instrumental parameters. The dependence of the spin polarization of the photoelectrons on the Gd overlayer thickness is shown in Fig. 16b: Obviously, spin depolarization is very effective. A mean free path for spin-exchange scattering of 3.8 Å is derived, a value much smaller than the mean free path for inelastic electron-electron scattering at comparable energies. Thus, the effectiveness of spin-exchange scattering of the photoexcited electron with localized magnetic moments is firmly established.

Clearly, the same experiment can be carried out with practically no restrictions on the type of the overlayer material. Particularly interesting is the case of cerium. At room temperature, Ce possesses one 4f-electron with ground state $^2F_{5/2}$. At first, one is tempted to predict that Ce depolarizes much less than Gd with its seven 4f-electrons. For the transition used, the opposite is true: The mean free

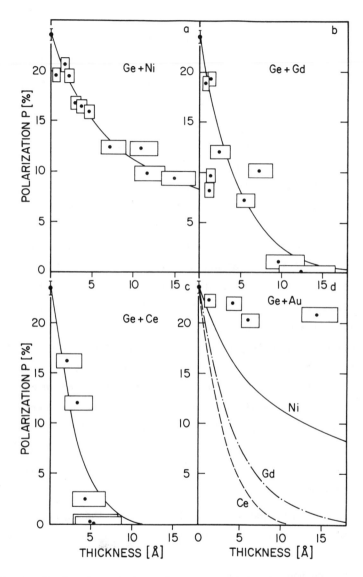

Fig. 16: Depolarization of photoelectrons after transversing an overlayer of thickness d. The initial polarization, P = 23 $\pm$ 1%,
of the excited electrons is obtained by optical spin orientation using the transition $\Gamma^{25'} \rightarrow \Gamma^{15}$ in germanium

path for spin-exchange scattering in Ce is found to be 3.2 $\overset{\circ}{A}$ only, see Fig. 16c. There is a qualitatively simple interpretation of this result which could serve as an incentive for future experiments:

The ground state of singly negatively charged Ce$^-$ with configuration 4f^2 is $^3H_4$. It lies 3.5 eV above the Fermi level (Lang et al, 1981). The energy of the excited electrons in p-Ge is 3.05 eV above E_F for the transition considered. Since the 4f-levels of Ce are probably not as localized as for the heavier rare earth, there will be energetic overlap of the 4f^2-states of Ce and the state of the excited photoelectron. Therefore, an incoming photoelectron will be easily caught into a 4f-orbital of Ce, where it interacts strongly via f-f exchange with the other electron. This mechanism will not hold for Gd: there, only s,p and d states of Gd are available at 4 eV above E_F. The (s-d)-f exchange, however, is much weaker than the f-f-exchange. By this mechanism the relatively strong depolarization by Ce finds a natural explanation. It would be worthwhile to measure the spin-flip cross section as function of the energy of the polarized, hot electrons. According to the argument given above, spin-exchange scattering should become particularly significant at those energies of the photoexcited electrons which correspond to a 4f$^{(n+1)}$-configuration of the rare earth, n being the number of f-electrons before interaction with the photoelectron. A model calculation on the resonance behaviour of the exchange interaction has recently been performed (O.L.T. de Menezes and J.S. Helman, to be published).

Apart from Ce and Gd, Fig. 16 shows also the depolarization by nickel and gold. The magnetic moment of gold is zero and no depolarization is observed, whereas for Ni a mean-free path of 12.5 $\overset{\circ}{A}$ is derived for spin-exchange scattering.

From depolarization curves of the type shown in Fig. 16 quantitative information on the exchange coupling between the photoelectron and the localized magnetic moment is obtained (Meier et al, 1984). For this purpose, use is made of the results derived in Ref. (Helman et al, 1973), where - in a simple model - the depolarization by a magnetic overlayer is calculated. For the exchange interaction $J \cdot \vec{s} \cdot \vec{S}$ ($\vec{s}$: spin of the photoelectron; $\vec{S}$: spin of the type of paramagnetic atom or ion making up the overlayer), the depolarization ρ by one monolayer of paramagnetic material is given by

$$\frac{(m^*aJ)^2 S(S+1)}{2m^* \hbar^2 V_o + (m^*aJ)^2} = \frac{1-\rho}{1+\rho} . \tag{20}$$

If the polarization of the photoexcited electrons is P_0 before reaching the monolayer, the polarization after transversing the monolayer will be $P_0 \cdot \rho$. The symbols in eqation (20) have the following meaning: V_0: energy of the vacuum level with respect to the bottom of the conduction band (in the present case, the photothreshold is ~2.6 eV and the energy gap of Ge is 0.7 eV; therefore: V_0 = 2.6 - 0.7 = 1.9 eV); $\hbar$: Planck's constant divided by 2π; m^*: effective mass of the excited electron; a: thickness of the paramagnetic monolayer.

For nickel, S is assumed to be 1/2. The monolayer thickness is taken to be 2.5 Å, and m^* is set equal to m, the free electron mass. With the value of the spin-flip mean free path, 12.5 Å, the exchange constant J is obtained from eq. (20): J = 0.85 $\pm$ 0.06 eV is derived This is not too far from the accepted value of 1.01 eV for nickel metal, defined by $\Delta E_{ex} = J(n_s \mu_B)$ where ΔE_{ex} is the exchange splitting of Ni (0.6 eV) and n_s is the number of Bohr magnetons per atom.

Similarly, for Gd the exchange constant is found to be 0.26 eV and for Ce 2.2 eV. The value for Gd in excellent agreement with results from other work (Wachter, 1972). For Ce the energy difference between the ground state $4f^2$-triplet level ($^3H_4$) and the respective singlet level ($^1G_4$) is 1.2 eV (Lang et al, 1981). Therefore, using $\Delta E = 2J |s| \cdot |S|$ with $|s|$ = 1/2 and $|S|$ = 1, the experimental value found for J may be slightly too large. However, for the resonance conditions described

above, a more refined theory should be applied (de Menezes et al, unpublished).

A basic aspect of these depolarization experiments deserves interest. Here, optical spin orientation is used to produce spin-polarized electrons in situ (= in the sample) in order to let them interact
subsequently with another system (the overlayer). This is a much
simpler procedure than producing polarized electrons externally - e.g.
by a GaAs source - with subsequent scattering on the surface of a
target. Using optical spin orientation the external source of polarized
electrons is replaced by a more conventient internal source.

References

Allenspach R, Meier F and Pescia D 1983 Phys. Rev. Lett. 51 2148
Allenspach R, Pescia D 1984 Phys. Rev. 1329 1783
Allenspach R, Meier F and Pescia D 1984 Appl. Phys. Lett 44 1107
    and references cited therein
Aspnes D E, Kelso S M, Olson C G and Lynch D W 1982 Phys. Rev. Lett 48
    1863
Bassani F and Pastori-Parravicini G 1975 in Electronic States and
    Optical Transitions in Solids, Pergamon Press
Bona G and Meier F 1985 Solid State Commun. (to be published)
Busch G, Campagna M, Cotti P and Siegmann H C 1969 Phys. Rev. Lett.
    22 597
Cardona M and Ley L 1978 in Photoemission in Solids 1, p 84ff: Topics
    in Applied Physics Vol. 26, Editors: Cardona M and Ley L
    (Springer-Verlag, Berlin)
Dresselhaus G 1955 Phys. Rev. 100 580
D'yakonov M I and Perel V I 1972 Sov. Phys.-Solid State 13 3023
Ekimov A I and Safarov V I 1970 JETP Lett. 12 198
Greuter F and Oelhafen P 1979 Z. f. Physik B34 123
Heine V 1969 in Physics of Metals 1. Electrons, Edited by Ziman J M,
    Cambridge University Press, p. 26ff
Helman S and Siegmann H C 1973 Solid State Comm. 13 891
Kastler A 1972 in New Directions in Atomic Physics Vol. II: Experiment,
    Editors: Condon E U and Sinanoglu O (Yale University Press)
Kisker E, Gudat W, Kuhlmann E, Clauberg R and Campagna M 1980 Phys.
    Rev. Lett 45 2053
Koster G F, Dimmock J O, Wheeler R G and Stolz H 1963 Properties of
    the Thirty-Two Point Groups (MIT Press, Cambridge, MA)
Kuhn H G 1969 Atomic Spectra, 2nd edition (Longmans, Green & Co. Ltd.)
    p. 129
Lampel G 1968 Phys. Rev. Lett 20 491
Lang J K, Baer A and Cox P A 1981 J. Phys. F Metal Phys. 11 121
Liebsch A 1976 Phys. Rev. B13 544
Meier F and Pescia D 1984 in Optical Orientation, Editors: Meier F and
    Zakharchenya B (North Holland Publ. Comp.) p 313ff.
Meier F, Bona G and Hüfner S 1984 Phys. Rev. Lett. 52 1152
de Menezes O L T and Helman J S, unpublished
Parsons R R 1969 Phys. Rev. Lett 23 1152
Pendry J B and Titterington D J 1977 Commun. on Physics 2 31
Perenboom J A A J, Wyder P and Meier F 1981 Physics Repts. 78 175
Pierce D T and Meier F 1976 Phys. Rev. B13 5484
Pikus G E and Titkov A N in Ref. (Meier F and Pescia D 1984)
Reyes J and Helman J S 1977 Phys. Rev. B16 4283
Riechert H, Alvarado S F, Titkov A N and Safarov V I 1984 Phys. Rev.
    Lett. 52 2297
Riechert H and Alvarado S F 1985 Festkörperprobleme XXV 1
Sattler K and Siegmann H C 1972 Phys. Rev. Lett 29 1565

Shevchik N J and Paul W 1972 J. Non-Crystalline Solids 8-10 381
Tinkham M 1964 Group Theory and Quantum Mechanics, McGraw-Hill Book
    Company p. 269
Wachter P 1972 Crit. Rev. Solid State Sci 3 189
Wöhlecke M and Borstel G in Ref.(Meier F and Pescia D 1984) p.434ff
Zürcher P, Meier F and Christensen N E 1979 Phys. Rev.
    Lett. 43, 54

Chapter 11

Spin-Resolved Photoemission from Nonmagnetic Metals and Adsorbates

U. Heinzmann and G. Schonhense

Fakultät für Physik, Universität Bielefeld,
4800 Bielefeld 1, FRG
Fritz-Haber-Institut der MPG, 1000 Berlin 33, FRG

11.1 Introduction

The concept of polarized photoelectrons ejected by circularly polarized light has been introduced by Fano (1969). One year later the "Fano-effect" was experimentally confirmed for free Cs atoms (Heinzmann et al. 1970). Spin-orbit interaction is the essential mechanism leading to a spin orientation of photoelectrons with degrees of polarization of up to 100 %. Optical transitions with σ-light induce an anisotropic distribution of the m_j sublevels of the final-state wavefunction, owing to the selection rules for electric dipole transitions (e.g. $\Delta m_j = +1$ for σ^+-light).

In solids, optical pumping using circularly polarized light has been studied by means of nuclear magnetic resonance by Lampel (1968). The first polarized photoelectrons emitted from nonmagnetic solids by σ-light have been observed for evaporated polycrystalline Cs films (Heinzmann et al. 1972) and a few years later the famous results for GaAs (Pierce and Meier 1976) initiated a number of systematic investigations of single-crystal semi-conductors (cf. chapter 10) and metals. A theoretical treatment for solid Cs was performed by Koyama and Merz (1975). Later, Reyes and Helman (1977), Feder (1977 and 1978), Borstel and Wöhlecke (1981) have established the theoretical framework, which was applied to a number of single-crystal faces.

Up to 1984 experimental analysis of the electron-spin polarization (ESP) in photoionization and photoemission using circularly polarized light was restricted to angle integrated measurements without resolution of the kinetic energy of the photoelectrons ejected; that type of experiment has been performed in detail for many free atoms and molecules (see for example Heinzmann 1980a and references therein, Heinzmann et al. 1981) as well as different solids. With the development of the new German dedicated electron storage ring for synchrotron radiation, BESSY, in Berlin, a light source of circularly polarized vacuum ultraviolet (vuv) radiation with sufficiently high intensity

has become available, making angle- and energy-resolved spin-polarization transfer studies from circularly polarized radiation onto
photoelectrons feasible. These measurements have been performed with
free atoms, with a solid-state system as well as with atoms adsorbed
on solid surfaces (as discussed in the following sections 11.2,
11.3 - 5, 11.6 - 7, respectively). The new technique extends the photon
energy range beyond 10 eV, where conventional methods for producing
circularly polarized radiation break down because no transparent or
even double refracting material exists.

The experimental technique used is briefly presented in section
11.2.1. The measured data of photoemission with subsequent ESP analysis
provides a quantitative experimental characterization of the symmetries
(angular momentum quantum numbers) of the electron bands (atomic or
ionic eigenstates) involved as discussed in sections 11.3 and 11.6.
This becomes especially important for critical points of the energy
bands and in regions, where hybridization effects play a dominant
role, as discussed in 11.4. Besides the presentation of experimental
results it is also the purpose of this chapter to demonstrate that
the angle- and spin-resolved photoelectron spectroscopy is becoming
a powerful tool to build a quantitative bridge from the atoms via the
molecules and via the adsorbates to the three-dimensional crystal
for the study of the electronic states and the structure of matter.
A special attempt is the study of atomic effects persisting in the
solid state and the adsorbate. In atomic photoionization similar
measurements yield a set of parameters which characterize the photoemission process quantummechanically completely: they allow the determination of all dipole matrix elements and all phase-shift differences
of the final-state wavefunctions (Heinzmann 1980a, b). To extend this
to more complicated systems like solid metals and adsorbates is the
pronounced goal of the running experiments.

11.2 Energy-, Angle- and Spin-Resolved Photoelectron Emission Technique

11.2.1 Experimental

The main components of the two apparatus built at BESSY in Berlin - one for studies of atomic photoionization (Heckenkamp et al. 1984) and one for photoemission experiments with solid surfaces (Eyers et al. 1984) and adsorbates (Schönhense et al. 1985) - are briefly discussed here. The synchrotron radiation is dispersed by a 6.5 m N.I. UHV monochromator of the Gillieson type (Eyers et al. 1983) with the electron beam in the storage ring being the virtual entrance slit. A spherical mirror and a plane holographic grating (1200 or 3600 lines/mm) form a 1 : 1 image of the tangential point in the exit slit. With a slit width of 2 mm a bandwidth of 0.5 nm has been achieved. Apertures movable in vertical direction are used to select radiation emitted above and below the storage ring plane, which has positive or negative helicity, respectively. In the plane the synchrotron radiation is linearly polarized. The optical degrees of polarization of the synchrotron radiation have been measured by Heckenkamp et al. (1984) by means of a rotatable four mirror analyzer (Heinzmann 1980b). Fig. 1 shows the results of the circular polarization P_{circ} and the linear polarization P_{lin} as functions of the vertical angle

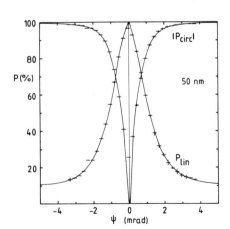

Fig. 1: Degree of circular and linear polarization P_{circ} and P_{lin}, respectively, of vuv synchrotron radiation emitted from the BESSY storage-ring plane as function of the vertical angle ψ ($\pm$ 0.1 mrad) as measured by Heckenkamp et al. (1984).

ψ (± 0.1 mrad). The solid lines which represent the theoretical predictions show excellent agreement with the experimental results demonstrating a complete linear polarization and a vanishing circular polarization of radiation emitted in the plane of the BESSY storage ring. Under the conditions of radiation accepted in the vertical angular range from 1 to 5 mrad out of the storage ringe plane, a photon flux of 10^{11} to 10^{12} photons s^{-1} with a degree of circular polarization P_{circ} = 93 % passes the monochromator exit slit and hits the phototarget under normal incidence producing photoelectrons in a region free of electric or magnetic fields.

As indicated in the schematic diagram of the apparatus shown in Fig. 2, the sample is cleaned by ion bombardment, heating in oxygen, and flashing; it is characterized by LEED and Auger-electron spectroscopy in a separate preparation chamber. The crystal on top of a manipulator moveable between preparation and photoemission chamber, can be cooled by use of a temperature-controlled liquid He-cryostat to temperatures of less than 40 K. Adsorbates are introduced via a doser nozzle which kept the background pressure below 10^{-9} mbar (base pressure $5 \cdot 10^{-11}$ mbar), allowing the continuous monitoring of the photoelectron spectra and LEED patterns as function of the coverage.

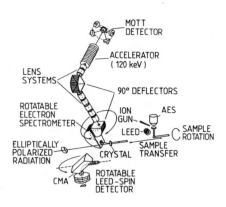

Fig. 2: Schematic diagram of the apparatus, built up at BESSY.

The photoelectrons are analyzed with respect to their kinetic energy either by a cylindrical-mirror analyzer followed by a *LEED spin detector* (Oepen et al. 1985) or by a simulated hemispherical spectrometer followed by a *UHV-Mott detector* (Eyers et al. 1984, Schönhense et al. 1985). Both electron spectrometers moveable up and down into the measuring and waiting position are rotatable about two axes which are perpendicular to the photon momentum at the crystal. These rotation axes being different with respect to the light-polarization ellipse will become important in studies of off-normal photoemission (cf. 11.5) where both spin analyzers shown in Fig. 2 complement each other.

In the upper system shown in Fig. 2 the photoelectrons analyzed with respect to their kinetic energy are directed by a 90° electrostatic deflection along the axis of rotation of the electron spectrometer. After a second deflection they are accelerated to 120 keV and scattered at the gold foil of the Mott detector for the spin-polarization analysis. Instrumental asymmetries have been eliminated by taking advantage of the reversal of the light helicity (typically each minute) as well as by use of four additional detectors in forward scattering directions (Heinzmann 1978) in the Mott detector (not shown in Fig. 2). Count rates of more than $10^3 s^{-1}$ have been obtained in the detectors for the spin analysis; the angular and energy resolution of the electron spectrometer was ±3° and 90 meV (Schönhense et al. 1985), respectively.

11.2.2 The "Complete" Photoionization Study

The reaction plane of symmetry for an angle- and spin-resolved photoionization process of an *unpolarized atom* or *unoriented molecule* using circularly polarized radiation is shown in Fig. 3. Because the momentum of the photon is negligibly small compared with the momentum of the photoelectron (valid in nonrelativistic approximation for photon energies ≤ 100 eV) there is a forward-backward symmetry in the reaction plane of Fig. 3. It also makes no difference whether right handed circularly polarized radiation comes from the left or left handed comes from the right. The rotational symmetry around the direction of the photon momentum causes both ESP components perpendicular to the photon spin to vanish for photoelectron emission angles $\theta = 0$, $\pi/2$, π. This is shown in Fig. 4, where the angle dependences of intensity $I(\theta)$ and spin-polarization components are shown for a certain atomic photoionization process (xenon), which has been simultaneously resolved with respect to all relevant variables i.e.: radiation wavelength 80 nm, radiation polarization σ^+, electron emission angle θ, electron kinetic energy corresponding to the final ionic state $Xe^+ \, {}^2P_{1/2}$, the 3 components of the ESP vector $\vec{P}(\theta)$: $P_\perp(\theta)$ perpendicular to the reaction plane, $A(\theta)$ parallel to the photon spin, $P_p(\theta)$ perpendicular to the photon spin but in the reaction plane.

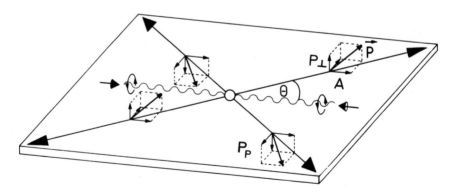

Fig. 3: Photoionization reaction plane in the case of circularly polarized radiation used.

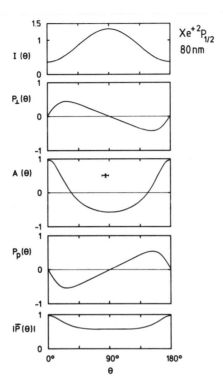

Fig. 4: Fit-curves of the experimental results (the size of a typical error-bar cross is given in the middle part) describing the angular dependences of the photoelectron intensity $I(\theta)$, of the 3 components and the length of the ESP vector for photoionization of Xe atoms at 80 nm; the photoelectrons leave the ion in the $^2P_{1/2}$ state.

The curves in Fig. 4 are fits to the experimental points (Hecken-kamp et al. 1984 and 1985) in accordance with the theoretical predictions by Cherepkov (1973) and Lee (1974).

$A(\theta)$ and $P_p(\theta)$ vanish, if linearly polarized or unpolarized instead of circularly polarized radiation is used (Heinzmann et al. 1979a, Schönhense 1980). All five curves in Fig. 4 show a reflection symmetry with respect to $\theta = \pi/2$, but $P_\perp(\theta)$ and $P_p(\theta)$ with changing sign. Thus, the polarizations of opposite sign cancel one another, if the photoelectrons ejected are extracted by an electric field regardless of their direction of emission. The only non-vanishing component of the spin polarization in an angle-integrated measurement is A (Fano effect). It is also worth noting that within the error limits the photoelectrons emitted into the forward direction $\theta = 0$ have been found by Heckenkamp et al. (1984) to be completely spin polarized (Fig. 4 middle part). This complete ESP in forward direction parallel to the photon spin as well as the fact that the electron polarization is proportional to the degree of photon polarization if partly polarized radiation is used, allows to characterize the process by the phrase "spin-polarization transfer" from spin polarized photons onto photoelectrons.

The bottom part of Fig. 4 demonstrates that the length of the ESP vector never vanishes as function of the emission angle θ. This can be generalized by the experimentally confirmed rule, that in an angular resolved photoemission experiment on atoms, molecules, adsorbates or solids it is very common rather than exceptional to get spin polarized photoelectrons.

Using the results in Fig. 4 and their energy dependences (Hecken-kamp et al. 1984 and 1985), all dipole matrix elements and phase-shift differences of the continuum wavefunctions describing the photoelectron emission from xenon atoms could be determined separately. Similar results of this quantummechanically complete experimental characterization have also been obtained in photoionization of mercury atoms (Schäfers et al. 1985).

11.3 Symmetry-Resolved Bandmapping of Pt(111)

Pt(111) was chosen in the experiments (Eyers et al. 1984 and 1985, Oepen et al. 1985) because of its high atomic number and its unreconstructed surface and the fact that several relativistic band-structure calculations exist.

11.3.1 Relativistic Bandstructure

Fig. 5 shows the non-self- consistent fully relativistic aug-mented plane wave (RAPW) bandstructure of Andersen (1970) extended to energies up to 22 eV above the Fermi level E_F by Borstel (Eyers et al. 1984). It is in excellent quantitative agreement with a recent corresponding self- consistent bandstructure calculation of Eckardt and Noffke (Leschik et al. 1984). In the following the symmetry proper-ties of the bands along the Λ-direction in $\vec{k}$ space are used, numbered at L from 1 to 10 with increasing energy. In normal emission the dominant direct interband transitions are those occuring from the

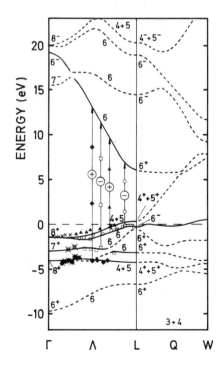

Fig. 5: Symmetry-resolved band-mapping of Pt in comparison with the calculated bandstructure. The mapping points (filled for positive and open for negative polarizations) have been obtained by a combination of intensity and polarization results of photo-electrons, partly shown in Fig. 6. For photon energies > 20 eV, bandmapping using a free-electron parabola as final state yields the crosses (from Eyers et al. 1984).

initial d-bands (Nos. 2 – 6) with the symmetries Λ_{4+5}^3, Λ_6^3, Λ_6^1, Λ_6^3, Λ_{4+5}^3 to the totally symmetric final-state band No. 7 of type Λ_6^1.   Since a transition $\Lambda_6^1 \to \Lambda_6^1$ is forbidden for normally incident light and without hybridization (11.4), only 4 direct transitions from the initial states Λ_{4+5}^3, Λ_6^3 occur (which evolve from the two nonrelativistic doublets Λ_3 and which are drawn as solid line in Fig. 5).

11.3.2 Spin-Resolved Photoelectron Spectra and Bandmapping

A few examples of photoelectron energy distribution curves (EDC) and corresponding photoelectron spinpolarizations (ESP) measured by Eyers et al. (1984) are shown in Fig. 6.

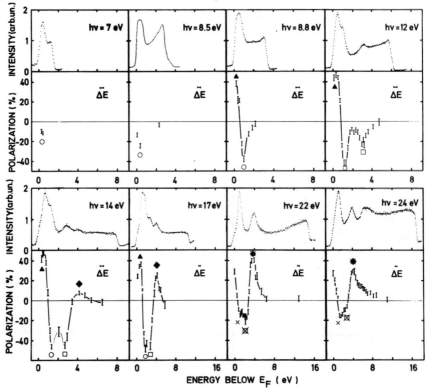

<u>Fig. 6:</u> Photoemission of Pt(111) (normal incidence and emission) using circularly polarized vuv radiation; photoelectron EDC (with respect to Fermi energy), upper parts; photoelectron polarization measured and normalized to a complete light polarization, lower parts. The symbols at the polarization peaks correspond to those in Fig. 5 (from Eyers et al. 1984).

The selection rules for the production of polarized electrons in cubic crystals have been discussed by Wöhlecke and Borstel (1981); they predicted a positive and a negative ESP for a transition of type $\Lambda_{4+5}^3 \rightarrow \Lambda_6^1$ and $\Lambda_6^3 \rightarrow \Lambda_6^1$, respectively.

The first direct transition (open circle) occurs near L from band No. 5 for a photon energy of about 6.5 eV as predicted in Fig. 5. Fig. 6 shows the corresponding spectrum for $h\nu = 7.0$ eV with the expected negative sign of the ESP. The low ESP may be explained by a small transition probability near L and a strong admixture of essentially unpolarized electrons from non $k_\perp$-conserving transitions possibly originating from band No. 5 along Q, which has a very high total density of states just below the Fermi energy (Mueller et al. 1971).

For increasing energy the probability of direct transitions $5 \rightarrow 7$ increase moving away from the L point. Thus the negative ESP is seen more clearly for a photon energy of 8.5 eV. Photoemission from the upper Λ_{4+5}^3 band (No. 6) sets in close to 8.8 eV with a positive polarization, as expected. If $h\nu$ is increased towards 14 eV, the two doublet bands 3 and 2 (Λ_6^3, Λ_{4+5}^3) produce an ESP of the expected sign. The sequence of spectra in Fig. 6 demonstrates how the signs of the spin polarization lead to an unequivocal characterization of the symmetries of the initial bands. For higher photon energies the interpretation of the data within the ground-state bandstructure becomes more and more inadequate because of the importance of self-energy corrections in the final states.

As shown in Fig. 6 less than ±100 % predicted ESP has been measured, this is due to theoretical as well as experimental effects; the reasons are discussed in the following section.

11.3.3 Temperature Dependence

Fig. 7.a shows EDC and ESP for a photon energy of 10.6 eV in normal photoemission of Pt(111) at room temperature. The spectra clearly show two peaks with positive and negative ESP arising due to the transitions Λ_{4+5} and Λ_6 to Λ_6, respectively. As discussed in the preceding section, the polarizations, however, never exceeded ±55 % in contradiction to the theoretical predictions of ±100 % in terms of application of the group theory. Fig. 7.c shows the ESP, measured by Eyers et al. (1985), in the first two peaks of the intensity spectrum (Fig. 7.b) as function of the temperature between 30 K

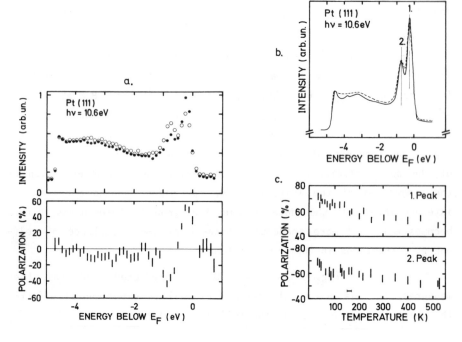

Fig. 7.a: Spin-resolved photoelectron spectrum of Pt(111) (normal incidence and emission at 300 K);Upper part: intensities measured directly in the two counters of the Mott detector, solid and open circles; Lower part: electron spin polarization
Fig. 7.b: Highly resolved EDC ($\Delta E \simeq 150$ meV FWHM) of Pt(111) at 300 K (dashed line) and 40 K (solid line)
Fig. 7.c: The temperature dependence of the electron polarizations of peaks 1 and 2 (Fig. 7.b) at 10.6 eV photon energy (all from Eyers et al. 1985).

and 550 K. With decreasing temperature the ESP increases from ±50 % to more than ±70 %. This effect might be explained by a decreased emission of unpolarized electrons from the non-$k_\perp$-conserving transitions as discussed in 11.3.2 or other phonon assisted non-direct transitions which enhance the inelastic contributions to the direct transitions. This is also seen in the widths of the intensity peaks shown in Fig. 7.b which become smaller by about 50 meV going from 300 K to 40 K. Taking into account that electrons in the inelastic background (Fig. 7.a) are unpolarized, the ESP analysis of both direct peaks alone yields data close to ±80 % for a temperature of 30 K, which is rather close to the theoretical prediction of a complete polarization as discussed in the preceding sections.

It is worth noting that each value shown in Fig. 7.c has been obtained by Eyers et al. (1985) just after the crystal has been cooled down to the corresponding temperature. This procedure was necessary in order to work with a clean crystal, because a marked depolarization seen as a function of the time demonstrated the influence of rest-gas adsorption on the crystal surface (11.7).

11.3.4 <u>Comparison with Quantitative Theoretical Calculations</u>

The general situation in the theoretical treatment of the spin-resolved photoemission is discussed in detail in chapter 4.5. There are two different kinds of theories to calculate spin effects in photo-emission: the first uses the 3-step model (Borstel and Wöhlecke 1982) handling the photoexcitation in the bulk, the tranportation of the photoelectrons to the surface and finally the transmission through the surface as independent processes. This way has the advantage that the sign of the ESP and thus the symmetry of the bands involved are easily correctly predicted, as discussed in detail in 11.3.2; but a quantitative prediction of the spin polarization is very rough, because the theory does not take into account finite lifetimes of the photoelectrons produced and of the holes left behind; it further uses only bound states as final states in the photoexcitation which neglects any influence of phases in coherent superpositions; it means that in the photoexcitation there is for example no ESP component $P_\perp$ perpendicular to the reaction plane as discussed in 11.2.2. Because the crystal has to be infinite in 3 dimensions, adsorbate-induced effects cannot be discussed quantitatively of course.

The second way to calculate spin-resolved photoemission is the relativistic one-step model. It uses LEED-states as final states and takes into account a priori the lifetime of the photoelectrons, coherent superpositions and a semi-infinite crystal. It has the dis-advantage that interpretations in terms of band structures are diffi-cult especially if finite lifetimes of the holes are also taken into account a priori. Very recently, two first one-step photoemission calculations including spin effects have been performed for Pt(111) and published (Ackermann and Feder 1985, Ginatempo et al. 1985) both taking into account the hole lifetimes a posteriori by convoluting the photocurrent density matrix with a Lorentzian. This treats the one-step photoemission exactly close to the Fermi energy, but might give some discrepancies to the experimental results at higher energies due to the approximation for the assumed energy dependences of the hole lifetimes.

The agreement between experiment (Eyers et al. 1984, 1985) and the calculations by Ginatempo et al. (1985) is satisfactory if one takes the experimental data of Fig. 6 obtained at room temperature but less satisfactory if one takes into account the temperature effects discussed in 11.3.3 and shown in Fig. 7. On the other hand the spin-polarization results calculated by Ackermann and Feder (1985) show good agreement with the experiment performed at low temperatures (30 K) where polarizations of more than ±70 % have been measured (Eyers et al. 1985). This is demonstrated in Fig. 8 upper part, where in the first peak A (middle part) corresponding to a transition from the Λ_{4+5} band the theoretical spin polarization is a little bit higher than the experimental one, whereas in the second peak B corresponding to a transition from Λ_6 it is vice versa: the polarization measured is higher than the calculated one (a not very often obtained behavior in spin-resolved photoemission). Usually, depolarization effects due to a non-zero energy resolution as well as an angular resolution and due to an unpolarized inelastic background in the photoelectron spectra are present in the experiment, while the theoretical results of Fig. 8 are the elastic currents only.

It seems nevertheless to be evident that the one-step theories have some freedoms of adjustable quantities (hole lifetime, imaginary potential part, inelastic contributions) in order to find a more quantitative agreement with the experimental data. To continue this procedure should be an important purpose of further theoretical investigations. Quantitative comparison of experimental spin polarizations with theory requires an exact knowledge of the experimental uncertainties, of the degree of circular light polarization and of the analyzing power and asymmetry function of the spin detector used. For the evaluation procedure of the data in Figs. 6 and 7 these "experimentally important" points have been taken into account carefully.

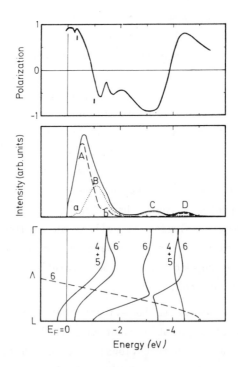

Fig. 8: Theoretical results for Pt(111) of Ackermann and Feder (1985). Lower part: Bulk band structure along ΓL for initial state (——) and for final state lowered by photon energy hν = 11 eV (---).
Middle part: Intensities I⁺ (---) and I⁻ (···) of elastic normal-emission photocurrent due to normally incident positive-helicity radiation (hν = 11 eV). (I⁺ and I⁻ peaks are labelled by letters A to D and a,b).
Upper part: Spin polarization for elastic current is compared with two experimental values (error bars from Eyers et al. 1985).

11.4 Hybridization Effects and Special Regions of Energy Bands

11.4.1 The W(100) Case

Hybridization, i.e. mixing of wavefunctions with different orbital symmetries may occur as a consequence of spin-orbit interaction if the wavefunctions belong to the same double-group representation. Only in special cases can such an admixture be seen in the photoelectron intensity spectra. This happens when transitions appear which should be dipole-forbidden if the states involved were of a pure single symmetry type. For dipole-allowed transitions that simple indicator of hybridization does not work. If, however, a state is admixed which produces spins of opposite sign, the ESP of the transition is diminished as compared to the value expected for non-hybridized bands.

The capability of ESP measurements to probe the hybridization properties of conduction bands has first been experimentally verified for the W(100) direction by Zürcher et al. (1979). The theoretical framework was established by Reyes and Helman (1977) who made quantitative predictions of the ESP for photoemission from tungsten. The band structure of W along Δ is shown in Fig. 9. Dashed lines indicate the

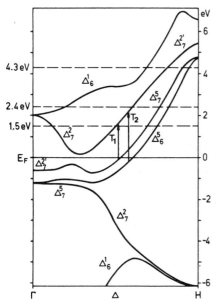

Fig. 9: Band structure of W(100) (Christensen and Feuerbacher 1974).

vacuum levels for the clean (4.3 eV) and two cesiated surfaces (2.4 and 1.5 eV). Since in this part of the Brillouin zone energy bands of different orbital symmetries are energetically rather close to each other, considerable orbital mixing due to hybridization is very likely to occur. From the dipole selection rules applied to *non-hybridized* orbitals one expects *completely polarized* photoelectrons of opposite sign for transitions T_1 and T_2 induced by circularly polarized light (Reyes and Helman 1977). However, hybridization will generally lower the ESP as illustrated in the table:

without hybridization	spins	*with* hybridization	spins
T_1: $\Delta_7^5 \rightarrow \Delta_7^{2'}$	$\downarrow$	T_1: $\Delta_7^5 + \Delta_7^{2'} + \Delta_7^2 \rightarrow \Delta_7^{2'} + \Delta_7^5 + \Delta_7^2$	$\downarrow + \uparrow$
T_2: $\Delta_6^5 \rightarrow \Delta_7^{2'}$	$\uparrow$	T_2: $\Delta_6^5 + \Delta_6^1 \rightarrow \Delta_7^{2'} + \Delta_7^5 + \Delta_7^2$	$\uparrow + \downarrow$

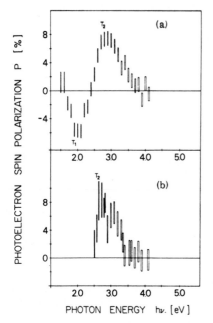

Fig. 10: ESP of the total photocurrent emitted by circularly polarized light from W(100) with photothreshold (a) 1.5 eV and (b) 2.4 eV. The extrema T_1 and T_2 arise due to the transitions indicated in Fig. 9 (Zürcher and Meier 1979).

Indeed, Zürcher et al. (1979) observed peak values of the ESP far off the theoretically predicted ±100 % as shown in Fig. 10. As discussed by the authors, depolarizing effects like contributions from off-normal emission due to the finite escape cone, a background of unpolarized electrons from surface photoemission or spin-exchange scattering at the Cs overlayer cannot explain the strongly diminished ESP. Assuming a reasonable contribution of surface emission and allowing for a small overlap of T_1 and T_2 due to lifetime broadening, the hybridization coefficients could be derived from the ESP data. For transition T_1 (T_2) the final state was estimated to be about 54 % (59 %) d-type of Δ_7^2 symmetry and 46 % (41 %) pd-type of Δ_7^5 symmetry. For the initial state corresponding to T_1 the d and pd contributions are interchanged, whereas for transition T_2 it was not possible to obtain the hybridization coefficients of the initial state from the experimental ESP data (Zürcher et al. 1979).

11.4.2 The Λ-Direction of Pt and Au

Closely related to the effects of hybridization is the question of crossings or anticrossings of energy bands. Owing to its potential for the experimental determination of bandstructure symmetries, spin-resolved photoemission is a unique tool for solving such questions. This has been demonstrated recently for platinum, where a discrepancy between theoretical bandstructure calculations could unequivocally be resolved by means of ESP measurements. The discrepancy concerns the Γ-L direction of Pt in the region 2 – 4 eV below E_F, where two relativistic calculations predicted a fundamentally different behavior: two band crossings appeared in the calculation of MacDonald et al. (1981) whereas Mackintosh (see Andersen 1970) found two anticrossings at the same points. Within the experimental uncertainty, bandmapping data (Mills et al. 1980) were compatible with both theories.

Spin- and momentum-resolved photoemission spectra of the corresponding energy regions were taken by Oepen et al. (1985) using the LEED spinpolarization detector (cf. chapter 5) and the set-up at BESSY. Fig. 11 shows for hν = 13 eV the measured ESP (a) and intensity curve (b). As illustrated in (d), this spectrum shows just the section (dashed curve) across one of the critical regions (band Nos. 3 - 4). The general features of the spectrum have been discussed in detail in section 11.3, here we focus attention on the region around 2 - 4 eV below E_F. While in the total intensity curve the feature C appears as a single broad peak, the ESP curve and even more clearly the partial intensity curve (-) reveal structures near -2.5 eV that have to be attributed to a second peak of minor intensity (marked by arrows). Since the ESP of both peaks has the same sign, the initial states must be of the *same* symmetry; hence the two bands in question cannot cross but must form an *anticrossing* as shown in (d). Note that away from the hybridization region transitions from band 3 to the final-state band are forbidden since both have Λ_6^1 symmetry. Corresponding spectra at several other photon energies (cf. section 11.3) gave

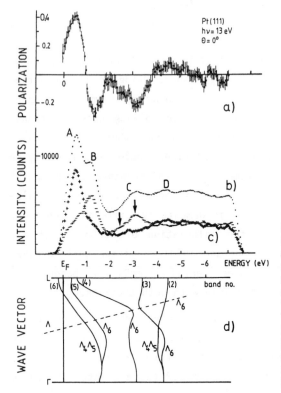

Fig. 11. Spin polarization (a), total intensity (b) and partial intensities (c) for spin parallel (+) and antiparallel (-) to the surface normal (Oepen et al. 1985).
Theoretical bandstructure of Pt along Λ (d) (after Borstel, Eyers et al. 1984) with the final- state band (dashed) being shifted by 13 eV.

evidence that band 2 must be of $\Lambda_4\Lambda_5$ symmetry while band 3 is of Λ_6 symmetry. Thus the *crossing* (at 4 eV below E_F) of the two bands of *unlike* symmetry is allowed.

The band structure of gold has attracted great interest because large discrepancies between experiments and calculations have been reported (see, e.g. Christensen 1981). Meier and Pescia (1981) studied transitions induced by circularly polarized light along the (111) direction near the L point. From the measured ESP of the total photo-current as function of photon energy they were able to precisely determine several interband energies. The experimental energy positions of the topmost bands agree well with the relativistic band-structure calculation of Christensen and Seraphin (1971), thereby clarifying the previous controversy. The symmetry character of the bands near E_F has been unambiguously assessed. Close to the L point the top occupied band is Λ_6^1 symmetry (equivalent to band 3 of Pt, Fig. 11). The forbidden transition from this band to the Λ_6^1 final state becomes allowed by hybridization with band Λ_6^3. This admixture to the initial state wavefunction could clearly be identified due to the negative ESP ($\Lambda_6^3 \rightarrow \Lambda_6^1$) at the photothreshold.

11.5 Off-normal Photoelectron Emission

11.5.1 Surface Transmission Effects

In contrast to the ESP induced by the dipole selection rules for optical transitions with circularly polarized light, there exists another spin phenomenon in solid-state photoemission which becomes only observable in fully energy-, angle- and spin-resolved measurements. Off-normally emitted photoelectrons experience a *spin-dependent diffraction* during their transmission through the surface. The physical origin of this effect is the spin-orbit interaction in the solid, therefore a strong analogy exists to spinpolarized LEED (see chapter 6). If the excitation process produces an *unpolarized* Bloch wave (for example after excitation with unpolarized light) these electrons may acquire a nonzero ESP due to spin-dependent phase-matching conditions at the solid / vacuum interface, i.e. when going from the Bloch-spinor regime to the free-electron regime. If the excited Bloch wave is already *polarized* the effect of the surface transmission becomes more involved: in addition to changes of the ESP even intensity asymmetries occur.

The surface transmission effect in its pure appearance – the nonzero ESP after excitation by unpolarized light – was observed by Kirschner et al. (1981). Photoelectrons in tungsten were excited by Ly_α radiation from a H-discharge lamp normally incident on the 001 surface. The electrons were sampled at a polar angle of $\theta = 70°$, energy analyzed by an electron spectrometer ($E \simeq E_F - 0.5$ eV) and spin analyzed by a LEED detector (see chapter 5). Fig. 12 shows the experimental data, taken in the form of azimuthal scans by rotating the crystal around its surface normal. The existence of the ESP and its structure connected to the crystal symmetry is a direct evidence of surface transmission effects (sometimes referred to as "final state ESP effects"). The quantization axis in Fig. 12 is the normal of the reaction plane spanned by incoming photon and outgoing electron. The same axis defines the well-known ESP component which occurs in photoionization of unpolarized atoms with unpolarized light (Heinzmann et

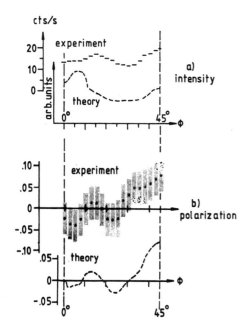

Fig. 12: Experimental and theoretical rotation diagrams (azimuthal scans) for (a) intensity and (b) ESP of photoelectrons released by *unpolarized* light from W(001); polar angle θ=70°, photon energy 10.2 eV (from Kirschner et al. 1981).

al. 1979a). Although arising under similar conditions and both being final-state effects, the physical nature of the atomic and solid-state effect is of course different: it is a consequence of quantum-mechanical interference between different photoelectron partial waves and between scattered partial waves, respectively.

The spin-dependent transmission through the surface may give rise to characteristic *intensity asymmetries* if the photoelectrons are excited by circularly (or elliptically) polarized light. In this case the electrons excited in the bulk are polarized and the transmission step depends explicitly on their spin-state. The surface can be viewed as a polarizing filter which weakens or enhances the transmitted intensity from a particular transition, depending on the sign of the ESP and hence on the helicity of the incident radiation. This is demonstrated in Fig. 13 for Pt(111). Part a) shows two intensity curves for σ^+ and σ^- light at $h\nu = 12$ eV yielding the asymmetry shown in b). The two peaks near E_F stem from transitions from the topmost occupied band split by spin-orbit interaction, i.e. from electrons

of opposite spin orientation. When the spin-dependence of the transmission step is known, this effect offers the possibility of an experimental determination of band-structure symmetries *without* explicit
ESP analysis.

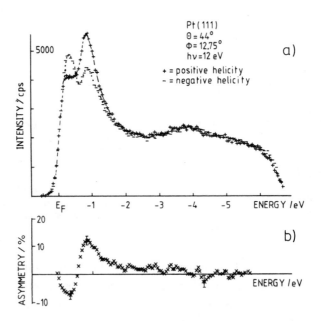

Fig. 13: Photoelectron intensity spectra and resulting asymmetry
$(I^+-I^-)/(I^++I^-)$ measured for σ-light of opposite helicity. ϕ is
measured from the ΓLUX mirror plane of Pt(111) (from Oepen et al.
1984).

11.5.2 Dependence of the ESP Vector on the Emission Angle

Similar to scattering of polarized electrons on free atoms (cf.
Kessler 1976), the spin dependent diffraction of polarized electrons
on surfaces generally changes the ESP vector in magnitude and direction. The scattering matrix depends on the electron energy and the
emission angle. Oepen et al. (1985) have measured the ESP vector in
the ΓLUX mirror plane of Pt(111) for different polar emission angles
and several photon energies. One example is shown in Fig. 14; P_{long}
and P_{trans} denote the longitudinal and transversal ESP components

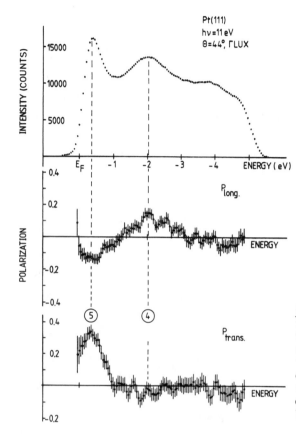

Fig. 14: Spectra of the intensity and the two inplane ESP components measured at normal incidence and a polar emission angle of 44° in the ΓLUX mirror plane of platinum (from Oepen 1984).

in the mirror plane. The two intensity peaks correspond to transitions from bands 5 and 4; obviously, the ESP vectors have different directions in the plane.

The variation of direction and magnitude of the ESP vector with emission angle and photon energy is illustrated in Fig. 15. The orientation of the ESP vector can deviate considerably from the direction of the photon spin (i.e. the surface normal). For the transition band 4 → 7 and an emission angle of 28° the electron and photon polarization vectors are nearly perpendicular to each other! There is evidence that this striking effect cannot be solely due to surface transmission, but the *excitation step* (within the three-step model) produces a spin component *perpendicular* to the photon spin, similar to the atomic case (see Fig. 3). If surface transmission were dominant, the energy dependence should be much more pronounced than displayed in Fig. 15. Very recently Borstel (1985) has outlined the capability of such measurements to extract, in principle, the relative phases and magnitudes of distinct transition matrix elements. This would be a major step towards the complete characterization of electronic states in solids – the counterpart of the complete photoionization experiments in the gas phase (cf. 11.2.2).

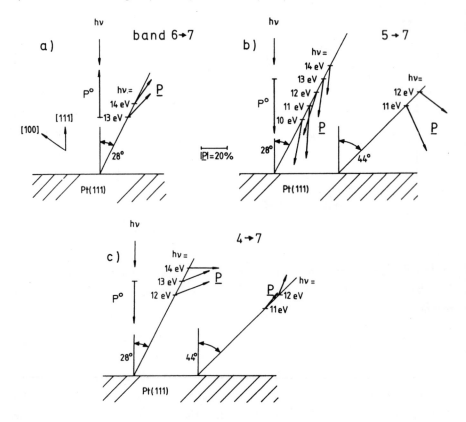

Fig. 15: Orientation and magnitude of the ESP vector in the ΓLUX mirror plane of Pt measured at two polar emission angles for various photon energies. Part a), b), and c) are transitions from bands no. 6, 5, and 4 (cf. Fig. 11), respectively. P° denotes the ESP vector for normal emission (from Oepen 1984).

11.6 Photoelectron Spinpolarization Spectroscopy of Physisorbed Rare Gases

The most recent progress in spin-resolved photoelectron spectroscopy was achieved for the case of adsorbate systems. This is a result of the development of electron storage rings which provide high photon intensities in the VUV range at very good ultrahigh vacuum conditions. Using the apparatus at BESSY (cf. 11.2) it has become possible to sample an ESP spectrum in less than one hour – a typical "lifetime" of a well-defined adsorbate overlayer. This opens up a wide field of possible investigations with a lot of fascinating new aspects.

11.6.1 Level Splitting

The symmetry operations which transform a wavefunction of an adsorbed atom into itself do no longer reflect the full symmetry of isotropic space but are restricted due to the adsorption geometry. Even for weakly bound (physisorbed) systems like the rare gases this reduction of symmetry by the presence of the surface may induce a splitting of energy levels. Consider, e.g., the $5p^6$ valence orbital of a free Xe atom (4p for Kr analogously). For these heavy atoms the spin-orbit interaction induces a strong mixing (hybridization) of the p_x, p_y and p_z orbitals such that the wavefunctions are eigenfunctions of j^2 and j_z. Starting from the atomic ground state ^1S_o (no spin-orbit interaction), photoemission of a valence electron leads to the p^5 configuration where angular- and spin momenta of the ion can couple to $^2P_{3/2}$ and $^2P_{1/2}$, briefly called the $p_{3/2}$ and $p_{1/2}$ hole states. Spin-orbit interaction in these final ionic states causes a splitting of 1.3 eV (0.6 eV for Kr).

Without magnetic interactions being present, the lowering of symmetry in the adsorbate system may result in a further splitting of $p_{3/2}$ into the $|m_j| = 3/2$ and $1/2$ sublevels. Indeed, Waclawski and Herbst (1975) observed a significant broadening of the corresponding peak in the photoelectron spectrum of xenon on tungsten, which they interpreted as the unresolved $|m_j|$ sublevel doublet split in the crystal field of the substrate-atom cores. On the other hand,

Matthew and Devey (1976) as well as Antoniewicz (1977) have pointed out that this model requires an unreasonably large positive charge on the surface atoms in order to produce the observed splitting. Instead, they proposed a splitting mechanism based upon image-charge screening which results in different relaxation shifts of the $p_{3/2}$ $|m_j| = 3/2$ and $1/2$ hole states. Finally, Horn et al. (1978) have measured a marked dispersion of the 5p-band in a Xe monolayer. From the measured dependence of the photoelectron intensities upon the angle of photon incidence they concluded that lateral interaction between adatoms is the dominant splitting mechanism (Scheffler et al. 1979). This explanation has been challenged repeatedly in recent years in the literature.

The nature of the *splitting mechanism* is closely related to the *energetic ordering* of the sublevels: substrate-induced relaxation effects cause the $|m_j| = 1/2$ level to have a lower binding energy than $|m_j| = 3/2$ (Matthew and Devey 1976), whereas the opposite is true if lateral interactions in the adlayer are responsible (Scheffler et al. 1979). Due to its unique capability of identifying orbital symmetries, ESP spectroscopy is a simple and reliable tool to clarify this question. On the basis of a theoretical model calculation for the photoemission from rare-gas adsorbates, Feder (1978) has shown that the ordering can be obtained from the ESP of the angle-integrated photocurrent if the dipole matrix elements are explicitly known. In the following, we outline a general group-theoretical treatment.

From LEED-patterns it is well known that at high coverage adsorbed rare-gas atoms form a hexagonally close-packed (hcp) overlayer. Hence, for normal emission (i.e. at the Γ point of the surface Brillouin zone) the symmetry group of the wave vector is C_{6v}. The wavefunctions are constructed of spatial and spin parts, with the first one transforming according to one of the representations of the single group. From the basis functions of C_{6v} tabulated by Koster et al. (1963, p.67) it is evident that p_z and $p_x p_y$ transform according to Δ_1 and Δ_5, respectively. Spin-orbit interaction is taken into account by reduction of the products with Δ_7 (generated by the

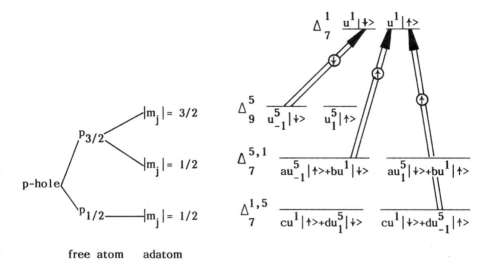

<u>Fig. 16:</u> Level scheme for transitions from the valence bands of ad-
sorbed rare gases. Basis functions are from Koster et al. (1963),
a–d being hybridization coefficients. Arrows indicate the transitions
having non-vanishing matrix elements.

spinors $|\uparrow>$ and $|\downarrow>$), yielding* $\Delta_1 \times \Delta_7 = \Delta_7^1$ and $\Delta_5 \times \Delta_7 = \Delta_7^5 + \Delta_9^5$
(Koster et al. 1963, p. 68). Thus the 6-fold degeneracy of the p-or-
bital is lifted and a splitting occurs into three sublevels of 2-fold
degeneracy (two of double group symmetry Δ_7 and one of Δ_9). In general,
hybridization will cause a mixing of Δ_7^5 and Δ_7^1 wavefunctions ($p_{xy}-p_z$
hybrids $\Delta_7^{5,1}$), whereas Δ_9^5 is of a single symmetry type (pure p_{xy}).

Fig. 16 shows the resulting level scheme. In order to derive
the ESP of photoelectrons emitted by circularly polarized light we
follow the way described by Meier and Pescia (1984). For normal incidence (propagation along −z) the radiation operator for circularly
polarized light (σ^-) transforms as x + iy (equivalent to $-i(s_x+is_y)$)

* According to common convention, the superscript (subscript) indicates the corresponding single (double) group representation.

and hence according to the basis function u_1^5 of Δ_5. The final state is assumed to be a plane wave with wave vector along the z direction transforming as Δ_7^1 (in the present geometry equivalent to a spherically symmetric s-wave). From the transition scheme one can easily see that all dipole matrix elements for optical transitions have the form $<f|Op.|i> = <u^1|u_1^5|u^1$ or u_1^5 or $u_{-1}^5>$ where it was used that the dipole operator acts only on the orbital part of the wavefunctions. Since the reduction of Δ_5 (light operator) $\times$ Δ_1 (initial state) does not contain Δ_1 (final state), the transition from a band with pure Δ_1 orbital symmetry is dipole-forbidden. The orbital part of the final-state wavefunction u^1 expressed in terms of products of basis functions $\Delta_5 \times \Delta_5$ reads $u^1 = \frac{1}{\sqrt{2}} (u_{-1}^5 v_1^5 + u_1^5 v_{-1}^5)$ (Koster et al. 1963, p. 69). Consequently the only non-vanishing matrix element is $<u^1|u_1^5|u_{-1}^5> \neq 0$. Finally, due to the orthogonality of the spin functions ($<\uparrow|\downarrow> = 0$) only the three transitions denoted by arrows are allowed for circularly polarized light, yielding the ESP values:

$$
P = \begin{cases} -1 \text{ for } P_{3/2} \; |m_j| = 3/2 \\ +1 \text{ for } P_{3/2} \; |m_j| = 1/2 \text{ and for } P_{1/2} \end{cases}
$$

These very general arguments are based upon *symmetry properties* only, they hold irrespective of the magnitude of the matrix elements, the hybridization coefficients a-d (as long as a and d $\neq 0$) and even irrespective of the physical nature of the splitting mechanism. It is thus not surprising that one gets the same results by using the angular part of the *atomic* wavefunctions Y_ℓ^m (analogously to the procedure described in Kessler 1976) (Schönhense et al. 1985a). Two spinresolved photoelectron spectra are shown in Fig. 17. For Xe and Kr the peak at lowest binding energy (peak 1) has nearly complete negative ESP which, according to the above arguments, corresponds to the $P_{3/2}$ $|m_j| = 3/2$ hole state, whereas peak 2 and 3 (here for Kr only) are highly positively polarized, i.e. $|m_j| = 1/2$. This result confirms the peak assignment given by Horn et al. (1978), which

indicates that the splitting is caused by *lateral adatom interactions*.
The high ESP of over ±85 % exceeds all values previously found for
other nonmagnetic solid-state systems. It is interesting, to compare
the results with the free-atom case: The degeneracy of the two $p_{3/2}$
sublevels with opposite sign of polarization *diminishes* the ESP to
$P \simeq -0.5$ since for free ions the intensity ratio between the $|m_j|$ =
3/2 and 1/2 channels is close to 3 (exactly 3 in the LS-coupling approximation). For the non-split $p_{1/2}$ level the complete positive polarization predicted by group theory has also been observed in a recent
gas-phase experiment (Heckenkamp et al. 1984) as shown in Fig. 4.

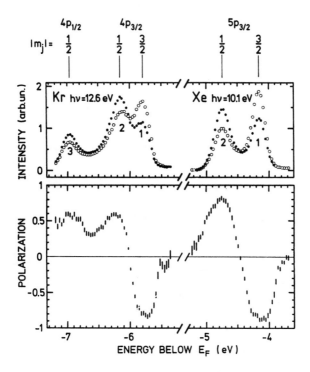

Fig. 17: ESP spectra of Kr and Xe monolayers. The measured intensity
asymmetry in the Mott detector (upper part) yields the ESP (lower
part) (from Schönhense et al. 1985a).

11.6.2 Resonance Behavior of the Spin Polarization

At photon energies just above the Xe 5p and Kr 4p thresholds the photoemission intensities are strongly enhanced so that for monolayer coverage the adsorbate photoemission can exceed the Pt d-band emission (close to E_F) by two orders of magnitude (cf. Schönhense et al. 1985a). In order to study the threshold behavior in more detail, the energy variation of intensity and ESP of the three Xe 5p-derived peaks has been measured as function of photon energy (summarized in Figs. 18-20) using circularly polarized synchrotron radiation and the apparatus shown in Fig. 2. Results have been obtained (i) for the commensurate √3 × √3 (R 30°) overlayer on the Pt(111) face, (ii) for the incommen- surate hexagonally close-packed (hcp) monolayer on Pt(111) and (iii) for the hcp monolayer on the basal plane of (natural single-crystal) graphite. Note that the intensity scales (in arbitrary units) are

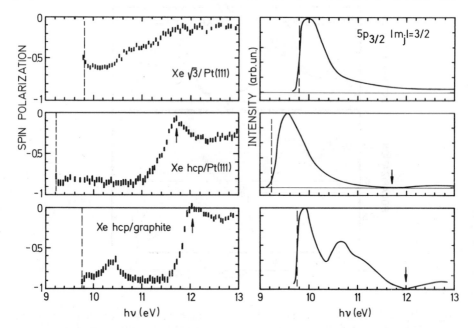

Fig. 18: Spectral variation of ESP (left) and corresponding intensity (right) for the Xe $5p_{3/2}$ $|m_j|$ = 3/2 hole state at normal incidence and emission (from Schönhense et al. 1985a and b).

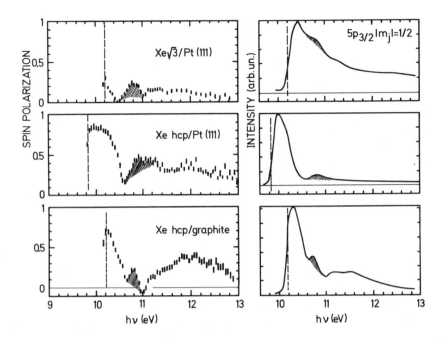

Fig. 19: The same as Fig. 18 but for Xe $5p_{3/2}$ $|m_j| = 1/2$

different from each other. It is advantageous to plot the data versus photon energy, because this representation is well-defined independent of workfunction changes. The figures reveal that the intensities of all peaks are strongly enhanced just above their thresholds (dashed lines, binding energies with respect to the vacuum level) but fall off rapidly towards higher energies within less than one eV. All ESP curves* show the signs predicted by group theory (see 11.6.1) but instead of having constant values ±1, they exhibit pronounced variations with photon energy! Evidently, there are important effects which were not included in the above symmetry considerations.

---

* It was found that the adsorbate layer strongly influences the substrate spectrum (cf. section 11.7). Therefore no reliable estimation of an underlying background of unpolarized secondary electrons could be made and hence the uncorrected ESP values are given.

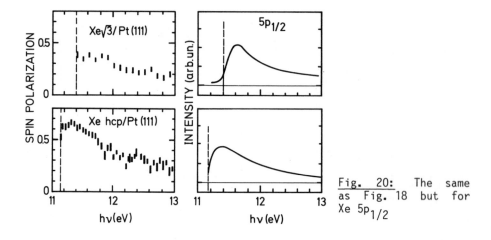

Fig. 20: The same as Fig. 18 but for Xe $5p_{1/2}$

One important mechanism which leads to strong variations of electron spinpolarizations and intensities (the so-called "dynamical" parameters) in near-threshold photoemission from *free* rare-gas atoms is *autoionization*. Autoionization resonances occur whenever discrete Rydberg-type excitations interact with a degenerate photoemission continuum leading to the well-known asymmetric Beutler-Fano profiles. In the particular case of xenon atoms the $5p \rightarrow ns$ and $5p \rightarrow nd$ series, converging to the Xe^+ $^2P_{1/2}$ ionization threshold give rise to strong ESP and intensity features between the fine-structure split $^2P_{3/2, 1/2}$ thresholds (cf. Heinzmann et al. 1979b). Such an autoionizing transition in Xe adatoms is most probably the origin of the hatched feature in the ESP and intensity curves of the $p_{3/2}$ $|m_j| = 1/2$ channel (Fig. 19), because it appears at nearly the same photon energy for all Xe layers studied independent of the substrate and the coverage.

The most surprising resonance features have recently been observed *below* the adsorbate photoemission thresholds. Very narrow Xe $5p \rightarrow 6s$ and Kr $4p \rightarrow 5s$ resonance transitions occur in the adsorbate phase. They show almost no relaxation shift compared to the gas phase and persist even in the submonolayer regime! Fig. 21 shows the results for Xe on graphite. The resonances were detected via spin-resolved spectroscopy of the electrons just above the low-energy cut-

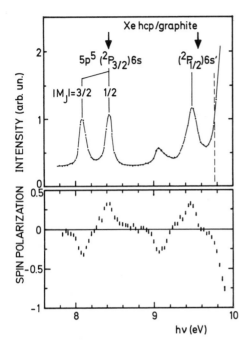

Fig. 21: Resonant electron emission from a xenon monolayer on graphite; arrows indicate the 6s and 6s' resonance-line positions for *free* atoms (from Schönhense et al. 1985b).

off of the spectrum, which are emitted due to a subsequent relaxation mechanism (probably a Penning-type decay). The ESP again provided a quantum-number labelling of the peaks according to 11.6.1 and furthermore clarified the nature of the decay mechanism observed (cf. Schönhense et al. 1985b). Note that the Xe* $5p^5(^2P_{3/2})6s$ excited level is split into $|M_J| = 3/2$ and 1/2 (core-) sublevels although the Xe* complex is electrically neutral, i.e. no image-charge screening should occur. The resonances in the two-dimensional overlayer are closely related to the surface excitons in rare-gas crystals observed by Saile et al. (1976). The feature at 9.1 eV might well be the n = 2 exciton of the atomic-line derived 8.1 eV-peak. Owing to its negative ESP it is evidently associated with the $|M_J| = 3/2$ core.

11.6.3 Influence of Substrate, Overlayer Structure and 2D Phase Transitions

An especially interesting point in adsorbate physics is the feasibility to vary the substrate crystal and in some cases also the adatom-adatom spacing via the coverage. For the weakly bound rare-gas atoms, which constitute a model case of matter in a state between gas and solid, one can thereby probe the influence of an increasing valence-orbital overlap on energy levels and photoemission dynamics.

Fig. 18 shows one striking difference between the commensurate Xe $\sqrt{3}$ phase on Pt(111) and the incommensurate hcp phases: whereas for $\sqrt{3}$ (Xe-Xe spacing 4.8 Å) the photoemission intensity is always different from zero, the $p_{3/2}$ $|m_j|$ = 3/2 peak *vanishes* for   both hcp layers (Xe-Xe spacing 4.4 Å), leading to sharp minima in the ESP curves (arrows). This different behavior has a remarkable effect on the photoelectron spectra. At $h\nu$ = 11.7 eV the spectrum of Xe $\sqrt{3}$ / Pt(111) clearly shows three Xe 5p-derived peaks; on adding a little more Xe, however, the first peak falls off until at monolayer saturation it has almost completely vanished (see Schönhense et al. 1985a). Hence, in the threshold region the peak heights may even depend reversely on the coverage! Another obvious difference between the two Xe phases on Pt(111) is the position of the photoemission thresholds (dashed lines in Figs. 18-20). This is essentially a consequence of the increased valence-orbital overlap in the hcp layer; the workfunction difference is only less than 100 meV. Furthermore, the ESP values for Xe $\sqrt{3}$ are considerably lower than for hcp – especially in the $p_{3/2}$ $|m_j|$ = 1/2 channel (Fig. 19). The unpolarized background of secondaries can only partly account for this effect.

On the other hand, Xe hcp monolayers on such different substrates as platinum and graphite show marked differences in the ESP and intensity curves, too. Additional features occur on the graphite surface: (i) a strong intensity structure between 10 and 11 eV in the $p_{3/2}$ $|m_j|$= 3/2 channel (Fig. 18), also visible in the ESP and (ii) weak features between 11 and 12 eV in the $|m_j|$ = 1/2 channel (Fig. 19).

A possible reason could be that the influence of the substrate on the Xe adatoms is weaker for graphite than for Pt(111). Consequently, more resonant transitions could persist on graphite. This interpretation is supported by the fact that the width of the hatched features is significantly smaller for graphite, i.e. less disturbed, than for Pt.

Besides the commensurate-incommensurate phase transition, also a 2D-liquid-solid phase transition is intensely studied at present (Poelsema et al. 1983, Wandelt 1985). It is to be expected that ESP spectroscopy can give important insights into the level splitting mechanism in the liquid phase at low coverage and hence for negligible valence-orbital overlap. Such investigations are in progress.

11.7 Adsorbate-Induced Changes in Substrate ESP Spectra

Besides the spinpolarization effects in photoemission from adsorbate levels, discussed in the preceding section, also the ESP of those photoelectrons originating from the substrate bands may show significant adsorbate-induced changes. Only very recently have such changes been observed and our knowledge of the interactions experienced by the photoelectron when travelling across an adsorbate layer is still rather sparse.*

Early experiments on cesiated GaAs never gave a clear indication for depolarization by spin-exchange scattering of photoelectrons at the Cs atoms on the surface (Pierce and Meier 1976). This result was no surprise, because for submonolayer coverages the Cs atom is ionized and thus has rare-gas electron configuration with zero spin leading to a very small spin-exchange constant. The situation becomes quite different if the adatoms carry magnetic moments. Then even a single paramagnetic monolayer may have a depolarizing effect as high as 50 % (Helman and Siegmann 1973). Meier et al. (1982b and 1984) performed a systematic study of the ESP of photoelectrons excited by circularly polarized light in Ge covered with variable amounts of Ni, Gd, Ce and Au. The mean free paths for spin-flip scattering turned out to lie between a few Ångstroms only for the magnetic materials (3.2 Å for Ce) and > 50 Å for Au, having no magnetic moment (cf. chapter 10).

A lot of details of the adsorbate-induced changes have been revealed in a recent energy- and angle-resolved ESP study (Eyers et al. 1985). Fig. 22 shows some selected intensity spectra of the adsorbate system xenon on platinum (111) for two different photon energies and various coverages. The spectra of the clean substrate (a) show the pronounced peaks due to the direct interband transitions along the Γ-L direction of Pt as discussed in section 11.3. The series of spectra

* The case of magnetic materials, where even nonmagnetic adsorbate overlayers strongly influence the magnetism of the surface (Meier et al. 1975 and 1982a) will not be discussed in this context.

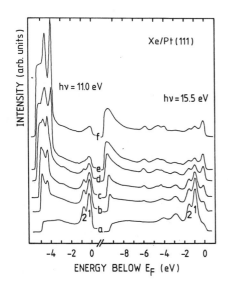

Fig. 22: Photoelectron spectra of Pt(111) for normal incidence and normal emission taken at various Xe coverages between (a) clean substrate and (f) complete incommensurate hcp monolayer (from Eyers et al. 1985).

at $h\nu = 11$ eV lies in the energy region of the anomalously intense photoemission from the Xe valence orbitals. Along with the strongly increasing $p_{3/2}$ $|m_j| = 3/2$, $1/2$ peaks on the left, the Pt photoemission intensity of peaks 1 and 2 (bands 6 and 5, Fig. 5) is drastically reduced with increasing Xe coverage. For the completed commensurate $\sqrt{3}$ overlayer (spectrum d) and the saturated incommensurate hcp mono-layer (spectrum f) the Pt peak heights have decreased by factors of about 2.5 and 4, respectively.

At $h\nu = 15.5$ eV (right series) photoemission from Xe is much weaker: even at full monolayer coverage (f) the Pt peaks still exceed the Xe peaks (located between -6 and -4 eV below E_F). Again, the Pt peak heights are reduced by roughly the same factors as observed for $h\nu = 11$ eV. However, in addition an adsorbate-induced new peak just below the Fermi level (0.25 eV below E_F) arises, which was only visible as a shoulder in the 15.5 eV spectrum of the clean substrate (a). The same behavior, i.e. a marked decrease in the d-band transition intensities and an enhancement of the peak just below E_F was also found for krypton overlayers. The main part of the observed decrease is clearly caused by inelastic scattering processes at the adatoms as is evident from the rise of the broad feature of secondaries at the low-energy cut-off of the 15.5 eV spectra.

One plausible explanation of the adsorbate-enhanced peak can be given in terms of photoelectron scattering, as well. The band just below E_F along the Q direction (cf. Fig. 5), which shows only a weak dispersion, gives rise to a maximum in the total density of states just below E_F. It is likely that part of the emission from this band is elastically scattered on the adatoms into the normal emission direction. This explanation is supported by the experimental facts that the new peak does not show a measurable dispersion (cf. Eyers et al. 1985) and its ESP is very low (< 10 %). Neglecting hybridization effects, interband transitions along the $Q-$line should yield unpolarized electrons.

Fig. 23 shows the ESP of peaks 1 and 2 as function of photon energy for the clean Pt(111) substrate and for Pt with a complete $\sqrt{3} \times \sqrt{3}$ (R 30°) overlayer of Xe atoms. Throughout the energy range studied a significant depolarization due to the Xe layer occurs. It must be pointed out that the effect appears even more pronounced if one takes the temperature dependence of the ESP into account: at T = 50 K (corresponding to the full circles) the ESP of the clean surface (open circles) lies close to ±70 % (cf. Fig. 7c). The depolarizations of both peaks show a similar systematic photon-energy dependence. Towards lower energies the ESP is reduced more and more. However, it is somewhat problematic to interpret this variation quantitatively in terms of an energy-dependent exchange constant, because below $h\nu = 12$ eV the new adsorbate-induced peak discussed above partly overlaps with peaks 1 and 2. Such an underlying contribution of essentially unpolarized electrons would appear as an additional depolarizing mechanism. For a complete Xe hcp monolayer the ESP was found to be less than ±15 % throughout the energy range of Fig. 23.

The first results of adsorbate-induced changes in substrate ESP spectra cannot be explained in terms of spin-flip scattering alone.

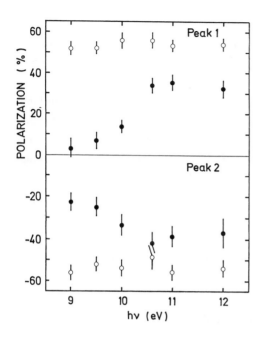

<u>Fig. 23:</u> Photoelectron polarization of peaks 1 and 2 of Pt(111) versus photon energy for clean Pt (T = 300 K, open circles) and for Pt covered by a complete √3 layer of Xe (T = 50 K, full circles) (from Eyers et al. 1985).

Even submonolayer coverages of Xe and Kr atoms can cause considerable changes which become observable in energy- and emission-angle resolving experiments. Elastic and inelastic photoelectron scattering processes, modifications of the substrate band structure in the surface region as well as the presence of surface states are strongly related to the structure and coverage of adsorbate overlayers. Systematic studies of high-resolution photoelectron intensity- and ESP-spectra together with quantitative theoretical calculations should give a deeper insight into these phenomena. Of particular interest are the chemisorbed (i.e. strongly bound) adsorbate systems, the study of which has just been started.

510

Acknowledgements:

We would like to thank A. Eyers, U. Frieß and F. Schäfers for their engagement and wealth of ideas in performing the experiments at BESSY. Thanks are due to N. Böwering for a critical reading of the manuscript. The friendly cooperation with G. Borstel, K. Hünlich, J. Kirschner, F. Meier, N. Müller, H.P. Oepen and with our colleagues from the Surface Science Department of the Fritz-Haber-Institut and the BESSY staff is gratefully acknowledged. The experiments were supported by the BMFT, the MPG, the KFA, and BESSY.

References:

Ackermann B and Feder R 1985, Solid State Commun. $\underline{54}$, 1077

Andersen O K 1970, Phys. Rev. B$\underline{2}$, 883

Antoniewicz P R 1977, Phys. Rev. Lett. $\underline{38}$, 374

Borstel G 1985, Solid State Commun. $\underline{53}$, 87

Borstel G and Wöhlecke M 1981, Phys. Rev. B$\underline{24}$, 2321

Borstel G and Wöhlecke M 1982, Phys. Rev. B$\underline{26}$, 1148

Cherepkov N A 1973, Zh. Eksp. Teor. Fiz. $\underline{65}$, 933
    (Sov. Phys. JETP $\underline{38}$, 463 (1974))

Christensen N E 1981, Solid State Commun. $\underline{37}$, 57

Christensen N E and Seraphin B 1971, Phys. Rev. B$\underline{4}$, 3321

Christensen N E and Feuerbacher B 1974, Phys. Rev. B$\underline{10}$, 2349

Eyers A, Heckenkamp Ch, Schäfers F, Schönhense G and Heinzmann U
    1983, Nucl. Instr. Meth. $\underline{208}$, 303

Eyers A, Schönhense G, Friess U, Schäfers F and Heinzmann U 1985,
    Surface Sience (in press)

Eyers A, Schäfers F, Schönhense G, Heinzmann U, Oepen H P, Hünlich K,
    Kirschner J and Borstel G 1984, Phys. Rev. Lett. $\underline{52}$, 1559

Fano U 1969, Phys. Rev. $\underline{178}$, 131

Feder R 1977, Solid State Commun. $\underline{21}$, 1091

Feder R 1978, Solid State Commun. $\underline{28}$, 27

Ginatempo B, Durham P J, Gyorffy B L and Temmerman W M 1985,
    Phys. Rev. Lett. $\underline{54}$, 1587

Heckenkamp Ch, Schäfers F, Schönhense G and Heinzmann U
    1984, Phys. Rev. Lett. $\underline{52}$, 421

Heckenkamp Ch, Schäfers F, Schönhense G and Heinzmann U
1985, Phys. Rev. A (in press)

Heinzmann U 1978, J. Phys. B11, 399

Heinzmann U 1980a, Appl. Opt. 19, 4087

Heinzmann U 1980b, J. Phys. B13, 4353 and 4367

Heinzmann U, Kessler J and Lorenz J 1970, Phys. Rev. Lett. 25, 1325

Heinzmann U, Schönhense G and Kessler J 1979a,
Phys. Rev. Lett. 42, 1603

Heinzmann U, Jost K, Kessler J and Ohnemus B
1972, Z. Phys. 251, 354

Heinzmann U, Osterheld B, Schäfers F and Schönhense G
1981, J. Phys. B14, L79

Heinzmann U, Schäfers F, Thimm K, Wolcke A and Kessler J 1979b,
J. Phys. B12, L679

Helman J S and Siegmann H C 1973, Solid State Commun. 13, 891

Horn K, Scheffler M and Bradshaw A M 1978, Phys. Rev. Lett. 41, 822

Kessler J 1976, "Polarized Electrons", Springer-Verlag, Berlin

Kirschner J, Feder R and Wendelken J F 1981, Phys. Rev. Lett. 47, 614

Koster G F, Dimmock J O, Wheeler R G and Statz H 1963,
"Properties of the Thirty-Two Point Groups", MIT Press,
Cambridge (Mass)

Koyama K and Merz H 1975, Z. Phys. B20, 131

Lampel G 1968, Phys. Rev. Lett. 20, 491

Lee C M 1974, Phys. Rev. A10, 1598

Leschik G, Courths R, Wern H, Hüfner S, Eckardt H and Noffke J
1984, Solid State Commun. 52, 221

MacDonald A H, Daams J M, Vosko S H and Koelling D D 1981,
Phys. Rev. B23, 6377

Matthew J A D and Devey M G 1976, J. Phys. C9, L413

Meier F and Pescia D 1981, Phys. Rev. Lett. 47, 374

Meier F and Pescia D 1984, in "Optical Orientation",
ed. by Meier F and Zakharchenya B P, North-Holland, Amsterdam

Meier F, Pierce D T and Sattler K 1975, Solid State Commun. 16, 401

Meier F, Pescia D and Schriber T 1982a, Phys. Rev. Lett. 48, 645

Meier F, Pescia D and Baumberger M 1982b, Phys. Rev. Lett. 49, 747

Meier F, Bona G L and Hüfner S 1984, Phys. Rev. Lett. 52, 1152

Mills K A, Davis R F, Kevan S D, Thornton G, and Shirley D A 1980,
Phys. Rev. B22, 581

Mueller F M, Garland J W, Cohen M H and Bennemann K H 1971,
    Ann. Phys. (N.Y.) $\underline{67}$, 19

Oepen H P 1984, Thesis Technische Universität Aachen (to be published)

Oepen H P, Hünlich K and Kirschner J 1984, BESSY Jahresbericht p. 239

Oepen H P, Hünlich K, Kirschner J, Eyers A, Schäfers F, Schönhense G
    and Heinzmann U 1985, Phys. Rev. B$\underline{31}$, 6846 - and to be published

Pierce D T and Meier F 1976, Phys. Rev. B$\underline{13}$, 5484

Poelsema B, Verheij L K and Comsa G 1983, Phys. Rev. Lett. $\underline{51}$, 2410
    and Surface Science (in press)

Reyes J and Helman J S 1977, Phys. Rev. B$\underline{16}$, 4283

Saile V, Skibowski M, Steinmann W, Gürtler P, Koch E E and Kozevnikov A
    1976, Phys. Rev. Lett. $\underline{37}$, 305

Schäfers F, Heckenkamp Ch, Schönhense G and Heinzmann U 1985,
    2. Europ. Conf. Atomic Mol. Physics, Amsterdam, Book of abstracts

Scheffler M, Horn K, Bradshaw A M and Kambe K 1979, Surf. Sience $\underline{80}$, 69

Schönhense G 1980, Phys. Rev. Lett. $\underline{44}$, 640

Schönhense G, Eyers A, Friess U, Schäfers F and Heinzmann U 1985a
    Phys. Rev. Lett. $\underline{54}$, 547

Schönhense G, Eyers A and Heinzmann U 1985b - to be published

Waclawski B J and Herbst J F 1975, Phys. Rev. Lett. $\underline{35}$, 1594

Wandelt K 1985, ECOSS 7, Aix-Marseille, Book of abstracts p. 343
    and private communication

Wöhlecke M and Borstel G 1981, Phys. Rev. B$\underline{23}$, 980

Zürcher P and Meier F 1979, J. Appl. Phys. $\underline{50}$(3), 2097

Zürcher P, Meier F and Christensen N E 1979, Phys. Rev. Lett. $\underline{43}$, 54

Chapter 12

Spin- and Angle-Resolved Photoemission from Ferromagnets

E. Kisker

Institut für Festkörperforschung, Kernforschungsanlage Jülich
D-5170 Jülich, FRG

12.1 Introduction

It has been a goal long-ago to actually map the electronic structure of ferromagnets by spin- and angle-resolved photoemission, i.e. to
extend the well-established technique of angle-resolved photoemission
by measuring the photoelectron spin. Previously, although it had been
speculated that the photoemitted electrons should be spin-polarized,
their spin character had been determined by comparing with calculated
band structures -which just were to be tested.

One of the difficulties with spin-resolved experiments is related
to the measurement of the spin polarization. There is no spin filter
comparable to optical polarizers, and measuring the spin has to be done
by some kind of scattering experiment which results in a loss of intensity by several orders of magnitude (for an introduction to polarized
electron physics, see for example Kessler 1976). Furthermore, the magnetic domains of the sample have to be aligned all parallel to the
spin-sensitive direction of the spin analyzer within the sampled surface area. This requires generally to immerse the sample into an external magnetic field which disturbs the electron trajectories and pre-
vented any good energy- and angle resolution.

Because of these difficulties, the spin polarization of electrons
photoemitted from ferromagnets remained somewhat mysterious until the
pioneering experiments of Eib and Alvarado (1976). The spin polarization of photoelectrons emitted from a Ni(100) surface by light of energy close to the photothreshold value was found to be in qualitative
agreement with the expectation based on the Stoner- Wohlfahrt- Slater

itinerant-electron model of ferromagnetism. In this picture, the majority-spin d-bands of Ni are fully occupied ("strong" ferromagnet), and
are separated from the Fermi energy E_F by the so-called Stoner gap   .
Only minority-spin d-bands cross through E_F, and a negative spin polarization is therefore expected at the photothreshold. With increasing
photon energy, the Stoner gap will be passed, and the spin polarization
has to become positive above some crossover photon energy (Wohlfarth
1971). Despite the qualitative agreement with this expectation, a quantitative evaluation of this data by Moore and Pendry (1978) and of data
on Ni(111) (Kisker et al. 1979) revealed that the exchange splitting
must be smaller by about a factor of two than predicted by self-consistent electronic-structure calculations. A similar conclusion had
been derived from angle-resolved photoemission (Himpsel and Eastman
1978).

In the above mentioned spin-resolved photothreshold experiments,
a magnetic field had been applied for saturating the samples magnetically, and direct angle- and energy resolution has therefore not been
obtained. Rather, the energy conservation law helped to identify the
initial states: With photon energy just equal to the work function,
only electrons from the Fermi energy are able to overcome the vacuum
barrier, and they have also to travel perpendicular to the surface for
escaping into vacuum. Therefore, with photon energy close to the
photothreshold, the initial photoemission state is well- defined. But
it is clear that with increasing photon energy a quantitative interpretation will become difficult because of the opening of the photoelectron escape cone and the uncertainty in initial state energy.

A major breakthrough occured when it was shown (Kisker et al. 1980a) that it is not necessary to apply a magnetic field during the photoemission experiment. Rather, it was possible to obtain nearly (-)100% polarized electrons from a Ni(110) sample without an applied magnetic field, making use of the magnetic remanence of geometrically favourably shaped samples (see fig.1). This opened the way for the future spin-, energy- and angle- resolved experiments.

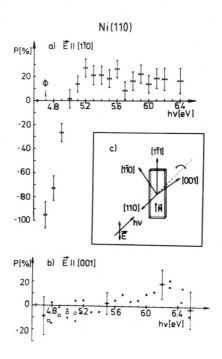

Fig.1: The measured spin polarization vs. photon energy in a photothreshold experiment on Ni(110) with a transverse magnetized sample (Kisker et al. 1980). a: Data with light electric vector EII(110). b: EII(100). c: Geometric configuration of the sample.

12.2 Experimental Set-ups for Spin- and Angle-Resolved Photoemission

As discussed above, it is possible to use ferromagnetic samples in their remanent state. To obtain high remanence, the samples have to be shaped suitably with a small demagnetization factor. As such are thin plates or evaporated films or so-called picture-frame shaped samples which consist of a frame the sides of which are parallel to easy magnetization directions. This kind of samples can be magnetically saturated by applying a short magnetic field pulse of about 200 Oe from a small unsupported coil of a few windings. Therefore, there is no re-striction to the access to the sample, and a conventional photoelectron spectrometer can be used for energy- and angle analysis.

Spin analysis can be performed in different ways (see chapter 5 of this volume). Up to now, for photoemission from ferromagnetic mate-rials, the Mott detector has been employed. It requires to accelerate the electrons to about 100 keV prior to scattering on a thin gold foil. Due to spin-orbit coupling, a left-right-asymmetry occurs when the electron beam is spin polarized perpendicular to the scattering plane (see, e.g. Kessler 1976). By making use of a large space angle, an efficiency of about 10^{-3} is obtained routinely.

The left and right count rates N_l and N_r in the Mott detector are accumulated as a function of binding energy, and the spin polarization $P(E)$ is calculated as

$$P(E)=1/S_{eff} \; (q(E)-1)/(q(E)+1) \tag{1}$$

where $q=N_l/N_r$. N_l and N_r have to be corrected for the so-called apparatus asymmetry if $N_1=N_r$ for unpolarized electrons. S_{eff} is called the effective Sherman factor, which has been calculated (see, e.g. Kessler 1976) for single elastic scattering. Multiple scattering in foils of finite thickness results in a smaller value, S_{eff}, and has to be determined by film thickness extrapolation. To make better contact with calculated band structures, it is more convenient to display spin-resolved energy distribution curves (SREDCs) I (E),I (E). These are related to the spin polarization P(E) and the spin-averaged intensity I(E) by the one-to-one correspondence

$$I^{\uparrow}(E)=0.5\ I(E)\ (1+P(E)),\quad I^{\downarrow}(E)=0.5\ I(E)\ (1-P(E)) \qquad (2)$$

Due to the surface-sensitivity, the samples have to be atomically clean during the photoemission experiment. Therefore, the apparatus has to be equipped with a surface preparation stage which consists generally of an ion-etching facility and a LEED- Auger system for determining the surface conditions.

As light sources, resonance lamps have been employed (Kisker et al. 1982, Raue et al. 1984a), and monochromatized synchrotron radiation in the energy range 20 to 70 eV (Clauberg et al. 1981a, Kisker et al. 1984). For details on the light sources, see chapter 5 of this volume. The beamline at the ACO storage ring which had been used for spin-resolved photoemission experiments has been described by Gudat et al. (1982). Due to the tunability and "cleanliness" (the light emerges from an ultrahigh vacuum system), synchrotron radiation is best suited for studying the band dispersions of the highly reactive transition metal surfaces. Energy resolutions of some tenths of an eV and angle resolution of about +- 3° are obtained.

A schematic of the experimental set-up which made explicit use
of the well-focused light spot as available with synchrotron radiation
is shown in fig.2. For more details on the electron optical lay-out,
see Kisker et al. (1982), Kisker (1983a). A similar apparatus, designed
for use with a high- efficient laboratory resonance lamp, is desribed
by Raue et al. (1984a).

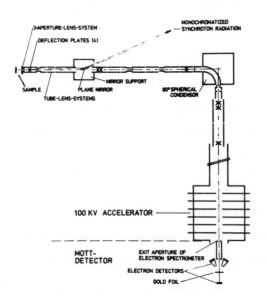

Fig.2: Schematic of an apparatus for spin-resolved photoemission.
(Kisker et al. 1985b)

12.3 Experimental Results and Discussion

12.3.1 Ni(110)

The electronic structure of Ni has for a long time been subject
to controversies (for a review see, e.g. Callaway 1980). As mentioned
already, early hints on the too small exchange splitting have been obtained from the spin-resolved threshold photoemission and from angle-
resolved photoemission. These findings have ultimately been confirmed
by spin- and angle- resolved photoemission by Raue et al. (1983). The
data, obtained on Ni(110) with the apparatus as described in detail
above, are shown in fig.3a,b for two orientations of the light electric

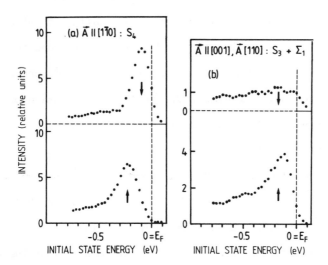

Fig.3: Spin- and angle-resolved EDCs for normal emission from
Ni(110) at room temperature for two orientations of the light electric
vector (EII(110) and EII(100)) (Raue et al. 1983).

vector and for normal emission. It is seen that the exchange-split states (fig.3a) are fully resolved and that the splitting amounts to about 0.18 eV. For the other azimuthal direction of the light electric vector (E∥(100)), only the ↓-spin state is below E_F (see fig.3b), and the exchange splitting cannot be determined.

The initial states for fig.3a are at X and of e_g-character. It has been found by spin-averaged angle-resolved photoemission (Heimann et al. 1981), that states of t_{2g}-symmetry have a larger splitting of about 0.3 eV .

A difference in exchange splitting between e_g and t_{2g} symmetry states is not inherent in the usually applied method for calculating the electronic structure of N-electron systems, the local-spin-density-functional (LSDF) theory. It has been shown by Liebsch (1979) that the small value of the exchange splitting, its different value for states of e_g and t_{2g} symmetry, the small d-band width, and the occurence of the valence band satellite are a consequence of correlations among d electrons which determine the spectral distribution of the created photoemission hole. It has been shown by Wang (1983) that by introducing non-local corrections to the LSDF approximation, an exchange splitting of only 0.4 eV is obtained, and $X_2\uparrow$ is found to be close to E_F. However, the d-band width is not reduced in this approach. Oles and Stollhoff (1984) using a model Hamiltonian and non- spherical exchange and correlation terms also were able to show that the exchange splitting depends on the e_g vs. t_{2g} symmetry. The values for the exchange splitting obtained by Oles and Stollhoff are considerably smaller than those obtained by LSDF theory.

Also, by inverse photoemission, disagreement between calculated band energies and peak positions in the EDCs has been observed by Borstel et al. (1985). Due to the discrepancies between peak positions in the photoemission data and calculated band energies, two band structures have to be kept in mind for Ni: The "experimental" and calculated ones, see fig. 4 (Martensson and Nielson 1984).

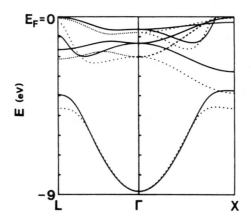

Fig.4: Comparison of an "experimental" (———) with a calculated (...) bandstructure of Ni along the Γ-L and Γ-X directions (Martensson and Nielson (1984).

12.3.2 Fe(100)

The Fe electronic structure has been tested along high-symmetry directions by angle-resolved photoemission by Eastman et al. (1980), Turner and Erskine (1982), Turner et al. (1984), Sakisaka et al. (1985), and others. The conclusion was that there is generally good agreement between experimental peak positions and calculated energy bands.

The first spin resolved energy distribution curves have been measured on an evaporated Fe film (Kisker et al. 1982). The data were in gross agreement with the expectations from the Fe DOS. In a spin- and angle- resolved experiment on Fe(100), the states $\Gamma_{25'}^{\uparrow}$, $\Gamma_{25'}^{\downarrow}$, and $\Gamma_{12}^{\uparrow}$ have been resolved by Feder et al. (1983) and interpreted quantitatively within a rigorous photoemission calculation, based on a ferromagnetic potential. Data, obtained at BESSY (Kisker et al. 1984) on the same crystal are shown in fig.5. The minority-spin EDC displays a peak at 0.4 eV binding energy, while the majority-spin EDC displays two broader peaks, one at 1.2 eV, and a second one at 2.6 eV. The initial state k-vectors are along Γ-H due to the fact of normal emission and predominantly normal-incident (s-polarized) light. From comparison with the band structure we conclude that the initial states are $\Gamma_{25'}^{\uparrow}$, $\Gamma_{12}^{\uparrow}$, and $\Gamma_{25'}^{\downarrow}$. The difference in binding energy between $\Gamma_{25'}^{\uparrow}$ and $\Gamma_{25'}^{\downarrow}$ is the exchange splitting at Γ, and it is determined as 2.2 eV. This value is slightly larger than that as obtained by Callaway and Wang (1977) with the van-Barth Hedin potential, but compares favourably with that obtained with the Kohn-Sham potential and that as calculated

by Moruzzi et al. (1978). It was shown recently by Hathaway et al.
(1985) that the position of $\Gamma'_{25}\!\uparrow$ depends somewhat on the chosen
lattice constant parameter. It is in best agreement with

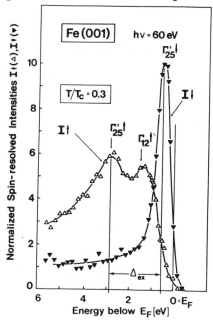

Fig.5: Spin- and angle-resolved EDCs from Fe(100) for emission centered
around the surface-normal direction with predominantly s-polarized light at 60 eV photon energy (Kisker et al. 1984). Indicated also are relevant critical point energies (Callaway and
Wang 1977).

experimental data at the minimum-energy lattice constant rather than
at the   experimental lattice constant. The problem of not yielding
the right lattice constant at the minimum energy is a recognized
problem of the LSDF approximation (Hathaway et al. 1985).

We note that in view of the dipole selection rules, emission from

Γ_{12} is forbidden. That it appears that strong in the spectrum might have one of three reasons: the limited angular resolution, crystal imperfections, or relativistic effects (Borstel 1985). Due to the high density of states of $\Gamma_{12}\uparrow$, even small experimental imperfections will result in a comparitively large contribution from this state.

The main features in the data have been well-reproduced in a recent rigorous photoemission calculation (Feder et al. 1984). However, it has been suggested that the good agreement between the peak positions and the energy bands is due to a compensation of self-energy effects (Feder et al. 1983) and is, therefore, somewhat fortuiteous.

For initial state identification one has to know the final states. We might mention here only briefly that the electronic states at energies between the vacuum level and about 100 eV have been determined recently for Fe(100) by the absorbed-current method (Kisker et al. 1985a, Tamura et al. 1985). It has been found that at 60 eV photon energy, direct phototransitions can occur near the Γ-point. It is expected furthermore that at lower photon energies, direct transitions throughout the Brillouin zone are likely to occur, enabeling to follow the dispersion of the Δ_5- symmetry bands across the Brillouin zone.

This dispersion is actually observed in spin-resolved EDCs taken at photon energies between 20 and 60 eV, see fig.6 (Kisker et al. 1985b). Below 33 eV photon energy, the minority-spin intensity drops suddenly, and the dominating peak in the SREDCs becomes a majority-spin peak. This indicates that at 33 eV photon energy, the initial state k-vector passes the first half of the Brillouin zone where the minority-spin Δ_5-band crosses E_F. The initial and final states corresponding to fig.6 are compiled in fig.7 (Feder et al. 1984).

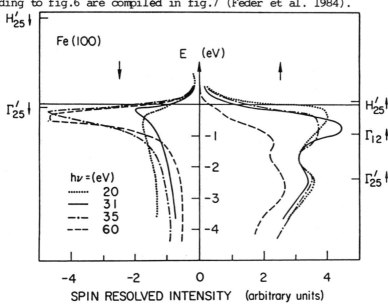

Fig.6: Spin- and angle-resolved EDCs from Fe(100) with the photon

energy varied between 20 and 60 eV (after Kisker et al. 1985b).

We note that the spin-summed EDCs, in contrast to the spin-
resolved ones, do not exhibit a strong dependence on photon energy.
They display a leading peak slightly below E_F, and a broader peak near
2.5 eV binding energy independent on photon energy. This, in the past,
has lead to the assumption that band dispersions are not observed in Fe
(Heimann and Neddermeier 1978, Schultz et al. 1979).

Recently, also the Fe(110) surface has been investigated with
spin- resolved photoemission (Schröder et al. 1985a). Similar good
agreement with calculated bands energies as for the (100) surface has
been found. Also, data obtained by the novel method of spin-polarized
inverse photoemission (Scheidt et al. 1984) have been interpreted satis-

factory in terms of the calculated electronic structure (see also chapter 13 of this volume).

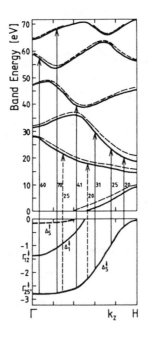

Fig.7: Identification of the k-vectors along Γ-H in a direct-transition model for the data of fig.8 (Feder at al. 1984).

12.3.3 Co(110)

Angle-resolved photoemission data by Himpsel and Eastman (1980) on hcp Co indicated strong discrepancies (a 20% too small occupied d-band width) as compared to self-consistently calculated fcc-Co band energies (Moruzzi et al 1978) (in lack of a selfconsistent hcp band-structure, which only recently has been calculated by Jarlborg and

Peter (1984). The "metastable" phase bcc Co has recently been grown by Prinz (1985) epitaxially on GaAs. Part of its electronic structure has been determined by spin-resolved photoemission at BESSY (Prinz et al. 1985). Remarkably, good agreement has been found between peak positions in the SREDCs and calculated band energies. This demonstrates the necessity for further spin-resolved studies of the hcp phase of Co and comparison with the calculated hcp bandstructure.

12.3.4 Oxygen Absorption Studies

We have studied previuosly the electronic structure of the 3d ferromagnets. It is known that many materials, consisting of non-magnetic constituents, are also ferromagnetic. A very interesting observation has, however, been made by the method of spin-resolved photo-threshold spectroscopy on Cr(110) by Meier et al.(1982). It was found, that the absorption of small quantities of Oxygen resulted in the appearance of spin- polarized electrons. It would be very interesting to study this phenomenon with spin- and energy resolved photoemission with synchrotron radiation, which allows to vary the escape depth of the photoelectrons by using light of different wavelength.

Interestingly, the opposite effect of O absorption has been observed for Ni(110) by Schmitt et al. (1985). A depolarization, accompanied by a reduction of the exchange splitting, has been found. For a

thorough discussion, see chapter 14 of this volume. In earlier spin-resolved photothreshold work, studying O absorption on Ni(110) (Clauberg
et al. 1981b), it was observed that oxygen absorption _prior_ to the development of overstructure merely leads to a change in work function.

12.3.5 Auger Electrons and Resonant Photoemission from the 3d-Transition Metals

Besides the group of primary photoelectrons which we have discussed so far there is another group of electrons which might contribute strongly to the photoelectron energy distribution curves: The
Auger electrons. They are observed when a core-hole is produced by an
energetic beam, as electrons, ions, or photons. The core hole is filled
"afterwards" by an electron from a higher shell, and a second, energetic, electron might be ejected (see also chapter 9 of this volume).
In the case of interest here, the two Auger electrons (the recombining
and the ejected one) are both from the valence bands (CVV-Auger process). Due to the two electrons and the core hole involved, the process
is more difficult to understand than primary photoemission. But the
phenomenon is of considerable interest since the interaction of the two
holes might lead to insight to correlation and screening effects.

It has been suggested by Lander (1953) that the CVV Auger
electron energy distribution should resemble the self- convolution of
the occupied valence band density-of-states if matrix element effects
are neglected (Feibelmann and McGuire 1977). Due to the Coulomb

correlation energy U_{eff}, a distortion and energetic shift of the Auger electron distributions is often observed (Antonides et al. 1977).

The 3p excitation threshold of Ni is at about 66 eV (Guillot et al. 1977). The usually weak structure at 6 eV binding energy ("6 eV satellite") has been found to be resonantly enhanced above 66 eV photon energy. The satellite has been interpreted as a $3d^8$ two-hole bound state (Penn 1979, Liebsch 1979, Feldkamp and Davis 1979) with the two holes on the same atom.

Feldkamp and Davis (1979) predicted that the resonant satellite should have a high spin polarization: In an atomic picture it arises from the transition

$$3p^6 \; 3d^9 = 3p^5 \; 3d^{10} = 3p^6 \; 3d^8 + el$$

The high spin polarization is due to the fact that Ni, as a strong ferromagnet, has only unoccupied minority-spin d-states above E_F. Accordingly, only a minority-spin core (3p) electron can be excited. If spin is conserved, only a minority-spin electron can recombine with the core hole. Since it is known that the ^1G-Term dominates the Auger electron distribution, a majority- spin electron has to be ejected to yield the singlett term ^1G.

It has actually been the first application of the method of spin- and energy-resolved photoemission with synchrotron radiation (from the ACO storage ring at Orsay) to determine the spin polarization of the resonant 6 eV satellite (Clauberg et al. 1981a), see fig.8. The measured spin polarization actually showed that the "satellite" is highly spin polarized, in almost quantitative agreement with the prediction of Davis and Feldkamp. For an illustration of the Auger transition see fig.9.

For Fe, resonant photoemission occurs similar as for Ni above the 3p excitation threshold, which is at 52 eV (Chandesris et al. 1983). But unlike as for Ni, a "satellite" in the sense of a feature at constant binding energy is not observed (Schröder et al. 1985b, Kirby et al. 1985, Walker et al. 1985). It has been shown that the spin-resolved

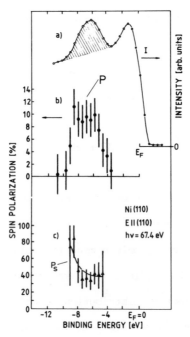

Fig.8: Spin polarization and energy distribution curve from Ni(110) at
67 eV photon energy (Clauberg et al.1981a).

Auger electron distribution is close to that as expected from the self-convolution of the spin-split DOS when exchange is taken into account (Schröder et al. 1985), see fig.10.

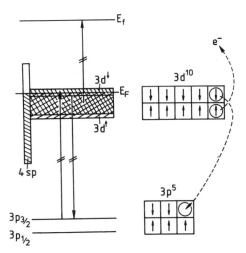

Fig.9: Model for the Super-Koster-Kronig Auger $M_{23}VV$ transition in Ni.

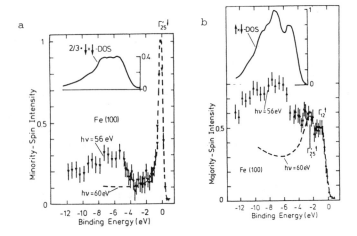

Fig.10: a: Majority-spin EDCs from Fe(100) at 56 eV photon energies. b: As above, but minority EDCs shown. For estimating the intensity of the resonant Auger distribution, hv=60 eV EDCs are indicated. Insets: Convolutions of the spin-split DOS (Schröder et al. 1985b).

12.3.6 Study of the Ferromagnetic to Paramagnetic Phase Transitions of Fe and Ni

12.3.6.1 Introduction

In the previous paragraphs a picture of the low-temperature electronic structure of the 3d ferromagnets has been obtained. We have seen that the theoretical models are adequate for an almost quantitative understanding. This is quite different for the so-called high-temperature ferromagnetism. At present, there are various controversial models concerning the electronic and the magnetic structure at high temperatures. With magnetic structure we mean the spatial distribution of the direction and of the value of the microscopic magnetization. A fundamental issue is: Is the long-range magnetization lost by gradually dissolving the magnetic moment with increasing temperature ("Stoner-model"), or is the long-range order lost by a fluctuation in direction of "local" microscopic magnetic moments? If the latter model applies, immediate questions arise on the magnitude of the local moments, or on their spatial size. It has been argued that regions strongly correlated in spin direction are as large as 20 $\overset{o}{A}$ in Fe above T_C (Capellmann 1982, Brown et al. 1983). If the correlated regions are as large as mentioned above, the clusters are large enough to sustain the low-temperature electronic structure. This model has been referred to as the "fluctuating band picture" (Capellmann 1974, Korenmann, Murray and Prange 1977).

Another extreme model is the so-called disordered local-moment picture, in which no correllation is anticipated between the magnetic moments which are assumed to exist on the lattice sites. In this model, the electronic structure has actually been calculated selfconsistently (Oguchi et al. 1983, Pindor et al. 1983, Hasegawa 1983, Gyorffy et al. 1985). But still the assumption is made of a random orientation of the local magnetic moments. There is yet no first- principles prediction on the magnetic structure at elevated temperatures.

As a bridge between the extreme models with no inherent short-range order and the huge amount of spin correllation, a new cluster-theory has been developed by Haines, Clauberg, and Feder (1985). With this theory, it is possible to calculate the electronic structure for an assumed amount of short-range order. In chapter 12 of this volume, this method will be outlined further in connection with a quantitative determination of the short-range- order parameter by comparison with the spin-resolved photoemission data to be reviewed below.

An important assumption in these models is that it is possible to "freeze" a certain spin configuration, to calculate its electronic structure, and to average over the possible configurations. This seems to be reasonable because of the time scales involved: The electron hop-ping times are much shorter than the spin-fluctuation times (see, e.g., Gyorffy et al. 1985). Because of the dynamics of the electron system at high temperatures, the question of timescales is also crucial for the interpretation of experiments testing the electronic structure. It might suffice here to point out that the photoelectron probing time is probably somewhere in between the electron hopping time and the spin fluctuation time (Kisker et al. 1985b). But this question has certainly to be addressed further in the future.

12.3.6.2 Fe(100)

We have shown in fig.5 the mapping of the initial states $\Gamma_{25}'\uparrow$, $\Gamma_{25}'\downarrow$ on the spin-resolved EDCs. Since the exchange splitting is fully resolved, this data appear well suited to study the variation of the electronic structure with increasing temperature. In fig.11, we show which changes occur to these SREDCs at higher temperatures. It is seen that the heights of the peaks from $\Gamma_{25}'\uparrow$ and $\Gamma_{25}'\downarrow$ are decreasing. A new ("extraordinary") peak emerges in the minority-spin EDC at the position of the majority-spin peak. This has been taken as evidence for the existence of regions which are magnetized in a different direction than the one of the spontaneous magnetization (Kisker 1984).

Since the photoemission experiment is very surface sensitive (the escape depth is of the order of a few atomic layers), it is interesting to have an indication for the importance of surface magnetic effects. It is clear that the relative energy- and angle-resolved spin polarization $P(T)/P(0)$ is closely related to the value of the relative magnetization $M(T)/M(0)$, although they must generally not be connected by a constant factor independent on temperature (Kisker 1983b). The relative spin polarization $P(T)/P(0)$ is, however proportional to the relative magnetization $M(T)/M(0)$ if it is measured at a binding energy where the spin-resolved EDCs remain stationary. This is the case at the position of $\Gamma_{25}^{\prime\,\uparrow}$ (2.6 eV), see fig.11. Data taken as a function of temperature

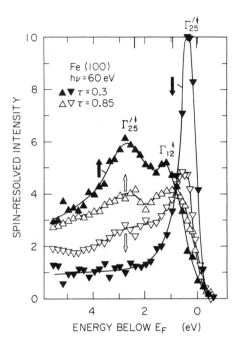

Fig.11: Spin- and angle- resolved energy distribution curves from
Fe(100) at two different temperatures (T=0.3 T_C and T=0.85 T_C).
The phototransitions occur near Γ (Kisker et al. 1984).

while the electron spectrometer is tuned to this energy are shown in
fig.12 and compared with the result of a model calculation on the
layer-dependence of the relative magnetization in the mean field approximation. It is seen that the spin polarization decreases almost
linearly with increasing temperature and that its dependence is close
to that as the calculated first to second layer magnetization curve.

From figs.11 and 12, we conclude furthermore that the peak separation of the exchange split peaks does not follow the reduction of the spontaneous magnetization which has decreased to about 1/3 of its value at room temperature at $T=0.85\ T_C$. This demonstrates clearly that the Stoner model is not valid.

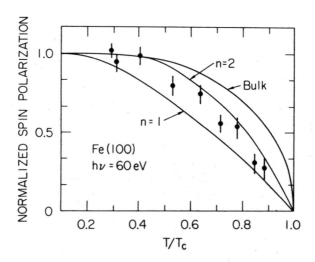

Fig.12: Spin polarization as function of temperature and comparison with a layer-dependent mean-field calculation. The electron spectrometer is set fixed at 2.6 eV binding energy ($\Gamma_{25}'\cdot\uparrow$). The photon energy is 60 eV. (Kisker et al. 1985b)

It will be shown in chapter 12 of this volume that the data (fig.11) have lead to the conclusion that short-range order of about 5 Å has to be present at $T=0.85\ T_C$ (Haines et al. 1985). A similar conclusion is obtained from an evaluation of X-ray photoelectron spectroscopy data on Fe(100) (Kirby et al. 1985), taken at temperatures up to $1.034\ T_C$ (Jarlborg and Peter 1985, Clauberg et al. 1985).

12.3.6.3 Ni(110)

Fe, because of its large value of the exchange splitting, allows
the observation of some of the effects on the electronic structure occuring at high temperatures most clearly. A much more difficult situation, both from the experimental and the theoretical point of view,
occurs in Ni due to its small value of the exchange splitting. Studies
on Ni(110) have been made somewhat earlier on Ni(110) by Hopster et al.
(1983). The data are shown in fig.13. The small value of the exchange

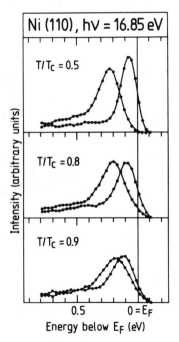

Fig.13: a: Spin- and angle-resolved energy distribution curves from
Ni(110) at 16.8 ev photon energy (NeI) for normal emission and
s-polarized light at three different temperatures (Hopster et
al. 1983). b: Spin-summed data. Initial states are near X.

splitting is already at room temperature comparable to the width
of the photoemission lineshapes. It is seen in fig.13 that the
exchange-split peaks are coalescing near T_C (see also fig.14a (Raue
et al. 1984b), where the separation of peak positions is shown as a
function of temperature). This might lead to the conclusion that the
Stoner model were valid. However, the width of the lines are increasing
with temperature (see fig.14b), and do not collaps above T_C to the low
temperature value. This fact is certainly beyond the Stoner model and
shows that local magnetic moments probably exist to temperatures larger

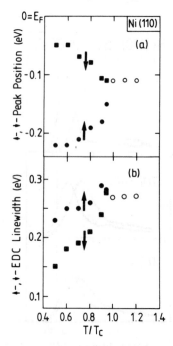

Fig.14: a: Initial state energies of the spin-split states of fig.13 as
function of temperature (Raue et al. 1984b). b: Temperature-
dependence of the width of the spin-split lines.

than the Curie temperature. It had been suggested (Kisker 1984) that unresolved "extraordinary" peaks as defined above are underlying the change in lineshape with temperature.

Further experiments taken at different emission angles and photon energies to test the influence of the wave vector on the changes of the electronic structure with temperature will probably lead to a better understanding of the high-temperature ferromagnetism and the microscopic nature of the ferromagnetic to paramagnetic phase transition.

12.4 Concluding Remarks

We have presented here a short review on the studies of the 3d transition metals by the new method of spin- , angle- and energy resolved photoemission. We have seen that the method is capable of mapping directly the spin-split electronic structure of ferromagnets up their Curie temperature. We have seen that the electronic structures of Fe and bcc Co seem to be reasonably well understood in the one-particle approximation at low temperatures, whereas Ni remains still a puzzle despite the fact that there are well-founded speculations why Ni is so different from Fe and Co. We expect that the source of the Ni problem will be enlighted further by a combination of more systematic studies by spin-resolved photoemission with synchrotron radiation and the novel method of spin-polarized inverse photoelectron spectroscopy.

The studies at elevated temperatures up to T_C suggest that for Fe, the truth might be in between the two extreme models of complete disorder and the fluctuating band picture, although the extrapolation of predictions made for temperatures above T_C to temperatures below T_C has to be done with care. Theories of Hasegawa (1983) and of Haines, Clauberg and Feder (1985) which take into account the long- range order below T_C are, for this reason, especially well suited for a determination of the short-range order parameter from existing data part of which has been reviewed here.

We mention only briefly here the work on epitaxial Gd on W by Weller et al. (1985). A magnetic reconstruction of the Gd surface has been inferred from the data. This might be seen in relation to an earlier observation in spin-resolved field emission from W/EuS system, where a   rotation of the spin polarization vector above the EuS bulk Curie temperature has been observed (Kisker et al. 1976).

The data which have been obtained so far are just a snapshot on the fascinating field. Certainly, this kind of work will be put on a broader basis in the near future, and systematic studies on the problems which have been reviewed here and those which are still awaiting for investigation will continue in several loboratories.

References

Antonides E, Janse E C and Sawatzki G A 1977 Phys. Rev. B15 1669

Borstel G, Thörner G, Donath M, Dose V and Goldmann A 1985 Sol. State Comm. 55 469

Borstel G 1985 Sol. State Comm. 53 87

Brown P J, Deportes D, Givord D and Ziebeck K R A 1983 Journ. Magn. Magn. Mat. 31-34 295

Callaway J and Wang C S 1977 Phys. Rev. B 16 2095

Callaway J 1980 Inst. Phys. Conf. Ser. 55 1

Capellmann H 1974 J. Phys. F4 1466

---- 1982 Journ. Magn. Magn. Mat. 28 250

Chandesris D, Lecante J and Petroff Y 1983 Phys. Rev. B27 2630

Clauberg R, Gudat W, Kisker E, Kuhlmann E and Rothberg G M 1981a Phys. Rev. Lett. 47 1314

Clauberg R, Gudat W, Kisker E, and Kuhlmann E 1981b Z. Phys. B43 47

Clauberg R, Haines E and Feder R 1985 Z. Phys. (in print)

Durham P, Staunton J and Gyorffy B L 1984 Journ. Magn. Magn. Mat. 45 38

Eastman D E, Himpsel F J and Knapp J A 1980 Phys. Rev. Lett. 44 95

Eib W and Alvarado S F 1976 Phys. Rev. Lett. 37 444

Feder R, Gudat W, Kisker E, Rodriguez A and Schröder K 1983 Sol. State Comm. 46 619

Feder R, Rodriguez A, Baier E and Kisker E 1984 Sol. State Comm. 52 57

Feibelmann P J and McGuire E J 1977 Phys. Rev. B15 3575

Feldkamp L A and Davis L C 1979 Phys. Rev. Lett. 43 151

544

Gudat W, Kisker E, Rothberg G M and Depautex C 1982 Nucl. Instr. Meth. 195 233

Guillot C, Ballu Y, Paigne J, Lecante J, Jain K P, Thiry P, Pinchaux R, Petroff Y and Falicov L M 1977 Phys. Rev. Lett. 39 1632

Gyorffy B L, Pindor A J, Staunton J, Stocks G M, Winter H 1985 J. Phys. F15 1337

Haines E, Clauberg R and Feder R 1985a Phys. Rev. Lett. 54 932

Hasegawa H 1983 J. Phys. F 13 1915

Hathaway K B, Jansen H J F and Freeman A J 1985 Phys. Rev. B31 7603

Heimann P and Neddermeyer N 1978 Phys. Rev. B18 3537

Heimann P, Himpsel F J and Eastman D E 1981 Sol. State Comm. 39 219

Hermanson J 1977 Sol. State Comm. 22 9

Himpsel F J and Eastman D E 1978 Phys. Rev. Lett. 41 507

---- 1980 Phys. Rev. B 21 3207

Hopster H, Raue R, Güntherodt G, Kisker E, Clauberg R and M. Campagna 1983 Phys. Rev. Lett. 51 829

Jarlborg T, Peter M 1984 Journ. Magn. Magn. Mat. 42 89

Kessler J 1976 Polarized Electrons, Springer Berlin

Kirby R E, Kisker E, King F K and Garwin E L 1985 Sol. State Comm., in print

Kirschner J, Glöbl M, Dose V and Scheidt H 1984 a Phys. Rev. Lett. 53 612

Kisker E, Mahan A H and Baum G 1976 unpublished

Kisker E, Gudat W, Campagna M and Kuhlmann E 1979 Phys. Rev. Lett. 43 966

Kisker E, Gudat W, Kuhlmann E, Clauberg R and Campagna M 1980 Phys. Rev. Lett. 45 2053

Kisker E, Clauberg R and Gudat W 1982 Rev. Sci. Instr. 53 507

Kisker E 1983a Rev. Sci. Instr. 54 1113

---- 1983b J. Phys. Chem. 87 3597

Kisker E, Schröder K, Campagna M and Gudat W 1984 Phys. Rev. Lett. 52 2285

Kisker E 1984 Journ. Magn. Magn. Mat. 45 23

Kisker E, Kirby R E, Garwin E L and King F.K. 1985a J. Appl. Phys. 57 3021

Kisker E, Schröder K, Gudat W and Campagna M 1985b Phys. Rev. B31 329

Kisker E, Prinz G A, Walker K H and Schröder K 1985c to be published

Korenman V, Murray J and Prange R E 1977 Phys. Rev. B16 4032

Lander J J 1953 Phys. Rev. 91 1382

Liebsch A 1979 Phys. Rev. Lett. 43 1431

Martensson H and Nilsson P O 1984 Phys. Rev. B30 3047

Meier F, Pescia D and Schriber T 1982 Phys. Rev. Lett. 48 645

Moore I D and Pendry J B 1978 J. Phys. C11 4615

Moruzzi V L, Janak J F and Williams A R 1978 Calculated Electronic Properties of Metals, Pergamon, New York

Oguchi T, Terakura K and Hamada N 1983 J. Phys. F13 145

Oles P and Stollhoff G 1984 Phys. Rev. B. 29 314

Penn D R 1979 Phys. Rev. Lett. 42 921

Pindor A J, Staunton J, Stocks G M and Winter H. 1983 J. Phys. F13 979

Prinz G A, Kisker E, Hathaway K B, Schröder K and Walker K-H 1985 J. Appl. Phys. 57 3024

Prinz G A 1985 Phys. Rev. Lett. 54 1051

Raue R, Hopster H and Clauberg R 1983 Phys. Rev. Lett. 50 1623

Raue R, Hopster H and Kisker E 1984a Rev. Sci. Instr. 55 383

Raue R, Hopster H and Clauberg R. 1984b Z. Phys. B54 121

Sakisaka Y, Rhodin T and Mueller D 1985 Sol. State Comm. 53 793

Schmitt W, H. Hopster, Güntherodt G 1985 Phys. Rev. B31 4035

Schröder K, Prinz G A, Walker K-H and Kisker E 1985a J. Appl. Phys. 57
    3669

Schröder K, Kisker E and Bringer A 1985b Sol. State Comm. 55, 377

Schultz A, Courths R, Schultz H and Hüfner S 1979 J. Phys. F9 L41

Staunton J, Gyorffy B L, Pindor A J, Stocks G M and Winter H 1985 J.
    Phys. F15 1387

Tamura E, Feder R, Krewer J and Kirby R E 1985 Sol. State Comm.,
    55, 543

Turner A M and Erskine J L 1982 Phys. Rev. B25 1983

---- 1984 Phys. Rev. B30 6675

Turner A M, Donoho A W and Erskine J L 1984 Phys. Rev. B29 2986

Walker K-H, Kisker E and Clauberg R 1985 to be published

Weller D, Alvarado S F, Gudat W, Schröder K, Campagna M 1985
       Phys. Rev. Lett. 54, 1555

Wang C S and Callaway J 1977a Phys. Rev. B15 298

Wang C S 1983 Journ. Magn. Magn. Mat. 31-34 95

Wohlfarth E P 1971 Phys. Lett. 36A 131

Chapter 13

Spin Dependent Inverse Photoemission from Ferromagnets

V. Dose and M. Glöbl

Physikalisches Institut der Universität
Am Hubland, D-8700 Würzburg, FRG

13.1 Introduction

The need for an experimental determination of the electronic structure of solids in terms of the energy versus momentum dispersion of electronic states has already been pointed out in previous sections of this volume. The additional complication from an analysis of the spin dependency of the electronic bands is the reason for the comparatively incomplete material available for the itinerant ferromagnets iron and nickel. The available data on occupied electronic states obtained with photoemission (PES) have been discussed by Gudat and Kisker in a previous chapter of this

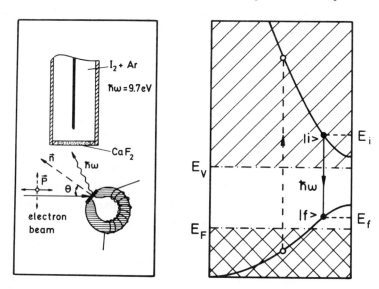

Fig.1 The right hand panel shows a schematic comparison of direct transitions in ordinary photoemission (dashed arrow) and inverse photoemission. The left hand panel shows a schematic of a spin polarized inverse photoemission experiment.

volume. They are restricted to electronic states below the Fermi level for the initial state and above the vacuum level for the final state. These ranges are indicated by the doubly and singly hatched area in the right hand panel of fig. 1. The broken "up" arrow indicates the direct transition preceeding the electron emission. It is apparent from fig. 1 that an important energy range namely that between the Fermi level E_F and the vacuum level E_V is inaccessible to ordinary photoemission. Inverse photoemission (IPE) is the only available tool to close this information gap offering exactly the same amount of detail as photoemission namely energy, momentum, and spin of unoccupied electronic states. (Dose 1983, 1984).

The technique consists of injecting electrons of well defined energy, momentum, and spin into a solid where they populate a previously empty band with energy E_i and reduced wave vector $\vec{k}_i$. If we observe the radiative decay of this excited state via emission of a photon of energy $\hbar\omega$ and momentum $\vec{q}$ to a final state of energy E_f and wave vector $\vec{k}_f$ we have by conservation of energy and momentum:

$$E_f = E_i - \hbar\omega \tag{1}$$

and

$$\vec{k}_f + \vec{q} = \vec{k}_i \tag{2}$$

Just as for ordinary photoemission $\vec{q}$ may be neglected for $\hbar\omega$ in the ultraviolet range and the energy as a function of momentum for the final state $|f\rangle$ derived. Fig. 1 shows that because E_i must exceed E_V the experimental analysis of electronic bands above E_V may be performed with either ordinary or inverse photoemission while the region between E_V and E_F is the exclusive domain of inverse photoemission. Consider the case where $E_f = E_F + \varepsilon$ and an initial state E_i in a corresponding photoemission experiment at $E_i = E_f - \varepsilon$. The matrix elements describing the two transitons are then equal. The cross section for the two processes, however, differ by a phase space factor of roughly $1/(2\alpha^2)$ where α is Sommerfelds fine structure constant. This relation between photo-ionization and radiative capture has been derived by Milne (1924) and applied to radiative capture in solids by Pendry (1980, 1981).

13.2 Experimental

The very low efficiency of the inverse photoemission process
leads to the requirement of a highly sensitive detector for the
emitted radiation. Such a detector was introduced to
Bremsstrahlungs spectroscopy by Dose (1977). It is a simple energy
selective Geiger Müller counter operating at a fixed quantum energy
of 9.7 eV with a resolution of 800 meV (FWHM). Such a radiation
detector constitutes the IPE analogue to the PES resonance lamp
since constant energy excitation in PES corresponds to constant
photon energy detection in IPE. Operation at fixed photon energy
and hence constant colour in IPE is also known as Bremssstrahlung
isochromat spectroscopy (BIS). The energy selective properties of
the Geiger counter result from a combination of the transmission
characteristics for it's CaF_2 entrance window and the molecular
photoionization efficiency of iodine which is used together with a
noble gas buffer, as a filling gas. The quantum efficiency of such
a counter has been calculated to be of the order of several percent
(Dose 1968). It has been shown recently that improved resolution of
the order of 400 meV can be obtained if the CaF_2 window is replaced
by a SrF_2 window (Goldmann et al. 1985).

The second ingredient for a spin dependent inverse photoemission
experiment, the spin polarized electron source does not present
problems since the advent of the negative electron affinity GaAs
source which offers spin polarization as an available option
(Pierce and Meier 1976, Kirschner et al. 1983). It is thus worth
pointing out that only small additional effort is necessary in
adding spin resolution to an IPE experiment in contrast to the
situation in PES. The first spin dependent IPE experiment was
performed on Ni(110) (Unguris et al. 1982). A schematic of their
experimental setup is given in the left hand panel of fig. 1.
Electrons with transverse (broken arrow) spin polarization impinge
on the Ni sample at a polar angle Θ in the <1$\bar{1}$1> azimuth. This is
also the direction of sample magnetization which was effected by a
small C-shaped electromagnet. The light emission from the sample

was observed with the band pass Geiger Müller counter discussed above. The observed spin dependent asymmetry in the photon production is

$$A = (N^+ - N^-)/(N^+ + N^-) \qquad (3)$$

where $N^{\pm}$ refer to the photon emission per unit incident charge in either spin direction. Let $n^{\pm}$ be the photon production for a hypothetical 100% spin polarized beam. We have then that

$$n^{\pm} = 1/2 \cdot \left\{ 1 \pm A/(P_o \cos\theta) \right\} \cdot (N^+ + N^-) \qquad (4)$$

where P_o denotes the actual spin polarization. Fig. 2 shows sample data for $\theta = 0^o$ and $\theta = 20^o$ extrapolated from the experimental numbers using equation (4). The most prominent result is that the peak seen just above the Fermi level E_F in the n^- spectra is

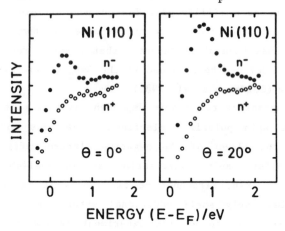

Fig.2 Inverse photoemission data from Ni(110) showing the 100% spin polarization of the empty d-band expected for a strong ferromagnet like Ni.

entirely absent in the n^+ spectra. This indicates 100% minority spinpolarization as expected for a strong ferromagnet like nickel. Comparison with a band structure calculation for ferromagnetic nickel shows that at normal incidence a direct transition into a minority d-band is possible. Selection rules would tend to suppress this transiton at strictly normal incidence. These rules are progressively relaxed for larger angles leading to the intensity increase in the minority d-band emission at $\Theta = 20^{\circ}$.

13.3 Room Temperature Data for Iron

Since the exchange splitting in nickel is only of the order of 300 meV a systematic spin resolved study of the energy versus momentum dispersion of empty bands in nickel would require an overall resolution exceeding 300 meV which is not yet available in spin polarized IPE experiments. Iron on the other hand offers an exchange splitting of about 2 eV which is safely larger than the resolution of the apparatus in fig. 1. The first measurements on iron where carried out by Scheidt et al. (1983). Electrons with longitudinal spin polarization impinged at a polar angle Θ in the <001> azimuth on an Fe(110) crystal. The relation between observed counting rates $N^{\pm}$ and counting rates $n^{\pm}$ for a hypothetical 100% spin polarization is in this case

$$n^{\pm} = 1/2 \cdot \left\{ 1 \pm A/(P_{o} \sin\Theta) \right\} \cdot (N^{+} + N^{-}) \qquad (5)$$

Since $P_{o} \sin\Theta$ enters into this relation the effective polarization vanishes for Θ versus normal incidence. This will show up in the larger statistical fluctuation of the spin resolved data for small angles Θ. From a schematic of the Brillouin zone for a bcc lattice shown in fig. 3 we find that variation of Θ in the <001> direction corresponds to the ΓNPH mirror plane of the crystal. Band mapping in mirror planes offers several advantages. The observed spectra are simplified due to degeneracies in the mirror plane. More-over the wave functions must have either even or odd parity under reflection in a mirror plane. The incident electron plane wave, however, has even parity and can consequently couple only to bulk states with even parity. Finally, for measurements in mirror planes

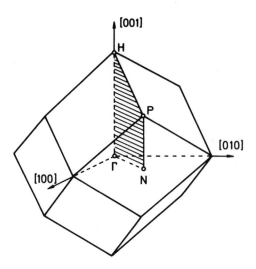

<u>Fig.3</u> Brillouin zone of the body centered cubic lattice. The
shaded plane indicates a mirror plane.

the parity of the final state wave function can be derived from an
analysis of the polarization of the emitted radiation. Light with
electric vector normal to the mirror plane results from final
states with odd parity, while polarization parallel to the mirror
plane signals final state wave functions with even parity.

A sample of spin resolved isochromat spectra for various angles
of incidence Θ is given in fig. 4. One upward dispersing transition
(B_1) is observed for majority final states while two branches
(B_2, B_3) show up for minority states. The larger statistical
fluctuation of the data for small Θ is due to the small effective
polarization. The gain of information from the spin analysis can be
demonstrated by a comparison of these data with the spin averaged
spectra given in fig. 5. These show two peaks at 0.5 eV and 1.5 eV
for $\Theta < 20^{\circ}$ and one single nearly angle independent structure for
$\Theta > 30^{\circ}$. Apart from these features which result from a
superposition of transitions B_1, B_2, and B_3 we find a step like
emission enhancement labeled S in fig. 5 which will be shown to be
due to empty surface states.

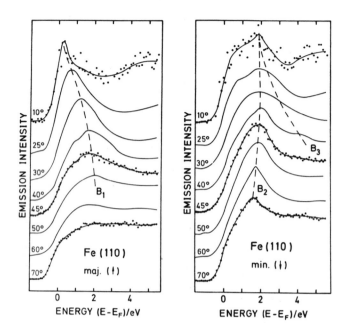

Fig.4 Inverse photoemission data from Fe(110) for various angles of incidence Θ of the primary electrons. Θ is varied in the mirror plane of fig. 3.

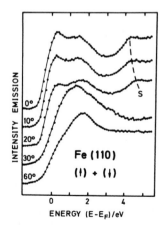

Fig.5 Spin averaged data on Fe(110) demonstrate the information loss compared to fig. 4. A new structure S which is difficult to identify in the spin resolved data due to statistical fluctuations results from image potential surface states.

The energy of the emission features in fig. 4 corresponds to the energy of the final band involved in the radiative transition. The goal of band mapping, the determination of the associated momentum can be reached only partly because electrons penetrating from the vacuum into a solid experience a force normal to the solid surface. As a result the momentum component $k_{\perp,i}$ in the solid differs from the momentum component $k_{\perp,a}$ by a usually unknown amount. The periodic forces parallel to the surface on the other hand result in diffraction such that

$$\vec{k}_{\parallel,i} = \vec{k}_{\parallel,a} + \vec{g}_{\parallel} \tag{6}$$

where $\vec{g}$ is a surface reciprocal lattice vector. Energy constraints favour strongly $\vec{g}_{\parallel} = 0$ at the rather low photon energies used in the work described here. For this case we have

$$k_{\parallel,a} = 1/\hbar \cdot (2mE_{kin,a})^{1/2} \cdot \sin\theta \tag{7}$$

which may be rewritten in terms of the sample work function Φ and the final state energy E_f as

$$k_{\parallel,a} = 1/\hbar \cdot \left\{ 2m(E_f + \hbar\omega - \Phi) \right\}^{1/2} \cdot \sin\theta \tag{8}$$

The data shown in fig. 4 may now be condensed in a plot of $E_f(k_{\parallel})$ using equation (8). This is presented in fig. 6 separately for the two spin directions. This figure shows also a projection of the bulk band structure and a theoretical prediction of $E_f(k_{\parallel})$ shown as the dashed lines near B_1, B_2, and B_3 (Kübler 1984). These were obtained from a search for two bulk bands 9.7 eV apart irrespective of the associated $k_{\perp}$. From a comparison with experiment we find that the calculated dispersion of B_1 and B_3 is more rapid than observed. B_2, showing only very weak dispersion results from a final d-band which extends with little energetic variation throughout the Brillouin zone.

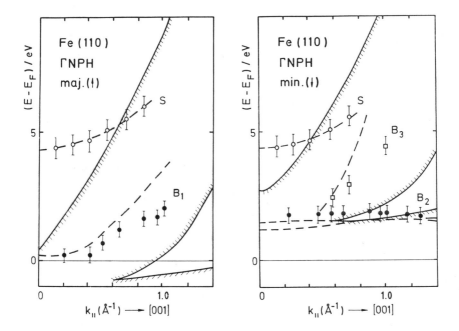

<u>Fig.6</u> Final state energy E_f as a function of $k_{\parallel}$, the component of
momentum parallel to the crystal surface in <001> direction
collected from the spectra in fig. 4. The dashed lines near
B_1, B_2, and B_3 are theoretical predictions.

Emission feature S starts for both spin polarizations in a band
gap at normal incidence which suggests a surface state as its
origin. Extended asymmetry measurements in the energy range of this
step structure have lead us to conclude that it is unpolarized. A
similar feature has been observed in the data from many crystal
surfaces and the accepted interpretation attributes it to radiative
transitions into bound states of the image potential (Johnson and
Smith 1983, Dose et al. 1984, Straub and Himpsel 1984, Reihl et al.
1984). Since the wave functions of such states extend far out into
the vacuum the absence of polarization effects is expected and
supports the assignment of the origin of this structure.

A more comprehensive interpretation of the observed spectra
which yields not only energetic positions but also intensities has

been given by Feder and Rodriguez (1984). They use a nonrelativistic one step model Green function formalism of photoemission to
calculate spin resolved ultraviolet Bremssstrahlung isochromat
spectra for Fe(110). Such calculations are theoretically more
satisfactory though physically less transparent than considerations
within the three step bulk direct transition model since they treat
penetration of the electron into the sample, transport in the
sample and radiative decay to a lower lying band coherently. More-
over bulk and surface emissions are treated on an equal footing.

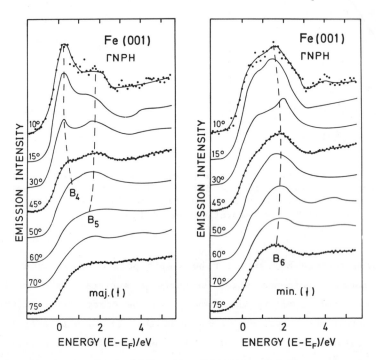

<u>Fig.7</u> Inverse photoemission data from Fe(001) for various angles
        of incidence Θ of the primary electrons. Θ is varied in the
        mirror plane of fig. 3

Isochromat spectra from Fe(001) are shown in fig. 7. The
variation of Θ is in the <110> direction in this case which means
that a different projection of the ΓNPH mirror plane is contained
in these spectra. The emission branches B_4 and B_5 are found for

majority final states in this case while in the minority system only one branch B_6, which is related to B_2 in fig. 4, shows up. B_4 and B_6 can be identified as direct transitions by an analysis of the theoretical bandstructure. The origin of B_5 remains unexplained.

13.4 Measurements at Elevated Temperature

While spin resolved band mapping of band ferromagnets provides already most useful bandstructure information at room temperature the extension of such measurements to elevated temperatures is eventually expected to provide the key to the discrimination between the various theoretical models for $T \neq 0$ band ferromagnetism. Measurements at elevated temperatures, however, cause complications due to sample contamination and interference

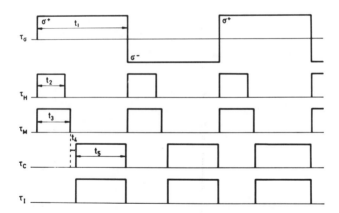

Fig.8 Timing diagram for temperature dependent measurements showing the relation between periodic change of polarization τ_σ, sample heating τ_H, sample magnetization τ_M, photon counting τ_c, and current integration τ_I.

between sample heating, magnetization, and current integration. A timing diagram of the only so far reported temperature dependent spin polarized inverse photoemission study of iron is given in fig. 8 (Kirschner et al. 1984). The basic period of the diagram is

560

the time t, of a given spin polarization indicated by the helicity $\sigma^{\pm}$ of the exciting light. The sample is heated for a time t_2 after each change of polarization. Since the magnetic field of the filament might influence the sample magnetization, the magnetization is renewed for a time $t_3 > t_2$ at each polarization change. After the heating and magnetization cycles are completed and a small interval t_4 has elapsed, the photon counting and current integration are started for a time interval t_5 which ends at the next polarization change.

Sample cleanliness is a problem with iron samples at elevated temperatures. Sulfur segregation limited the useful temperature range to $T/T_C \leqslant 0.86$ for the measurements described here.

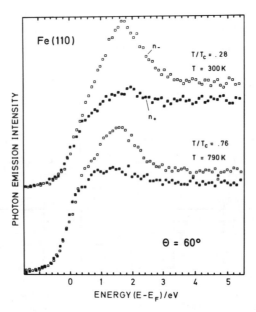

Fig.9 Temperature dependent spin resolved spectra from Fe(110) at $\Theta = 60°$.

Temperature dependent sample spectra from Fe(110) at a polar angle of 60° are shown in fig. 9. An angle of 60° brings us close to the P point in the Brillouin zone at 1.5 eV final state energy. At this large angle of incidence the effective polarization is

close to its maximum resulting in a small statistical fluctuation
of the data. Fig. 9 shows that the minority emission peak does not
shift in energy within the limits of experimental resolution. The
observed asymmetry, however, decreases clearly as the temperature
rises. It is worth noting that the majority counterpart of this
band has been observed by conventional photoemission to lie about
0.5 eV below the Fermi energy at the P point (Eastman et al. 1980).
An energetic shift of this occupied band upon heating has not been
reported. The for this pair of bands at this particular point in
the Brillouin zone temperature independent exchange splitting
amounts then to (2.1 ± 0.2) eV. This kind of temperature dependence
is in line with predictions of the short range magnetic order model
(Korenman et al. 1977, Capellmann 1979, Korenman and Prange 1980).

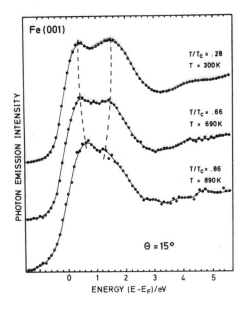

Fig.10 Temperature dependent spin averaged spectra from Fe(001) at
$\theta = 15°$.

Quite a different behaviour as a function of temperature is
observed for a pair of states near the Γ - H line. Fig. 10 shows
spin averaged data from Fe(001) as a function of temperature. The

562

spin character of the observed two peaks may be inferred from an inspection of fig. 7, which shows that the peak at 0.2 eV results from a majority band while that at 1.8 eV is due to a minority band. The exchange splitting of these bands is then (1.6 ± 0.2) eV at room temperature. Upon temperature increase these two bands appear to merge into one single peak about midway between the room temperature peaks, indicating a collapsing band state at this particular point in $\vec{k}$ space. This behaviour is in line with a prediction of the disordered local moment theory (Hasegawa 1980, Hubbard 1981, Heine et al. 1981).

13.5 <u>Outlook</u>

Spin dependent inverse photoemission studies of empty bands are certainly in their infancy. A considerable impetus to this field of research can be expected in the near future especially if we succeed in improving the experimental resolution at an acceptable expense in sensitivity. The experiments reported in this chapter are so far the only ones which have come to our attention.

Our own work on iron has only been possible through a most fruitful cooperation with Dr. J. Kirschner in the Institut für Grenzflächenforschung und Vakuumphysik of the Kernforschungsanlage Jülich. We are indebted to Prof. H. Ibach for encouraging this cooperation and acknowledge financial support from the Deutsche Forschungsgemeinschaft.

<u>References</u>

Capellmann H 1979 Z. Phys. **B34** 29
Dose V 1968 Rev. Sci. Instru. **39** 1055
-- 1977 Appl. Phys. **14** 117
-- 1983 Progress in Surface Science **13** 225
-- 1984 J. Phys. Chem. **88**, 1681
Dose V, Altmann W, Goldmann A, Kolac U, and Rogozik J 1984
    Phys. Rev. Lett. **52** 1919
Eastman D E, Himpsel F J, and Knapp J A 1980 Phys. Rev. Lett. **44** 95

Feder R and Rodriguez A 1984 Solid State Commun. **50** 1033
Goldmann A, Donath M, Altmann W, and Dose V 1985
        Phys. Rev. **B32** xxxx
Hasegawa H 1980 J. Phys. Soc. Jpn. **49** 178 963
Heine V 1981 J. Phys. F **11** 2645
Hubbard J 1981 Phys. Rev. **B23** 5974
Johnson P D and Smith N V 1983 Phys. Rev. **B27** 2527
Kirschner J, Oepen H P, an Ibach H 1983 Appl. Phys. A **30** 177
Kirschner J, Glöbl M, Dose V, and Scheidt H 1984
        Phys. Rev. Lett. **53** 612
Korenman V, Murray T L, and Prange R E 1977
        Phys. Rev. **B16** 4032 4048 4058
Korenman V and Prange R E 1980 Phys. Rev. Lett. **44** 1291
Kübler J private communication
Milne E A 1924 Phil. Mag. **47** 209
Pendry J B 1980 Phys. Rev. Lett. **45** 1356
--   1981 J. Phys. C. Solid State Phys. **14** 1381
Pierce D T and Meier F 1976 Phys. Rev. **B13**, 5484
Reihl B, Frank K H, and Schlittler R R 1984 Phys. Rev. **B30**, 7328
Scheidt H, Glöbl M, Dose V, and Kirschner J 1983
        Phys. Rev. Lett. **51** 1688
Straub D and Himpsel F J 1984 Phys. Rev. Lett. **52** 1922
Unguris J, Seiler A, Celotta R J, Pierce D T, Johnson P D,
        and Smith N V 1982 Phys. Rev. Lett. **49** 1047

Chapter 14

Photoemission and Bremsstrahlung from Fe and Ni: Theoretical Results and Analysis of Experimental Data

R. Clauberg [+] and R. Feder[*]

Institut für Festkörperforschung, Kernforschungsanlage Jülich,
D-5170 Jülich, FRG

[+] permanent address: IBM Zürich, Research Laboratory,
                       CH-8803 Rüschlikon, Switzerland

[*] permanent address: Theoretische Festkörperphysik, FB 10,
                       Universität Duisburg GH, D-4100 Duisburg,
                       FRG

14.1 Introduction

As shown in Chapters 12 and 13 of this book (and references therein), recent experimental achievements have established spin-resolved ultraviolet photoemission and its inverse (bremsstrahlung) as powerful tools for probing ferromagnetism of transition metals and of their surfaces. Since magnetic properties are coded in measured spectra in a rather complicated manner (cf. Chapter 4), the amount of information one actually obtains from experimental data depends on the level of theoretical sophistication one uses in the analysis. It therefore seems pertinent to present and discuss in some detail results calculated by means of the currently most advanced theories, and to illustrate how these results are employed to extract physical information from experimental data. With these aims, we focus in the following on photoemission and its inverse from Fe and Ni firstly at temperatures well below the Curie temperature T_c, where ferromagnetism is well understood in terms of a band (Stoner-Wohlfahrth-Slater) model, and secondly at temperatures near T_c, where the nature of ferromagnetism has until recently been a subject of intense study and controversy (cf. Chapter 2 of this book).

14.2 Analysis at Low Temperature

A quantitatively adequate theoretical treatment of spin-resolved photoemission and its inverse from 3d transition metal ferromagnets at temperatures well below T_c is achieved by means of a one-step model multiple scattering formalism (cf. Chapter 4.5 and references therein). As found by recent relativistic calculations (Ackermann 1985, Ackermann and Feder 1985), spin-orbit coupling may be neglected except if one is interested in certain special "spin hybridization" features. One then has to perform two successive Schrödinger-equation-based calculations, for spin-up and spin-down effective potentials $V^S(E,\underline{r})$. The choice of these self-energies is at present still in a rather empirical stage, and will be specified below for Fe and Ni individually. While one-step theory results are capable of quantitatively reproducing experimental results and thereby verifying assumed potentials V^S, we emphasize that they are not entirely satisfactory by themselves but should be complemented by calculations of the (upper and lower state) bulk band structure and - with a view to surface effects - also the layer-projected density of states in order to reveal the physical origin (like bulk interband transitions or surface effects) of individual spectral features.

14.2.1 Fe(001) and Fe(110)

We first review a theoretical study of spin- and angle-resolved ultraviolet photoemission from Fe(001) (Feder et al. 1984). The central input quantities to the calculations, the effective potentials $V^S(E,r)$, are constructed as follows. Their real parts for the initial (lower) and final (upper) states include a local-density exchange-correlation approximation $\propto \alpha^S(E,r)\ [\rho^S(r)]^{1/3}$ (with $s = +/-$) (cf. Chapter 4.2.2 of this book, eq.(11)), where ρ^+ and ρ^- are spin-up and spin-down charge densities taken from a self-consistent ground state calculation (Gloetzel 1979) (very similar to that of Moruzzi et al. 1978). (Very recently, Wang et al. (1985) pointed out some deficiencies in the local-spin-density approximation used in these ground state calculations, but results based on a non-local approach are not yet available). The exchange parameter $\alpha^S(E,r)$ is chosen spin-independent as $\alpha = 2/3$ (i.e. the Kohn-Sham value) for $E = E_F$. For the initial states ($E \leq E_F$), electron gas results suggest an increase of α; This is, as the simplest assumption, taken as linear in $(E_F - E)$ with the slope as a parameter to be adjusted by comparison with experiment. (For the data discussed below, a value $\alpha = 0.676$ is thus found at 3 eV below E_F). For the final states (with E between 16 and 70 eV above E_F) $\alpha = 2/3$ and an $\alpha(E)$ decreasing with increasing E (cf. Chapter 4.2.2 eq.(12)) are used. The imaginary self-energy part for the final state is taken as a spatially uniform function $V_i(E)$ increasing with E as suggested by (SP)LEED (cf. Chapter 4.2.2 Fig. 4) (neglecting the refinements of localized absorption and of a slight spin dependence). The initial-state imaginary self-energy part (corresponding to the hole lifetime), which must vanish at E_F and increase in going to lower energy, have been assumed either as linear functions with adjustable slopes or as the monotonically increasing functions obtained by Hubbard model calculations (Tréglia et al. 1982). The surface potential barrier is taken as a step (of height $V_{or} = 14$ eV) with an adjustable position above the topmost atomic plane. Variation of the barrier position permits a distinction between bulk spectral features and those arising from surface states (or resonances), since only the latter are substantially affected.

For several photon energies (from 20 to 70 eV), spin-resolved photoemission intensity versus initial-state energy spectra calculated with the above model assumptions are compared (in Fig. 1) with their experimental counterparts (due to Kisker et al. 1985; cf. also Chapter 12 of this book). The interpretation of individual spectral features is aided by the (initial and final state) bulk band structure (calculated with the same real potential part as used for photoemission) (Fig. 2) and layer-by-layer densities (Fig. 3) (calculated with the same complex potential as the photoemission spectra, using a Green function method due to Hora and Scheffler (1984)).

The calculated minority spin spectra exhibit a prominent peak (labelled A in Fig. 1) just below the Fermi level, which remains at the same position in the entire photon energy range from 20 to 70 eV. As can be seen in Fig. 4, the top layer density of states (for $k_{\parallel} = 0$) is rather small, while the bulk layer density of states has a pronounced peak corresponding to the flat $\Delta_5^{\downarrow}$ band (cf. Fig. 2). The latter thus clearly provides the initial state, which leads to A via transition into a bulk band (for $\hbar\omega = 20$ eV) or into a strongly attenuated LEED state (for 31 eV and above). This is consistent with the additional finding that neither position nor height of A are affected by changes in the surface barrier position. Calculations for off-normal emission angles (not shown in Fig. 1) reveal a strong sensitivity in height and position, the latter shifting to lower energies (by up to

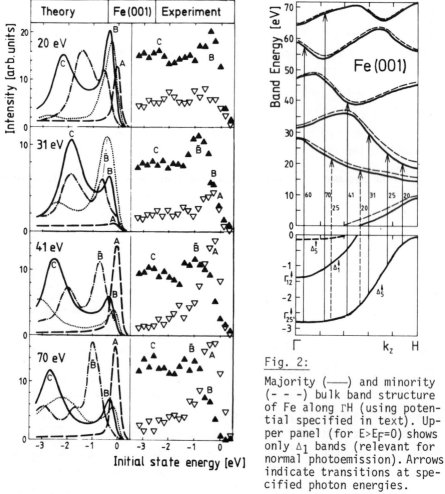

Fig. 2:

Majority (——) and minority (- - -) bulk band structure of Fe along ΓH (using potential specified in text). Upper panel (for $E > E_F = 0$) shows only Δ_1 bands (relevant for normal photoemission). Arrows indicate transitions at specified photon energies.

Fig. 1.

Photoemission from Fe(001) for radiation with $\underline{A} \parallel$ [010] and energies as indicated. Left column: calculated majority (——) and minority (- - -) spectra for polar emission angle $\Theta_e = 0$, and majority spectra for $\Theta_e = 8°$ with $\underline{k}_\parallel \parallel \underline{A}$ (–·–·) and $\perp \underline{A}$ (·····). Right column: experimental majority (▲) and minority (▽) data due to Kisker et al. (1985).

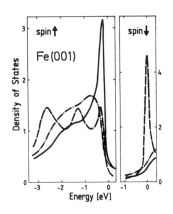

Fig. 3:

Layer-projected majority (left) and minority (right) density of states of Fe(001) for $\Theta_e=0$ ($k_{\parallel}=0$): bulk (- - -) and topmost surface layer with barrier position 0.20 (——) and 0.25 a_0 (lattice constant above topmost internuclear plane).

0.7 eV for $\Theta_e = 8°$ and $k_{\parallel} \perp E$). This implies a left-shift (by about 0.2 - 0.3 eV) and a significant broadening of finite-acceptance cone data compared to the ideal normal-emission case, as is confirmed by Fig. 1. The measured strong increase of A (relative to the other features) between 31 and 41 eV photon energy is reproduced by the calculations.

The theoretical majority spin spectra for $\Theta_e = 0$ (Fig. 1) display two peaks, labelled B and C. The binding energy of B (~ 0.35 eV) is independent of the photon energy. While the band structure does not provide an explanation, the strongly enhanced top layer density of states at this energy (cf. Fig. 3) and the strong sensitivity of the height of B to the surface barrier position identify this peak as associated with a surface state resonance. In both theory and experiment B is dominant at $h\nu = 20$ eV and decreases at higher photon energies (appearing in the data only as a shoulder of a new peak $\tilde{B}$ to be discussed below). The fact that recent spin-averaged photoemission work (Turner et al. 1984) also reports the surface feature B but interprets it as due to a minority spin state, stresses the need for spin analysis. Peak C arises from direct transitions from bulk $\Delta_5^{\uparrow}$ to $\Delta_1^{\uparrow}$ states, as is seen from the arrows in Fig. 2 (dashed for 20 eV and solid for 31, 41 and 70 eV). At photon energies 60 and 70 eV, C originates from the vicinity of $\Gamma_{25'}$.

The majority peak $\tilde{B}$, which is very prominent in the experimental spectra for $h\nu = 31$ eV and above, is absent in the normal-emission theoretical spectra in accordance with non-relativistic dipole selection rules, but appears for off-normal emission (see $\Theta_e = 8°$ in Fig. 1), mainly due to even-symmetry initial states ($\underset{\sim}{k}"\|A$) at higher $h\nu$ and odd ones at lower $h\nu$. An alternative mechanism, hybridization of $\Delta_5^{\downarrow}$ and $\Delta_1^{\uparrow}$ initial states by spin-orbit coupling, is too weak to be observable, as was found by recent fully relativistic photoemission calculations (Ackermann 1985, Ackermann and Feder 1985).

The assumed self-energy approximations are thus seen to lead to theoretical photoemission spectra, bulk band structure and layer densities of states, which permit a fairly quantitative interpretation of the experimental spectra. But how precisely does this actually determine the self-energy and thereby also the quasi-particle bulk band structure? This obviously depends on the sensitivity of the calculated spectra to variations in the assumed self-energy ingredients. For the final-state self-energy, calculations (Feder et al. 1984) showed that the relative peak heights at a given photon energy are hardly affected by changes within physically reasonable limits (e.g. taking $\alpha = 2/3$ or an energy-dependent form). On the other hand, positions of peaks respond considerably to changes in the initial-state self-energy. This applies in particular to the majority peak C (associated with $\Gamma_{25'}^{\uparrow}$). If one uses $\alpha = 2/3$ throughout, instead of the linearly increasing form (with $\alpha = 0.676$ at -3 eV), C and $\Gamma_{25'}^{\uparrow}$ are shifted towards E_F by about 0.5 eV. Since a rising imaginary potential part V_i also produces a right-shift of C (by about 0.2 eV for $V_i^{\uparrow}$ of about 0.5 eV at -3 eV), an increase in α can (beyond some $\alpha > 2/3$) be compensated by an increase in $V_i^{\uparrow}$. This implies some uncertainty in α and V_i. This uncertainty is, in the present case, overruled by an experimental uncertainty arising mainly from the width of the angular acceptance cone: as is suggested by the off-normal theoretical results in Fig. 1, and the above discussion, these contributions broaden and shift features (like A and C) in a manner dependent on the acceptance angle. (Note especially for peak C

the width and scatter in the experimental data). Despite these uncertainties, initial-state real potentials as used in ground state local-
density functional theory (e.g. Callaway and Wang 1977, Moruzzi et al.
1978, Gloetzel 1979) can be ruled out, since they lead, by themselves
and even more so together with the $V_i(E)$- induced right-shift, to a
binding energy of peak C too small to be compatible with the experimental data. As for $V_i(E)$, a distinction between the linear approximation and that due to Trëglia et al. (1982) cannot be made on the
grounds of the data, but a form of such or similar nature is needed.
This is also supported by other experiments and plausible from phase
space arguments for hole-lifetime-determining Auger processes involving the unoccupied d-band parts (cf. e.g. Eberhardt and Plummer 1980,
and Eastman et al. 1980). To draw more definite conclusions, spin-resolved experimental data with improved angular and energy resolution
are needed.

In summary, the main results of the above analysis for Fe(001)
are: (1) one-step-model calculations reproduce experimental features
with regard to existence, position and relative intensity; (2) it is
important to use a quasi-particle self-energy (including an energy-
dependent imaginary part) with a real part different from the local-
density functional ground state potential (cf. discussion and references in Chapter 4.2.2); (3) off-normal acceptance angle effects have
to be identified before making contact with a bulk band structure
along a high-symmetry line; (4) majority and minority features (C and
A) are due to bulk interband transitions near $\Gamma_{25'}$, with an "exchange
splitting" between 2.2 and 2.5 eV, and (5) a majority feature (B) originates from a surface-state resonance.

We now turn, more briefly, to an analogous analysis (due to Feder
and Rodriguez 1984) of spin-dependent ultraviolet inverse photoemission
(bremsstrahlung) data from Fe(110) (cf. Chapter 13 and refs. therein),
which probe the empty part of the d-band. The theoretical model assumptions are similar to the one described above. The analysis rests again
on the three pillars: one-step theory spectra, (quasi-particle) bulk

band structure and layer-projected density of states.

Typical theoretical results are shown in Fig. 4. Most of the bremsstrahlung spectral features can be understood as due to bulk

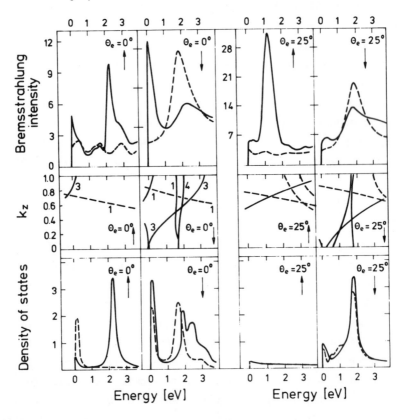

Fig. 4: For spin up (↑) and spin down (↓) electrons incident on ferro-magnetic Fe(110) in a (1Ī0) plane at polar angles Θ_e=0 and 25°: Top row: Bremsstrahlung isochromats as calculated for p- (——) and s-polar-ized (- - -) photons of 9.7 eV exiting in (1Ī0) plane at an angle of 67° with respect to incident electron direction. The intensity is given in photons per electron per Hartree per steradian. Energy E (with re-spect to E_F) of lower (final) state. Second row: corresponding bulk band structure along the line ($k_\parallel$, k_z) with $k_\parallel$=√2(E+9.7-∅) cos (Θ_e) for upper (initial) state (shifted downwards by $\hbar\omega$=9.7 eV)(- - -) and lower (final) state (——). For Θ_e=0, the numbers next to the bands indicate the Σ symmetry type. Third row: corresponding $k_\parallel$-resolved local densi-ties of states in top (——) and bulk (- - -) layer for the lower state.

interband transitions (see crossing points between lower (final state) and upper (initial state) bands, with the latter shifted downwards in energy by the photon energy 9.7 eV). In accordance with dipole selection rules, the emitted s- and p-polarized light intensities strongly differ from each other. The peak at 2.2 eV in the p-polarized spectrum due to $\Theta_e=0$ spin-up electrons cannot be associated with a bulk transition, but is identified as a surface-state feature from the layer density of states (bottom row of Fig. 4). This interpretation is corroborated by the strong sensitivity of the position of this peak to variation of the position of the surface potential barrier. In addition, there are several minor peaks, which do not correspond to the limiting cases of bulk or surface state transitions.

Calculated spectra are compared with experimental data (due to Scheidt et al. 1983) in Fig. 5, bearing in mind: (1) the data dots indicate only the centres of vertical statistical error bars of a length of 3 to 8 times the dot diameter; (2) there is a substantial spread in experimental photon detection energy ($\pm$ 0.35 eV) and angle; (3) the experimental s to p-polarized photon acceptance ratio is not known; (4) inelastic scattering contributions are present in the measured spectra but neglected in the calculations. Under these conditions, one cannot expect agreement in line shape, width and height of features. However, good agreement is seen to be reached with regard to existence and energetic position of features. The dominant spin-down peak near 1.8 eV, which is practically dispersionless with Θ_e (i.e. $k_{\parallel}$), is identified by the band structure (Fig. 4) as due to a bulk interband transition. The same holds for the leading spin-up peak, which disperses towards higher energy with increasing Θ_e. In contrast, the smaller spin-down peak near 0.3 eV cannot be ascribed to a bulk interband transition, but on the grounds of the surface-enhanced density of states (cf. Fig. 4, bottom rightmost panel) possibly to a transition into a surface resonance state.

As for conclusions on the self-energy model and the associated quasi-particle band structure, we note that the assumptions made are

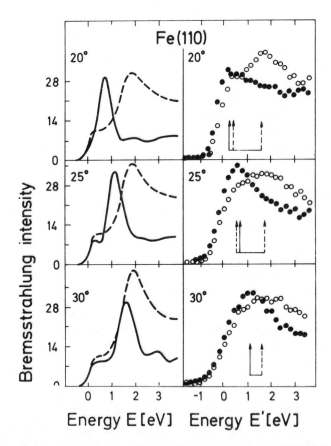

Fig. 5: Bremsstrahlung isochromats due to polarized electrons incident
on ferromagnetic Fe(1 1 0 in a (1 1 0) plane at polar angles (with respect
to the surface normal) as indicated. Theoretical results: sum of s-
and p-polarized 9.7 eV photon spectra convoluted by a Gaussian of width
0.3 eV, for spin up (——) and spin down (- - -) electrons; energy E of
lower (final) state with respect to E_F. Experimental data (Scheidt et
al. 1983) for $\hbar\omega = 9.7 \pm 0.35$ eV for spin up (●●●) and spin down (○○○);
energy $E' = E_{kin} + \emptyset - 9.7$ eV, where E_{kin} is the kinetic energy of the
incident electrons. Intensity units: as in Fig. 1 for theoretical, arbitrary for experimental results. The solid and dashed vertical arrows
indicate the majority and minority peak positions assigned to the data
in Scheidt et al. (1983).

broadly consistent with the available bremsstrahlung data on Fe(110),
but higher-resolution experiments (discriminating also between s- and
p-polarized photons) are required for detailed information.

14.2.2 Ni(110)

Since the exchange splitting in Ni is by about a factor of 6 smaller than in Fe, spin-resolved spectroscopies demand even more re- fined experimentation to permit a theoretical analysis. Inverse photo- emission (for a first experiment, on Ni(110), see Unguris et al. 1982; for calculations, on Ni(001), see Thörner and Borstel 1984) has not yet reached such a stage. We therefore focus in the following exclus- ively on spin-resolved photoemission. A theoretical analysis is pre- sented for data on Ni(110), measured at normal emission and photon energy 16.85 eV by Raue et al.(1983). This case interests for three reasons. Firstly, it demonstrates the advantage of spin-resolution, since the lifetime broadening of the states is of similar size as the exchange splitting, requiring fitting procedures in conventional pho- toemission even for the exact determination of the peak positions. Secondly, the analysis of these data builds a basis for a chemisorp- tion study on the same material (section 14.2.3). And thirdly, the analysis of these low-temperature data is of fundamental importance for the understanding of the corresponding high-temperature data, which reveal the changes in the electronic structure at the ferromagnetic to paramagnetic phase transition (section 14.3.2).

The theoretical investigation of the photoemission spectra (Clauberg 1983) is done with a one-step model of the photoemission process based on the Bloch wave method (Liebsch 1978, Feibelman and Eastman 1974, Spanjaard et al. 1977). In this model, the sample is treated as a semi-infinite crystal with a constant spacing between the atomic layers parallel to the surface and the surface barrier is des- cribed by a truncated image potential. The initial and final states of photoemission in this semi-infinite system are given as linear combi- nations of Bloch waves, including those whose amplitudes exponentially decrease towards the interior of the crystal. This allows a detailed analysis of the calculated spectra in terms of the complex band struc- ture and the layer density of states, going far beyond the reach of the old three-step model of photoemission.

Since the presently existing photoemission programs are all based on the "golden rule" or equivalent expresssions for the photoemission current and a single-(quasi-) particle description, many-body effects must be included empirically. The main many-body effects are the inelastic scattering of the excited photoelectron and the self-energy corrections for the hole-state generated in the photoemission process. Especially for Ni it is well known that the latter correction is important for explaining the differences between the experimental photoemission spectra and the ground state calculations in the local-density approximation. In section 14.2.1 the many-body effects were approximated by the inclusion of imaginary parts into the final- and initial-state potentials as well as by changing the real part of the muffin-tin potential in the initial-state. Here, the same approximation is used for the final-state, but the self-energy corrections for the hole-state are treated differently. In the one-step-model "golden-rule" expression for the photocurrent I (cf. Chapter 4.5) the energy-conserving $\delta(E_f-h\nu-E_i)$ is replaced by the diagonal part A_i of the hole-state spectral function, i.e.

$$I \propto (E_f-E_v)^{1/2}\sum_i|<\Psi_f|\underset{\sim}{A}\cdot\underset{\sim}{p}|\Psi_i>|^2 A_i(E_f-h\nu) \tag{1}$$

where the spectral function of the hole-state $A_i(E_f-h\nu)$ is related to the complex self-energy $\Sigma(E_f-h\nu)$ by

$$A_i(E_f-h\nu) = \frac{1}{\pi}\frac{\mathrm{Im}\Sigma(E_f-h\nu)}{(E_f-h\nu-E_i-\mathrm{Re}\Sigma(E_f-h\nu))^2+(\mathrm{Im}\Sigma(E_f-h\nu))^2} \tag{2}$$

(Hedin and Lundqvist 1969, Hedin et al. 1971), and $\Psi_{i(f)}$ is the initial (final) state in photoemission, E_v the vacuum energy, $\underset{\sim}{A}$ the vector potential of the exciting light, and $h\nu$ the photon energy. In the limit of a vanishing imaginary part of the self-energy, equation (1) becomes

$$I \propto (E_f-E_v)^{1/2}\sum_i|<\Psi_f|\underset{\sim}{A}\cdot\underset{\sim}{p}|\Psi_i>|^2\delta(E_f-h\nu-E_i-\mathrm{Re}\Sigma(E_f-h\nu)) \tag{3} .$$

This means the "original" energy E_i is replaced by a quasi-particle energy $\epsilon = E_i + \mathrm{Re}\Sigma(\epsilon)$. In the limit of the three-step-model, this corresponds to calculating the photoemission spectra with a so-called correlated band structure (Davis and Feldkamp 1980, Treglia et al. 1982). At present there is no theory which gives an accurate descrition of the self-energy (cf. Igarashi 1983, 1985). The most important approaches to the problem are those by the three-body correlation (Igarashi 1983, 1985), the t-matrix (Liebsch 1979, 1981; Penn 1979), and the second order perturbation (Treglia et al. 1982) theories.

Fig. 6 shows a part of the one-dimensional ferromagnetic band structure of Ni along the $\Gamma(\Sigma)K(S)X$ direction. The initial states include self-energy corrections in the low-density limit of the t-matrix approach (for details see Clauberg 1983, 1984) superposed on a ground state muffin-tin potential (Moruzzi et al. 1978). The band structure along $\Gamma(\Sigma)K(S)X$ determines the normal photoemission from Ni(110). A detailed analysis of the transition matrix elements between Bloch waves in the initial and in the final state of the photoemission calculation (Fig. 7) shows that for Ne I radiation ($h\nu=16.85$ eV) the spectra are mainly due to emission from the region near the X point

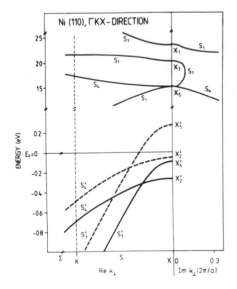

Fig. 6:

One-dimensional ferromagnetic band structure along the $\Gamma(\Sigma)K(S)X$ direction including self-energy corrections for the initial states. Only initial states are shown which are allowed in normal emission. For the final states at the X point the imaginary parts of the wave vectors are shown in addition.

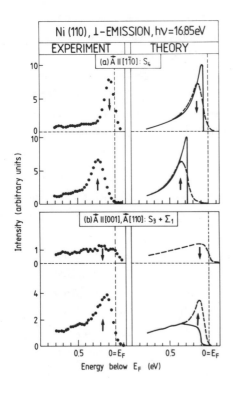

Fig. 7:

Experimental and calculated spin-resolved energy distribution curves for normal emission from Ni(110) and different directions of the electric field vector. The dashed curves in the theory include the lifetime broadening of the hole state and the experimental resolution. The proportionality factor between the experimental count rates and the units in the calculations is chosen to give equal peak heights for the measured and calculated $S_4^{\uparrow}$ spectrum.

into the final state band gap between the X_5' and the X_3 point. The imaginary part of the wave vector of the complex continuation of the S_1 band between these two points determines the mean escape depth of the photoelectrons and is given in Fig. 6. The mean escape depth is about 8 a_0 (a_0 is the Bohr radius) or about 3.5 atomic layers parallel to the (110) surface. This should be compared with the mean free path of about 25 a_0, which is the mean escape depth for propagating final states. The resulting relaxation of $k_\perp$ conservation, i.e. the increased probability for indirect transitions, gives rise to spectra with quite broad structures, even without the inclusion of a lifetime broadening of the initial states. The asymptotic line shapes in Fig. 7 are a direct consequence of these indirect transitions, since the band edges at $X_2^{\uparrow\downarrow}$ ($X_5^{\uparrow}$) give upper energy limits to the spectra. The spectra calculated according to eq.(3) are given by the solid lines in Fig. 7. To approximate the spectra which also include the imaginary part of the self-energy according to eq.(1.b) and eq.(2), the results of eq.(3) are convoluted with a Lorentzian of half-width $Im\Sigma(\epsilon)$ at half maximum. This gives exactly the results of eq.(3) if one considers constant-final-state spectra or usual energy distribution curves with a final state wave function constant over the range of energies smeared out by $Im\Sigma(\epsilon)$. In general this procedure is expected to give a very good approximation to eq.(3), since the final state wave function is a quite smooth function of energy if the final state broadening is included by using a complex final state potential. The dashed lines in Fig. 7 include this approximation to the effect of $Im\Sigma(\epsilon)$ and additionally have been convoluted by a Gaussian of 0.1 eV full width at half maximum to include the experimental resolution. The comparison of these calculated spectra with the experimental spectra (Raue et al. 1983) shows good agreement not only for the peak positions, but more importantly for the relative intensities, peak widths and line shapes, thereby demonstrating again the advantage of the one-step model in photoemission. In particular the use of the complex band structure is a necessity for including emission into evanescent final states (band gaps) which determines the spectra discussed here.

There is only one large discrepancy between the measured and the calculated spectra, this is the missing $S_3^{\uparrow}$ peak in the calculations (solid line). The analysis of the transition matrix elements shows that this reduction of the emission intensity at the expected peak position is related to a destructive interference between the main Bloch wave propagating to the surface and the reflected one propagating into the solid, leading to a minimum in the layer density of states (directly extracted from the photoemission calculation) of the first few layers (counted from the surface). Because of the very small escape depth of the photoelectrons in band gap emission, this minimum is reflected in the photoemission spectra. The destructive interference in the first layers is a general phenomenon which occurs at special points in the band structure of a perfect semi-infinite crystal (Lee and Holzwarth

1978). Removal of the destructive interference will restore the $S_3^\uparrow$ peak in the spectra. A corresponding spectrum, where the different contributions to the transition matrix element are added incoherently is also shown in Fig. 7. This spectrum agrees fairly well with the experiment. There are at least two different effects which may soften or even remove the destructive interference. Firstly, in the same spectra calculated with the Green function method including the imaginary part of the self-energy into the initial state potential, the $S_3^\uparrow$ peak appears for a finite imaginary part (cf. section 14.2.3) and its intensity is found to decrease when reducing the imaginary part. This indicates an agreement between the two theoretical methods in the case of vanishing imaginary part. For finite imaginary part, however, the above discrepancy exists. The exact reason still has to be examined. Secondly, the interference may be weakened by the well-known relaxation of the Ni(110) surface, which disturbs the truncated bulk geometry in the first layers, thereby disturbing also the Bloch condition for the wave functions, which is fundamental for the destructive interference.

Finally, there is one result to be mentioned, which is important for the understanding of the corresponding photoemission data at high-temperatures: The exact location of the S_4 transition in the Brillouin zone. Although, the S_4 peak itself is coupled to the transition directly at the X point, the small escape depth of the photoelectrons leads to a strong contribution from indirect transitions. As a result, the intensity at half maximum of the lower energy side of the peak is given already by transitions from a point midway between K and X. Therefore the high-temperature data cannot be related simply to the X point but must include this broadening in $\underset{\sim}{k}$ space.

In the next section the influence of adsorbates on the spectra discussed above will be investigated.

14.2.3 Ni(110)(2 x 1)-O

Chemisorption of "foreign" atoms on 3d transition metal surfaces generally modifies the electronic structure of the topmost surface layers in such a way that the magnetic moment per atom and thereby the average surface magnetization is reduced (cf. Göpel 1979, Rau 1982; Chapter 1, 7, and 12 of this book and references therein). In the following we illustrate for the case of oxygen on Ni(110) how spin-resolved photoemission measurements and calculations can reveal details of chemisorption-induced changes in the electronic and magnetic structure of the surface.

Oxygen chemisorbs dissociatively on Ni(110). At a coverage of 0.3-0.35 monolayers, it forms a (2 x 1) overlayer and induces a massive reconstruction of the substrate. The geometrical arrangement of the atoms in the surface region has been determined by ion-scattering (Schuster and Varelas 1983) and scanning tunneling microscopy (Baro et al. 1984) as a "saw-tooth" model (cf. Fig. 8), in which the periodicity of the two topmost Ni layers is doubled in the [1$\overline{1}$0] direction

Fig. 8:

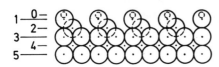

Saw-tooth reconstruction model of Ni(110)(2x1)-O (Schuster and Varelas 1983): projection onto (001) plane perpendicular to the surface, with labels 0 and 1-5 referring to oxygen overlayer and monoatomic Ni layers, respectively.

and the oxygen atoms reside in long bridge positions 0.25 Å above the topmost [001] Ni rows. (The bare Ni "ribbons" reported by Baro et al. (1984), which account for the coverage less than 0.5, are not shown in Fig. 8).

Spin-resolved photoemission spectra from this Ni(110)(2x1)-O system have recently been measured (Schmitt et al. 1985) and calculated (by means of Pendry's multiple scattering one-step theory (cf. Chapter 4.5))(Feder and Hopster 1985). As can be seen in Fig.9, experiment and theory are in close agreement. The spectra exhibit the following changes due to oxygen adsorption: (1) an overall intensity

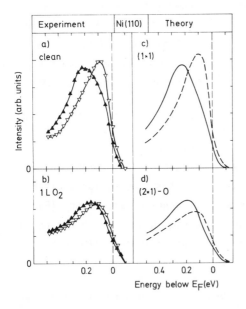

Fig. 9:

Normal photoemission spectra from Ni(110) excited by unpolarized light (hν=16.85 eV) at normal incidence. Experimental majority (▲▲▲) and minority data (▽▽▽)(Schmitt et al. 1985) for clean surface (panel (a)) and after exposure to 1 L O_2 (panel (b)). Calculated results (Feder and Hopster 1985) for relaxed (1x1) surface (c), and for (2x1) saw-tooth reconstruction model with oxygen and four magnetically dead Ni layers (d).

(The intensity scales were chosen such as to match the heights of the majority peaks in parts (a) and (c). The heights of the other peaks relative to these two are the actual measured and calculated ones.)

reduction, which is stronger for minority than for majority-spin electrons; (2) a strong reduction of the "exchange splitting" between majority and minority peaks.

The reasons for these changes are elucidated by calculations of the photocurrent for s- and p-polarized light for a sequence of geometrical and magnetic models going from clean Ni(110) to the final Ni(110)(2x1)-O model. The spin-dependent effective potentials V^σ for Ni were constructed in essentially the same way as described above (section 14.2.2). Magnetically dead layers were mimicked by using the spin-average $(V^+ + V^-)/2$. An enhancement of the clean Ni(110) surface magnetization M_1 at zero temperature by about 15 % with respect to the bulk, as was found by self-consistent ground state calculations (Freeman et al. 1982 and Chapter 1 of this book), was not included, since at T=300 K it is roughly compensated by the stronger decrease of M_1 with increasing temperature. For oxygen, a renormalized-atom potential with an Xα exchange part (cf. Chapter 4.2.2 eq.(11))(with α=0.7) was employed.

A selection of calculated spectra is shown in Fig. 10. The results for clean unrelaxed Ni(110)(i.e. a truncated bulk geometry)

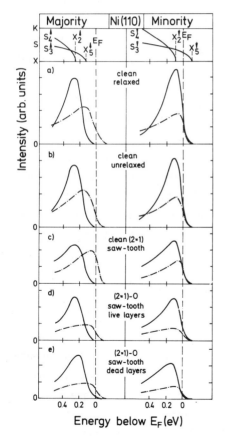

Fig. 10:

Theoretical majority (left column) and minority (right column) normal photoemission spectra from Ni(110) due to normally incident light (hν=16.85 eV) with $\underline{E}||[1\bar{1}0]$ (——) and $\underline{E}||[001]$ (-·-·) as obtained for the following surface models: a) clean with multilayer relaxation (δ_{12}=-8.4%, δ_{23}=3.1%) (cf. Gauthier et al. 1984) and surface magnetization M_1=M_b, b) clean truncated bulk with M_1=M_b, c) clean (2x1) saw-tooth reconstruction with M_1=M_b, d) (2x1)-O saw-tooth model with M_1=M_b, and e) (2x1)-O saw-tooth model with M_1=M_2=M_3=M_4=0, i.e. four magnetically dead Ni layers. Topmost panel: self-energy-corrected bulk band structure along KX direction. (Feder and Hopster 1985).

have already been discussed in Section 14.2.2. The $S_3^\uparrow$ peak appears in Fig. 10 as a consequence of including the initial-state lifetime in the multiple scattering treatment. Surface relaxation, with an 8.4 % contraction of the first and a 3.1 % expansion of the second inter-layer spacing (as determined by means of LEED by Gauthier et al. (1984) and, similarly, by Xu and Tong (1985)), is seen in Fig. 10a to have little influence on the spectra. The more massive reconstruction of a (fictitious) clean saw-tooth model (as in Fig. 8, but without oxygen) produces the following changes: the $S_4^{\uparrow\downarrow}$ peak decreases by about 40 %, while the $S_3^\uparrow$ peak is reduced by only about 10 % but shifted towards E_F by about 0.5 eV; the 30 % decrease of the $S_3^\downarrow$ peak is largely due to the cut-off at E_f after a change similar to that of the $S_3^\uparrow$ peak. This selective drastic decrease of the S_4^σ peaks, obtained by the full pho-toemission calculation, is qualitatively plausible from a simple tight-binding consideration of the initial states: the d-wave basis functions of S_4 symmetry (x^2-y^2) on each site strongly overlap with those on the neighbouring sites along the [1$\bar{1}$0] direction parallel to the surface. Removal of these sites in layers 1 and 2 (cf. Fig. 8) therefore strong-ly alters the corresponding tight-binding states in these layers, lea-ving only bulk-type layers to contribute to the S_4 spectra. In contrast, the S_3 basis functions z(x+y) mainly overlap along the [001] direction and the tight-binding states are consequently less affected by the re-construction. Adsorption of oxygen (Fig. 10d), on the other hand, has little effect on the S_4 spectra, but reduces the S_3 peak heights by about 50 %. This is again plausible from the tight-binding picture: the S_3 basis function z(x+y) on the top layer Ni atoms are most invol-ved in bonding with oxygen atoms in long bridge positions along [001].

The surface models considered so far involve changes in geome-try, but all Ni layers have the same (bulk) magnetization. Consequent-ly, the "exchange splitting" between majority and minority peaks has not been affected. Calculations with varying numbers of magnetically dead layers (as extreme cases of reduced magnetization) gave the fol-lowing results (Feder and Hopster 1985). Quenching the magnetization in layers 1 and 2 of the clean saw-tooth model and the (2x1)-O model had almost no effect. For the dominant S_4 peaks, this is plausible from the above discussed alteration of the S_4 initial states in (2x1) Ni layers. A significant reduction of the exchange splitting, as observed experimentally (cf. Fig. 9), was only achieved by quenching the magnetization of the bulk-like Ni layers 3 and 4 (cf. Fig. 10e) (irrespective of the magnetization of layers 1 and 2). The selective $S_4^{\downarrow}$ attenuation is explained by the fact that the nonmagnetic (1x1) top layers support propagating S_4 states at energies near $X_2^{\uparrow}$, but not near $X_2^{\downarrow}$.

Do these findings actually imply the existence of magnetically dead deeper layers ? It is important to note that quenching of the magnetization of layers 3 and 4 is sufficient for reducing the observed exchange splitting, but not necessary. It would already suffice to selectively quench the S_4 splitting in these layers, in line with the tight-binding argument that the absence of alternate rows in the adjacent layer 2 changes the overlap with (x^2-y^2) orbitals and thereby S_4 states, whereas S_3 states are much less affected. The latter appears consistent with spin-dependent inverse photoemission (bremsstrahlung) data (Seiler et el. 1985), which show no shift of the empty minority band upon oxygen chemisorption.

In conclusion, the above theoretical analysis for Ni(110)(2x1) -O supports the saw-tooth reconstruction model and indicates that bulk-like Ni layers 3 and 4 exhibit a drastic reduction of the exchange splitting between $S_4^{\uparrow}$ and $S_4^{\downarrow}$ states. More generally, it recommends spin-resolved photoemission and its inverse as promising tools for studying $\underset{\sim}{k}$- and symmetry-dependent details of surface magnetism in chemisorption systems, but also indicates the need for complementary methods (like spin-polarized electron scattering, cf. Chapter 7) to deduce more integral magnetic surface properties, like the total magnetic moment per atom.

14.3 Analysis Near the Curie Temperature

The theoretical understanding of transition metal ferromagnetism near the Curie temperature T_c in terms of the underlying electronic structure has been a matter of vivid debate for a long time (cf. Chapter 2 of this book). General consensus was reached with regard to the inadequacy of the Stoner model near T_c and the importance of spin-density fluctuations at finite temperatures. While the modern theories of finite temperature ferromagnetism are based on the assumption of local magnetic moments existing even above T_c, the correlation between these moments, i.e. the extent of short-range magnetic order was still a matter of controversy, with a "disordered-local-moment picture" with no short-range order at all on the one side, and a "local-band picture" with massive short-range order on the other side. Progress in this matter has most recently been made by theoretical analysis of spin- and angle-resolved photoemission data (Haines et al. 1985a). One of the important findings in the theoretical analysis of finite temperature photoemission and inverse photoemission (Clauberg et al. 1985, Korenman and Prange 1984) is the dependence on the wave vector $\underset{\sim}{k}$ at which the photoemission process takes place, as well as on other effects leading to a different temperature dependence for different states in the Brillouin zone. It is therefore not possible to establish a universal correspondence between a particular temperature behaviour (e.g. merging of peaks) and a particular model (e.g. the disordered-local-moment limit). In fact, it may well happen that, for a given surface, data associated with some $\underset{\sim}{k}$ points naively appear to be in line with the disordered-local-moment model, whilest the data for other $\underset{\sim}{k}$ points seem compatible with the local-band limit (see e.g. the inverse photoemission data on Fe discussed in Chapter 13). Information on magnetic order can therefore not be read off the data directly, but has to be extracted by means of a detailed theoretical analysis involving calculations for a range of assumed degrees of short-range order.

In the following we review such analyses for Fe and Ni at some special points of the Brillouin zone, where experiments have been done or are expected to give interesting results.

14.3.1 Fe(001)

All the $\underset{\sim}{k}$ points we want to discuss for Fe can be investigated
by photoemission or ultrviolett bremsstrahlung from the (001) face of
Fe. The investigations are based on a bulk interband transition model
with a plane wave final state (initial state in inverse photoemission)
and a tight-binding approximation for the initial state (see Chapter
4.5.7 and references therein). Since the local magnetic moment in Fe
near T_c is likely to be reduced by only about 10 - 15 % from its value
at T = 0 (Heine et al. 1981, Pindor et al.1983), it is assumed that
$\underset{\sim}{\Delta}_{i1}$ has the same magnitude at all sites i. Fig. 11 shows the part of
the nonmagnetic band structure (derived from the tight-binding parame-
ters of h) relevant for the spectra to be discussed.

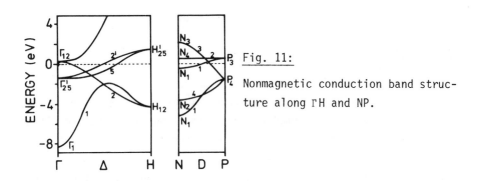

Fig. 11:

Nonmagnetic conduction band structure along ΓH and NP.

The bcc bulk lattice of Fe is partitioned into cubic clusters
with periodic boundary conditions. The size of the cluster is given
by a cube-edge length of 10 lattice constants, i.e. 2000 atoms. The
exchange field directions $\underset{\sim}{\Delta}_{i2}$ ($\underset{\sim}{\Delta}_{i0}$ = 0) at the atomic sites i, are
set up to vary randomly from site-to-site subject to a prescribed cor-
relation function (cf. Chapter 4.5.7) which consists of a long-range
part corresponding to the average magnetization of the cluster and a
Gaussian short-range part (of full width Λ at half maximum) (Haines

1985). Fig. 12 shows one configuration of local exchange fields $\{\underset{\sim}{\Delta}_{i2}\}$ in a (001) central plane of the cluster for $T>T_c$ and a short-range order given by $\Lambda=5.4$ Å, corresponding to a nearest-neighbour correlation coefficient of 0.55. A variety of short-range order constellations is visible in this figure.

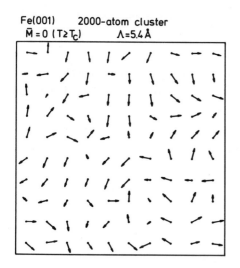

Fe(001)   2000-atom cluster
$\bar{M}=0$ $(T \geq T_c)$ $\Lambda=5.4$ Å

Fig. 12:

Configuration of exchange field directions $\underset{\sim}{\Delta}_{i2}$ (with $|\underset{\sim}{\Delta}_{i2}|$=const) in an (001) central plane of a 2000 atom bcc Fe cluster, for short-range order $\Lambda=5.4$ Å and average magnetizazion zero. The arrows represent the projections of the $\underset{\sim}{\Delta}_{i2}$ onto the plane. (Clauberg et al. 1985)

In calculating physical quantities an average over a set of configurations must be performed to ensure that the randomness resulting from this restricted-random generation of the configuration $\{\underset{\sim}{\Delta}_{i2}\}$ is averaged out. The spin-resolved photoemission or inverse photoemission intensity in the direction $\underset{\sim}{k}$ is calculated via the diagonal elements of the one-electron Green function (associated with the Hamiltonian) with respect to the initial state $\underset{\sim}{A} \cdot \underset{\sim}{p} |ks\rangle$. To evaluate these elements, a recursion method is used (Haydock 1980).

We will discuss here four special points of the Brillouin zone to demonstrate the main effects influencing the spectra at finite temperature and to extract information about the degree of short-range order by comparison with experiment.

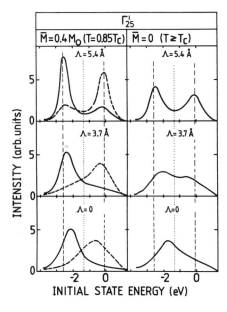

Fig. 13:

Theoretical spin-up (solid lines) and spin-down (dashed lines) spectra for photoemission from the Γ_{25}' state. The vertical dashed lines mark the energies of $\Gamma_{25}^{\uparrow}{}'$ and $\Gamma_{25}^{\downarrow}{}'$ at T=0, and the dotted line indicates the nonmagnetic Γ_{25}' energy. The exchange splitting is Δ_2=2.5 eV. (Clauberg et al. 1985)

The point for which we compare with experiment is the Γ_{25}' point in the centre of the Brillouin zone. Fig. 13 shows spectra corresponding to a temperature below T_c and to a temperature at or above T_c. A value of Δ_2=2.5 eV is used according to the analysis of the low-temperature data (cf. section 14.2.1). These spectra reveal a strong dependence on the degree of short-range magnetic order. Without short-range order the $\Gamma_{25}^{\downarrow}{}'$ and $\Gamma_{25}^{\uparrow}{}'$ peaks get broadened, shift towards each other, and finally collapse into a broad central peak when the temperature increases up to T_c. This "collapse" of the Γ_{25}' line splitting occurs despite the constant exchange splitting used in the calculations and indicates that most of the electrons in this initial state are so fast, or so delocalized, that they are hardly affected by the local exchange field of a single site, but "feel" an exchange field which has averaged to zero. For increasing short-range order the shifts of the peaks are reduced as well as the broadening. For a short-range order described by Λ=5.4 Å the two peaks appear still at their T=0 positions. For larger values of Λ only the widths of the peaks decrease. In addition, for $T<T_c$ the spectra for $\Lambda \geq 5.4$ Å reveal the emergence of two peaks of

opposite spin-direction at the energies of the dominating peaks. These new peaks are a consequence of a reversed average magnetization in regions of an extension Λ, which is large enough to build up a well-defined local band structure for states like $\Gamma_{25'}$. This local band structure is also the explanation for the appearance of the two dominating peaks still at their T=0 positions.

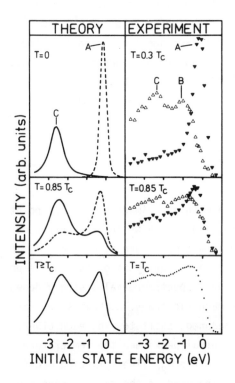

Fig. 14:

Theoretical spin-up (solid lines) and spin-down (dashed lines) spectra for photoemission from $\Gamma_{25'}$ with $\Lambda=5.4$ Å at temperatures as indicated, with an energy-dependent initial state lifetime (Feder et al. 1984) and convolution by a Gaussian of 0.35 eV full width at half maximum. Experimental spin-up (solid triangles, solid circles) and spin-down (open triangles, solid circles) spectra from Kisker et al. (1985). (Clauberg et al. 1985)

Fig. 14 shows experimental spin-resolved photoemission data (Kisker et al. 1985) in comparison to the calculated spectra with $\Lambda=5.4$ Å. Peaks A and C have been found to be interband transitions from Δ_5 states near $\Gamma_{25'}^{\downarrow}$ and $\Gamma_{25'}^{\uparrow}$, repectively (cf. section 14.2.1). We therefore focus on these peaks and disregard the extra experimental peak B, which is presumably due to off-normal emission from a neighbourhood of Γ_{12}. Considering only the peaks A and C the temperature dependence of the experimental data is in good agreement with

that of the model spectra for Λ=5.4 Å after inclusion of an energy-de
pendent lifetime broadening in the initial state and the experimental
resolution. In particular, the emergence of opposite-spin features at
the positions of the dominating peaks is observed experimentally.
Comparison wit Fig. 13 shows that this latter effect requires a short-
range magnetic order of at least Λ=4 Å. By including also the final
state broadening connected with the finite escape depth of the photoelectrons, the agreement between theory and experiment is increased
even more, since this reduces the dip between the two main peaks in
the model spectra (Clauberg et al. 1985).

A similar temperature dependence as for the Γ_{25}' point is expected
for the H_{25}' point which belongs to the same symmetry as Γ_{25}'. This point
does not appear in photoemission spectra, but in inverse photoemission
spectra, since both, the $H_{25}'^{\uparrow}$ and $H_{25}'^{\downarrow}$ peak are above E_F. Fig. 15 shows

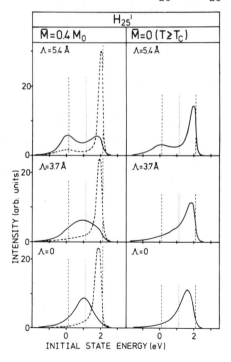

Fig. 15:

Theoretical spectra for inverse
photoemission from the H_{25}' state.
The exchange splitting is Δ_2=2.0eV.
The presentation is the same as
in Fig. 13.
(Clauberg et al. 1985)

the theoretical spectra as a function of long-range and short-range magnetic order, calculated with $\Delta_2 = 2.0$ eV. Obviously, in contrast to $\Gamma_{25'}$, the dependence on temperature and short-range order is very different for the two spin-directions. While for $\Lambda = 0$, the low-energy peak is strongly broadened and continuously shifted towards the high-energy peak by approaching T_c, the high-energy peak is only slightly shifted and broadened. Considering the dependence on short-range order, the high-energy peak is nearly unaffected, and the low-energy-peak shifts back to its $T=0$ position with increasing short-range order, indicating the establishment of a local band structure. Simultaneously, the appearance of peaks with opposite spin-directions is observed as for the $\Gamma_{25'}$ point. So the behaviour of the low-energy peak in the majority-spin spectrum shows the same dependence as for $\Gamma_{25'}$, and it is the deviating behaviour of the high-energy peak in the minority-spin spectrum which must be explained. The important point is, that the $H_{25'}$ point is on the top of the whole d-band. At zero temperature, in the whole Brillouin zone there are no majority-spin d-electrons at the energy of the minority-spin $H_{25'}^{\downarrow}$ point, except the few due to hybridization with the s-band, but inversely there is a high density of minority-spin states with the right symmetry (e.g. Δ_5 as a sub-symmetry of $H_{25'}$ or $\Gamma_{25'}$) at the energy of the majority-spin $H_{25'}^{\uparrow}$ point. Going to finite temperatures, i.e. introducing magnetic disorder, spin and wave vector are no longer good quantum numbers and the majority-spin electrons are strongly scattered into minority-spin states by the magnetic disorder. Oppositely, the minority-spin electrons can not be scattered into majority-spin states at nearby energies. This results in the observed fact, that the electrons which belong to $H_{25'}^{\downarrow}$ at $T=0$ are nearly unaffected by the disorder.

The strong dependence of the majority-spin spectra on short-range order makes the $H_{25'}$ point a promising candidate to gain further information about the actual degree of short-range magnetic order in Fe near T_c by spin-resolved inverse photoemission.

While the H_{25}' point demonstrates the effect of the energetic position in the band structure, the Γ_{12} point shows the effect of a symmetry different from Γ_{25}' for a state energetically close to Γ_{25}' and also at the same point of the Brillouin zone. The Γ_{12} peak whose majority-spin part already appeared as peak B in the experimental data in Fig. 14, can not be observed in total in a photoemission or inverse photoemission experiment, since $\Gamma_{12}^{\uparrow}$ is below E_F and $\Gamma_{12}^{\downarrow}$ above. But the density of states projected onto the corresponding $\underset{\sim}{k}$ point, which is given in the model calculations, reveals very nicely the influence of the symmetry of the initial state on the temperature behaviour. In contrast to the Γ_{25}' point there are no collapsing peaks in Γ_{12} in the disordered-local-moment limit. There is a two-peak structure and the appearence of new peaks with opposite spin character at the energies of the dominating peaks even without short-range order. This demonstrates that the electrons at Γ_{12} are mainly influenced by the local exchange field at a single site and hardly feel the magnetic disorder of the environment. Accordingly, the overlap integrals to the neighbour atoms are found to be larger for Γ_{25}' than for Γ_{12} symmetry.

As a last example Fig. 16 shows the theoretical spectra for a point in the D_3 band ($\underset{\sim}{k}=(1/2)NP = (1/4)(122)(2\pi/a)$ as a pure d-band with a large group velocity. Without short-range order the spin-resolved spectra below T_c for both directions, each reveal a peak close to the nonmagnetic D_3 energy and a second peak close to its T=0 position, i.e. a three-peak structure in the spin-integrated spectra, very similar to those calculated for a special point with high group velocity in the local-band theory (Korenman and Prange 1980). In addition there are small shoulders at the T=0 positions for the opposite spin-directions whose weight increases with increasing short-range order. But in contrast to Γ_{25}', for D_3 there still remains the peak close to the nonmagnetic energy which is only slightly shifted towards the T=0 position. The spectra seem to reflect a strong superposition of the averaging of the disordered exchange field towards a zero-field and the determination of the

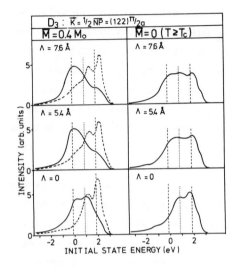

Fig. 16:

Theoretical spectra for the D_3 band $\underset{\sim}{k}=(1/2)(122)(2\pi/a)$. The exchange splitting is $\Delta_2=2.0$ eV. The presentation is the same as in Fig. 13.

(Clauberg et al. 1985)

spectra by a single site field. As a result of the high group velocity even for large Λ there is still a large probability that part of the electrons feel an exchange field strongly reduced by averaging.

Beside the effects discussed here the spectra are influenced by the bonding and anti-bonding character of the wave functions (Clauberg et al. 1985). In general all the effects can be understood qualitatively but the actual spectra strongly depend on the relative importance of the different effects and can not be derived without a quantitative model calculation. Finally we want to mention two facts. Firstly, all the calculations in the disordered-local-moment limit discussed above are in good agreement with recent alloy-type coherent-potential approximation (CPA) calculations without short-range order (Durham 1985). Secondly, up to now the experimental data from spin-resolved photoemission and inverse photoemission allow only the extraction of a lower limit of about 4 $\overset{o}{A}$ for the short-range magnetic order near T_c in Fe, for the determination of an upper limit further experimental data are needed. An indication for an upper limit

of about 6 - 8 Å has been extracted from most recent valence band
X-ray photoemission data (Kirby et al 1985) by Haines et al.(1985b).

14.3.2 Ni(110)

For Ni we want to focus on one single point in the Brillouin
zone for which temperature-dependent spin-resolved photoemission data
are available. These data originate from a region around the X_2 point.
The corresponding low-temperature data have been discussed in section
14.2.2. Fig. 17 shows a model calculation for the X_2 point using the
same theory as for Fe in section 14.3.1. The calculations are done for
an fcc cluster with a cube-edge length of 6 lattice constants, i.e.
864 atoms, and an average over 5 to 10 configurations is performed
(Haines et al. 1985c). The assumed exchange splitting of 0.31 eV is
larger than the experimentally observed value of 0.18 eV (Raue et al.
1983), and the effect of final state broadening (cf. section 14.2.2)
is not included. Nevertheless the spectra reveal the general dependence
on temperature and short-range order for the X_2 point.

For $\Lambda=5.4$ Å, which equals the corresponding Λ for Fe not only in
units of Å but also in units of the nearest-neighbour distance, there
is still a collapsing peak behaviour, as for $\Lambda=0$ at the Γ_{25}' point in
Fe. Haines et al. (1985c) showed that with decreasing exchange splitting the tendency for a collapsing line splitting increases. This ex-
plains the difference in comparison with Fe, since the exchange splitting is smaller by about a factor of 6 in Ni than in Fe. The spectra
for $\Lambda=5.4$ Å are in quite good agreement with the experimental data
(Hopster et al. 1983, Raue et al. 1984) given in Fig. 18, considering
the strong shift and broadening of the majority-spin peak as a function
of temperature as well as the smaller effect on the minority-spin peak.
It is also obvious, that a short-range order of $\Lambda \geq 10$ Å is in disagreement with the experiment. The exact value of the upper limit for the
short-range magnetic order can not be extracted from the present results, since this would require a calculation with the correct exchange splitting of 0.18 eV, and also the inclusion of the final state

598

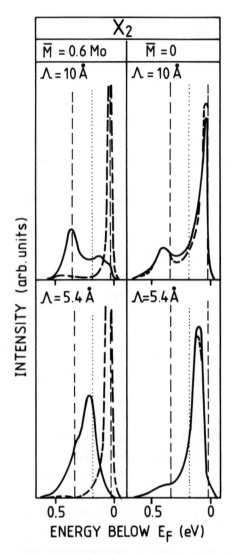

Fig. 17:

Theoretical spectra for photo-emission from the X_2 point. The theory is the same as in section 14.3.1 for Fe. The exchange splitting is $\Delta_2=0.31$ eV. The presentation is the same as in Fig. 13.
(Haines et al. 1985c)

broadening which is important for these experimental data. Korenman and Prange (1984) have shown that, despite the observed collapse of the line splitting, these experimental data are compatible with local band theory. Their calculated model spectra are presented in Fig. 19. The calculation is not exactly for the X point, but very close to it, relying on the fact that the band gap emission leads to a considerable

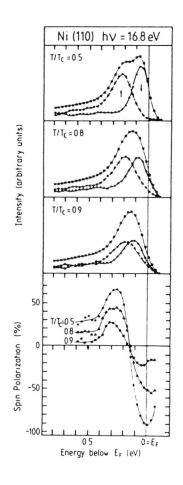

Fig. 18:

Spin-integrated and spin-resolved
energy distribution curves for three
different temperatures. The lower
panel shows the corresponding spin-
polarization curves $(\uparrow-\downarrow)/(\uparrow+\downarrow)$ at
the temperatures indicated.

(Hopster et al. 1983)

contribution of states close to the X point to the measured spectra
(cf. section 14.2.2). So, oppositely to Fe, there is a clear evidence
for an upper limit of short-range order in Ni and it is the lower
limit which has still to be confirmed.

In conclusion of the analysis of finite temperature ferromagnetism, it has been demonstrated that photoemission and inverse photoemission spectra, and especially their spin-resolved versions, contain
information about the short-range magnetic order near T_c in ferromagnets. A lower limit for this degree of short-range order could be ex-

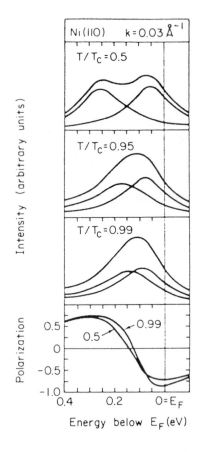

Fig.19:

Majority, minority, and total spectral function at three temperatures, for an S_4 symmetry band near the X point in Ni. The three upper panels correspond to relative magnetizations c=0.90, 0.45, and 0.25, respectively. The lower panel shows the polarization $(\uparrow-\downarrow)/(\uparrow+\downarrow)$, as a fraction of c, at the temperatures indicated. (From Korenman and Prange 1984).

tracted for Fe, and an upper limit for Ni. To complete the information on short-range order in Fe and Ni, further experimental data are needed. In particular, it is also important to investigate to what extent bulk and surface short-range order differ from each other. (For the case of long-range order cf. Chapter 3). In addition we want to emphasize that the theoretical calculations showed collapsing peak structures, fixed two-peak structures, and also three-peak structures in the disordered-local-moment limit as well as in the local-band limit. There is no universal correspondence between the observed features and one of the models. The temperature dependence of the spectra depends strongly on the particular electronic state which is investigated, and the degree of short-range magnetic order can not be deduced from the experimental data without quantitative model calculations.

References:

Ackermann B 1985, Ph.D.-Thesis, Universität Duisburg/Kernforschungs-
    anlage Jülich
Ackermann B and Feder R 1985, J. Phys. C
Baro AM, Binnig G, Rohrer H, Gerber Ch, Stoll E, Baratoff A, and
    Salvan F 1984, Phys. Rev. Lett. 52, 1304
Callaway J and Wang CS 1977, Phys. Rev. B16, 2095
Clauberg R 1983, Phys. Rev. B28, 2561
────── 1984, Jülich-Report No. 1926 (ISSN 0366-0885)(Kernforschungs-
    anlage Jülich, FRG)
Clauberg R, Haines EM, and Feder R 1985, Z. Phys. B
Davis LC and Feldkamp LA 1980, Solid State Commun. 34, 141
Durham P 1985, J. Magn. Magn. Mater. 45, 38
Eastman DE, Himpsel FJ, and Knapp JA 1980, Phys. Rev. Lett. 44, 95
Eberhardt W and Plummer EW 1980, Phys. Rev. B21, 3245
Feder R and Rodriguez A 1984, Solid State Commun. 50, 1033
Feder R, Rodriguez A, Baier U, and Kisker E 1984, Solid State
    Commun. 52, 57
Feder R and Hopster H 1985, Solid State Commun.
Feibelman PJ and Eastman DE 1974, Phys. Rev. B 10, 4932
Freeman AJ, Wang CS, and Krakauer H 1982, J. Appl. Phys. 53, 1997
Gauthier Y, Baudoing R, Joly Y, Gaubert C, and Rundgren J 1984,
    J. Phys. C17, 4547
Gloetzel D 1979, private communication
Göpel W 1979, Surface Sci. 85, 400
Haines EM 1985, Comp. Phys. Commun.
Haines EM, Clauberg R, and Feder R 1985a, Phys. Rev. Lett. 54, 932
────── 1985b,Solid State Commun.
Haines EM, Heine V, and Ziegler A 1985c, J. Phys. F
Haydock R 1980, in "Solid State Physics: Advances inResearch and
    Applications", edited by F. Seitz, D. Turnbull, and H. Ehrenreich,
    (Academic, New York) Vol. 35
Hedin L and Lundqvist S 1969, in "Solid State Physics", vol. 23, eds.

F. Seitz, D. Turnbull, and H. Ehrenreich, (Academic, New York)

Hedin L, Lundqvist BI, and Lundqvist S 1971, in "Proceedings of the 3rd Materials Research Symposium: Electronic Density of States", edited by L.H. Bennett, Natl. Bur. Stand. (U.S.) Spec. Publ. No. 323 (U.S.G.PO., Washington, D.C.)

Heine V, Samson JH, and Nex CMM 1981, J. Phys. F$\underline{11}$, 2645

Hopster H, Raue R, Güntherodt G, Kisker E, Clauberg R, and Campagna M 1983, Phys. Rev. Lett. $\underline{51}$, 829

Hora R and Scheffler M 1984, Phys. Rev. B$\underline{29}$, 692

Igarashi J-I 1983, J. Phys. Soc. Japan $\underline{52}$, 2827

—— 1985, J. Phys. Soc. Japan $\underline{54}$, 260

Kirby RE, Kisker E, King FK, and Garwin EL 1985,

Kisker E, Schröder K, Gudat W, and Campagna M 1985, Phys. Rev. B$\underline{31}$, 2285

Korenman V and Prange RE 1980, Phys. Rev. Lett. $\underline{44}$, 1291

—— 1984, Phys. Rev. Lett. $\underline{53}$, 186

Lee MJG and Holzwarth NAW 1978, Phys. Rev. B$\underline{18}$, 5365

Liebsch A 1978, in "Electron and Ion Spectroscopy of Solids", edited by L. Fiermans, J. Vennik, and W. Dekeyser (Plenum, New York)

—— 1979, Phys. Rev. Lett. $\underline{43}$, 1431

—— 1981, Phys. Rev. B$\underline{23}$, 5203

Moruzzi VL, Janak JF, and Williams AR 1978, "Electronic Properties of Metals" (Pergamon, New York)

Penn DR 1979, Phys. Rev. Lett. $\underline{42}$, 921

Pindor AJ, Staunton J, Stocks GM, and Winter H 1983, J. Phys. F$\underline{13}$, 979

Rau C 1982, J. Magn. Magn. Mater. $\underline{30}$, 141

Raue R, Hopster H, and Clauberg R 1983, Phys. Rev. Lett. $\underline{50}$, 1623

—— 1984, Z. Phys. B$\underline{54}$, 121

Scheidt H, Glöbl M, Dose V, and Kirschner J 1983, Phys. Rev. Lett. $\underline{51}$, 1688

Schmitt W, Hopster H, and Güntherodt G 1985, Phys. Rev. B$\underline{31}$, 4035

Schuster M and Varelas C 1983, Surface Sci. $\underline{134}$, 195

Seiler A, Feigerle CS, Pena JL, Celotta R, and Pierce DT 1985

Spanjaard D, Jepsen DW, and Marcus PM 1977, Phys. Rev. B$\underline{15}$, 1728

Thörner G and Borstel G 1984, Solid State Commun. $\underline{49}$, 997

Treglia G, Ducastelle F, and Spanjaard D 1982, J. Phys. (Paris) $\underline{43}$, 341

Turner AM, Donoho AW, and Erskine JL 1984, Phys. Rev. B$\underline{29}$, 2986

Unguris J, Seiler A, Celotta R, Pierce DT, Johnson PD, and Smith NV 1982, Phys. Rev. Lett. $\underline{49}$, 1047

Wang CS, Klein BM, and Krakauer H 1985, Phys. Rev. Lett. $\underline{54}$, 1852

Xu ML and Tong SY 1985, Bull. Am. Phys. Soc. $\underline{30}$, 460

Chapter 15

Polarized Electrons in Surface Physics: Outlook

M. Campagna

Institut für Festkörperforschung der KFA Jülich,
D-5170 Jülich, FRG

and

II. Physikalisches Institut der Universität zu Köln,
D-5000 Köln 41, FRG

The contributions to this monograph document the recent successes of experiments and theories involving polarized electrons in solid state and surface physics. The task of the editor in choosing the topics was clearly not easy since rather different problems are currently being studied using polarized electrons. Both spin-dependent interactions, i.e. exchange and spin-oribt coupling are being exploited to investigate problems ranging from critical phenomena at surfaces to relativistic effects in the electronic structure of heavy metals like Pt. The progress of this field till 1984 has been succinctly reviewed recently with different accents. We would like to address the reader to these original papers (Siegmann et al 1984, Siegmann 1984, Campagna 1984, Campagna 1985a). We would also like to mention a personal view of the present author on the development of spin-polarized photoelectron spectroscopy with applications to magnetic solids since the first successful experiment 1969 (Busch et al 1969) and with emphasis on the research performed in the Jülich-Cologne area (Campagna 1985b).

Driving forces for the current research in polarized electron spectroscopies are:

1. The substantial shift in research activity in magnetism from bulk to surfaces, interfaces and thin film

phenomena. This is promoted by the simultaneous progress in thin film preparation techniques involving increasing sophistication. They include, for example, molecular or atom beam epitaxy, various forms of sputtering, chemical vapor deposition (CVD) or even metal-organic chemical vapor deposition (MO-CVD). As we know all these techniques - although far from being fully understood - are playing a crucial role in present day solid state technology. This is going to be so also for the foreseeable future. "Magnetic Information Technology" is the appealing name that (Kryder and Bortz 1984) used for indicating the new needs in research in magnetism. Techniques quite familiar in semiconductor technology are being transferred to surface and interface magnetism. In an even broader sense the colleagues in Japan - many of them working in industry - call this line of research "Magnetics for Electronics" (Sakurai 1983). The continuity in research projects and applications in Japan in the past 25 years has indeed made the field of thin film magnetism a very lively (and rewarding) one in this country. And the Japanese colleagues - we have to admit - are quite alert in following also the most recent developments up to the use of polarized secondary electrons to view magnetic domains (Koike and Hayakawa 1984).

2. Besides the tremendous activity with III-V also the preparation of high quality II-VI or II-IV-V$_2$ compounds and their interfaces - for example with metals - is progressing. Relativistic effects in the electronic structure of these materials via spin-orbit coupling can be investigated by the optical orientation technique, well establish for III-V compounds (Meyer and Zakharchenya 1984). Recently (Riechert and Alvarado 1985) have discovered that the precession of the spin polarization vector of excited photo-carriers in the band-bending region of GaAs can be used to study the electronic parameters of this important region of the semiconductor near its

surface. The extension of these investigations to these other materials should be matter of future investigations. This is relevant also for further progress in polarized electron sources.

3. The recent technological advances involving electron beams, primarily optics and two-dimensional detectors. We can easily predict that both advances will further revolutionize the field for producing and analyzing polarized electron beams.

4. The progress in synchrotron radiation and laser research. The properties that make synchrotron radiation especially appealing to polarized electron research are the same that make it so promising to the research community in general, primarily

a) wavelength tunability
b) the expected 2 to 3 order of magnitude increase in brightness when 3rd generation sources will be available (wigglers and undulators in connection with low emittance storage rings)
c) polarization (both linear and circular) and time structure

Photon sources like (tunable) lasers in the visible are becoming increasingly cheaper and realible. The use of lasers operating in the vacuum ultraviolet is starting to be a real alternative to the one of conventional sources for special applications (e.g. multiphoton-processes at metal surfaces) in electron spectroscopy. Even further in the future we see also the application of Free Electron Lasers.

5. Progress in ion and excited-atom sources. This techniques generating polarized electrons by means of neutralization via Auger processes of primarily at clean or adsorbate covered magnetic surfaces is still in its infancy (Onellion et al 1984).

The experimental work involving polarized electrons, which made the various techniques recently become spectroscopies, greatly profited from the input from both theory and calculations like computer simulations. This symbiosis will remain the basis for the next investigations to come. The premises for a large amount of systematic spectroscopic work involving e.g. clean and adsorbate covered non-magnetic surfaces or magnetic transition metal compounds are now favourable. On the other hand the phantasy and steadiness of the researcher, as usual, will be limiting factor in providing new, original applications and uses of polarized electrons. The electron spin as a differential probe will further attract the attention of many alert spectroscopies. This 1985 volume on polarized electrons attesting this factor coincides amusingly with the 60th birthday or discovery of the electron spin. It makes the 75th birthday of the electron spin in the year 2000 an interesting date to look for the next critical perspective.

References

Siegmann H C, Meier F, Erbudak M, and Landolt M 1984 "Advances in Electronics and Electron Physics" 62 p. 1

Siegmann H C 1984 Physica 127 B 131

Campagna M   1984 Physica 127 B 117

Campagna M 1985 J. Vac. Sci. Tech. A3 (3) 1491

Busch G, Campagna M, Cotti P, and Siegmann H C 1969 Phys. Rev. Lett. 22, 597

Campagna M 1985 Z. Phys. B 60,   Special edition upon the occasion of the 60th birthday of Bernd Mühlschlegel

Kryder M H and Bortz A B 1984 Phys. Tod. 37 (12) 20

Sakuray Y ed. 1983 "Recent magnetics for electronics", Jap. Annual Reviews in Electronics, Computers and Telecommunications, 10, North Holland Publishing Company

Koike K and Hayakawa K 1984 Jap. J. Appl. Phys. 23 L187; Appl. Phys. 45 585

Meyer F and Zakharchenya B 1984 Modern Problems in Condensed Matter Sciences, Vol. 8: "Optical Orientation", North Holland Publishing Company, Amsterdam

Riechert H and Alvarado S F 1985 Proceedings of the 4th General Conference of the Condensed Matter Division of the EPS, Berlin, March 18-22, 1985, Festkörperprobleme XXV, ed. P. Grosse, Vieweg and Sun

Onellion M, Hart M W, Dunning F B, and Walters G K   1984 Phys Rev. Lett. 52 380